PIPE FITTING AND PIPING HANDBOOK

Prentice-Hall Books by Louis Gary Lamit

Piping Systems: Drafting and Design (1981)
Piping Systems: Drafting and Design Workbook (1981)
Industrial Model Building (1981)
Descriptive Geometry (1983)
Descriptive Geometry Workbook (1983)
Pipe Fitting and Piping Handbook (1984)

PIPE FITTING AND PIPING HANDBOOK

Louis Gary Lamit

Illustrations By
Patrick Joseph Scheetz

PRENTICE-HALL INC., *Englewood Cliffs, NJ 07632*

Library of Congress Cataloging in Publication Data

Lamit, Louis Gary. (date)
   Pipe fitting and piping handbook.

   Includes index.
   1. Pipe-fitting--Handbooks, manuals, etc. 2. Pipe lines--Handbooks, manuals, etc. I. Title.
TH6711.L36 1984   696'.2   83-23121
ISBN 0-13-676602-1

Editorial/production supervision
   and interior design: **Karen Skrable**
Cover design: **Edsal Enterprises**
Manufacturing buyer: **Tony Caruso**

© 1984 by **Prentice-Hall, Inc.,** Englewood Cliffs, New Jersey 07632

All rights reserved. No part of this book may be reproduced, in any form or by any means, without permission in writing from the publisher.

Printed in the United States of America

10 9 8 7 6 5 4 3 2

ISBN 0-13-676602-1

Prentice-Hall International, Inc., *London*
Prentice-Hall of Australia Pty. Limited, *Sydney*
Editora Prentice-Hall do Brasil, Ltda., *Rio de Janeiro*
Prentice-Hall Canada Inc., *Toronto*
Prentice-Hall of India Private Limited, *New Delhi*
Prentice-Hall of Japan, Inc., *Tokyo*
Prentice-Hall of Southeast Asia Pte. Ltd., *Singapore*
Whitehall Books Limited, *Wellington, New Zealand*

Dedication

For **Margie**

CONTENTS

PREFACE *xi*

1 PIPE DATA *1*

Fig. 1-2 Relationship of nominal size to inside diameter (ID), outside diameter (OD), and schedule number *5*
Table 1-1 Allowable (S) values for seamless alloy, carbon steel, and wrought iron pipe (in 100 psi units)
Table 1-2 ASTM carbon steel pipe and flange specifications *9*
Table 1-3 Schedule numbers and pipe sizes for steel and stainless steel pipe *10*
Table 1-4 API standard line pipe, threaded. Weights and dimensions are nominal *16*
Table 1-5 Cast iron flanged pipe for water. ASA B16.1 class 125 or ASA B16.2 class 250 *16*
Table 1-6 Approx. weight of round seamless cold-finished carbon-steel mechanical tubing (pound per foot). Carbon, 0.25% maximum standard sizes for warehouse stocks random lengths *17*
Table 1-7 Sizes and weights of SPS copper and 85 red brass pipe *17*
Table 1-8 Weights and dimensions of lead pipe *17*
Table 1-9 Sizes and weights for aluminum pipe *18*
Table 1-10 Polyethylene plastic pipe dimensions *19*
Table 1-11 Corrosion-resistance data, polyethylene pipe *19*
Table 1-12 Cast iron soil pipe *20*
Table 1-13 Heat loss from bare pipe and fittings *20*
Table 1-14 Pipe size metric conversion *21*
Table 1-15 Pipe and water weight per line foot *22*
Table 1-16 Weights of metals *22*
Table 1-17 Heat loss from horizontal bare steel pipe (BTU per hour per linear foot at 70° room temperature) *22*
Table 1-18 Total thermal expansion of piping materials (inches per 100 ft. above 32°F) *22*

2 FITTINGS *23*

Fig. 2-2 Common pipe fittings *26*
Table 2-1 Dimensions for welded fittings *28*
Table 2-2 Forged steel sock-welding fittings *30*
Table 2-3 Dimensions of steel flanged fittings *31*
Table 2-4 Dimensions of cast iron flanged fittings *32*
Table 2-5 Cast steel flanged fittings *33*
Table 2-6 Forged steel screwed fittings *34*

3 FLANGES *35*

Fig. 3-2 Selection and application of forged steel companion flanges *36*
Fig. 3-3 Selection and application of facings and gaskets for steel flanges *37*
Table 3-1 Facing dimensions for steel flanges (ANSI B16.5-1973) *38*
Table 3-2 American standard ring joint facing and rings *39*

Table 3-3   Number for ring-joint gaskets and grooves *41*
Table 3-5   Weldneck flanges *42*
Table 3-6   Slip-on, threaded, and socket-type flange dimensions *42*
Table 3-7   Lap joint flanges *43*
Table 3-8   Blind flange dimensions *43*
Table 3-9   Templates for drilling. Classes 400, 600, and 900 steel *44*
Table 3-10 Templates for drilling. Classes 1500 and 2500 steel *45*
Table 3-11 Templates for drilling. Classes 25 and 125 cast iron *46*
Table 3-12 Templates for drilling. Classes 250 and 800 hydraulic cast iron *47*
Table 3-13 Bolting dimensions for 150-lb flanges *48*
Table 3-14 Bolting dimensions for 300-lb flanges *48*
Table 3-15 Bolting dimensions for 400- and 600-lb flanges *49*
Table 3-16 Standard cast iron companion flanges and bolts (for working pressures up to 125 PSI steam, 175 PSI WOG) *49*
Table 3-17 Templates for drilling. Bronze and corrosion-resistant standards *50*

4   THREADS *51*

Table 4-1   Normal thread engagement *52*
Table 4-2   American national standard taper pipe threads (NPT) *53*
Table 4-3   Normal engagement for tight joints *54*
Fig. 4-2   Making up screwed joints *55*
Fig. 4-3   Wrench types and proper use *56*
Table 4-4   American standard straight threads *57*
Table 4-5   British standard taper pipe threads *58*
Table 4-6   Drill sizes for pipe taps *58*
Table 4-7   Tap and drill sizes (American standard course) *58*

5   VALVES *59*

Fig. 5-2   Principal functions of valves *60*
Fig. 5-3   Valve flow characteristics *60*
Table 5-1   Flow resistance *61*
Fig. 5-4   Valve stem variations *64*
Table 5-2   Valve trim selection guide *65*
Table 5-3   Service conditions for gate valves *65*
Table 5-4   Selection guide for disk variations for globe valves *65*
Table 5-5   Selection guide for lubricated versus nonlubricated plug valves *66*
Fig. 5-5   How to assemble a flanged joint *67*
Table 5-6   Selection of valves and valve designs *68*

Fig. 5-6   Typical materials on a cast iron gate valve *68*
Fig. 5-7   Split-wedge gate valve *68*
Fig. 5-8   OS & Y globe valve *69*
Fig. 5-9   Angle globe valve *69*
Fig. 5-10   Gear-operated butterfly valve *69*
Fig. 5-11   Composition disk globe valve *69*
Fig. 5-12   Swing check valve *70*
Fig. 5-13   Lift check valve *70*
Fig. 5-14   Strainer installation *70*
Table 5-7   Steam trap selection *71*
Fig. 5-15   Installation of safety and relief valves *72*

6   PIPE SYMBOLS *73*

Fig. 6-2   Piping drafting line key *74*
Fig. 6-3   Special pipeline line symbols *75*
Fig. 6-4   Basic piping fitting symbols *76*
Fig. 6-5   Comprehensive guide to piping drawing symbols *78*
Fig. 6-6   Flow diagram symbols *79*
Fig. 6-7   Instrumentation symbols *80*

7   PIPE SETUP *82*

Fig. 7-2   Alignment of pipe *83*
Fig. 7-3   Branch and header layout *83*
Table 7-1   How to cut odd-angle elbows *84*
Table 7-2   Cutback at 90 degrees for ID nozzle of OD of header. Standard weight pipe *85*
Table 7-3   Cutback at 90 degrees for ID of nozzle to OD of header. Extra heavy pipe *85*
Table 7-4   ID of nozzle to OD of LR ELL on centerline. Standard weight pipe *86*
Fig. 7-4   Branch and header layout for 45-degree laterals *86*
Table 7-5   Reinforcing pads (arc of nozzle diameter and circumference of header) *87*
Table 7-6   Reinforcing pads (arc for large OD pipe) *87*

8   OFFSETS *88*

Fig. 8-2   45-degree common offset *89*
Fig. 8-3   45-degree rolling offset *90*
Fig. 8-4   Simple offsets *90*
Fig. 8-5   Rolling offsets *92*
Table 8-2   30-degree offsets *92*
Table 8-3   45-degree offsets *94*
Table 8-4   45-degree triangles—base to hypotenuse *95*

9 BENDS 97

Fig. 9-2   Standard pipe bends *98*
Table 9-2   Length of pipe in bends *99*
Fig. 9-3   Calculation of pipe bends *100*
Table 9-3   Center-to-end (CE), back center-to-end (B/CE), and arc length for 30-degree bends of varying radii and pipe sizes *102*
Table 9-4   Center-to-end (CE), back center-to-end (B/CE), and arc length for 45-degree bends of varying radii and pipe sizes *104*
Table 9-5   Center-to-end (CE), back center-to-end (B/CE), and arc length for 60-degree bends of varying radii and pipe sizes *106*
Table 9-6   Arc for 90-degree bends *108*
Table 9-7   True angle of bends in two planes *109*
Table 9-8   Dimensions for 30-degree and 45-degree company bends to bending and elbows *110*

10 MITERS 111

Fig. 10-2   One weld, two-piece miter. Angle of cut equals ½ angle of turn. Angle of cut equals angle of turn/divided by 2. See Table 10-1 for dimension A (cut back) *112*
Fig. 10-3   90-degree miter using two welds. See Table 10-1 for setback and dimensions *112*
Table 10-1   Miter welding dimensions for one-miter and two-miter bend *113*
Table 10-2   Miter welding dimensions for three- and four-miter bends *114*

11 DEVELOPMENTS AND PATTERNS 115

Fig. 11-2   Template layout *116*
Fig. 11-3   Multipiece turn *117*
Fig. 11-4   Template for 90-degree branch *117*
Fig. 11-5   Pattern for wraparound template *117*
Fig. 11-6   Marking cuts without templates *117*
Fig. 11-7   90-degree, two-piece turn. Development and pattern *118*
Fig. 11-8   90-degree, three-piece turn. Development and pattern *118*
Fig. 11-9   90-degree, multipiece welded turn. Development and pattern *119*
Fig. 11-10   Branch and header development and pattern (equal diameters) *119*
Fig. 11-11   Branch and header development and pattern (unequal diameters) *120*
Fig. 11-12   Lateral development and pattern (equal diameters) *120*
Fig. 11-13   Lateral development and pattern (unequal diameters) *121*
Fig. 11-14   60-degree WYE connection development and pattern *121*
Fig. 11-15   Blunt end development and pattern *122*
Fig. 11-16   Orange peel head development and pattern *122*

12 WELDING 123

Fig. 12-2   Submerged arc welding *124*
Fig. 12-3   Manual shielded metal arc welding *125*
Fig. 12-4   Gas-metal arc welding *125*
Fig. 12-5   Location of elements for standard welding symbols *126*
Fig. 12-6   Gas and arc weld symbols *127*
Fig. 12-7   Weld symbols *127*
Table 12-1   Electrode data *127*
Table 12-2   Troubleshooting arc welding equipment *128*
Fig. 12-8   Welding joints *129*
Table 12-3   Welding and brazing temperatures *130*
Table 12-4   Temperature data chart *130*

13 PIPE FLOW 131

Table 13-1   Flow resistance for valves and fittings *133*
Table 13-2   Flow conversion chart *134*
Table 13-3   Water pressure to feet head *135*
Table 13-4   Feet head of water to PSI *135*
Table 13-5   Properties of liquids and gases *135*
Table 13-6   Boiling points of water at various pressures *135*
Table 13-7   Flow of water-equalization of pipe discharge rates for pipe and copper tubing *136*

14 INSULATION 137

Table 14-1   Thermal insulation *138*
Table 14-2   Properties of insulating materials *141*
Table 14-3   Recommended thickness of hydrous calcium silicate pipe insulation *142*
Table 14-4   Recommended thickness of 85 percent magnesia pipe insulation *143*
Table 14-5   Recommended thickness of block insulation *143*

15 PIPE SUPPORTS 144

Fig. 15-2 Pipe support symbols *145*
Fig. 15-3 Spacing of pipe supports, deflection of horizontal pipelines (based on standard pipe filled with water) *146*
Fig. 15-4 Thermal expansion of pipe *146*
Table 15-1 Suggested maximum spacing between pipe supports for horizontal straight runs of standard and heavier pipe (at maximum operating temperature of 750°F) *147*
Fig. 15-5 Rod swing angle of acceptability *147*
Fig. 15-6 Determining lengths of structural braces *148*
Fig. 15-7 Stanchion length determination *148*
Table 15-2 Thermal expansion of pipe materials *148*
Table 15-3 Spacing of pipe supports, stresses calculated for standard weight pipe *149*

16 RIGGING AND HOISTING 150

Table 16-1 Safe load (in pounds) on improved plow steel wire rope (6 strands, 19 or 37 wires per strand, hemp core) *152*
Table 16-2 Safe working load of wire rope in tons of 2000 lbs (safety factor 5-OSHA) *152*
Fig. 16-2 Clamp fastenings *153*
Fig. 16-3 Knots for rigging and hoisting *153*
Table 16-3 Safe working load (SWL) of new fiber ropes (rope used six months will have a SWL of 50 percent) *155*
Table 16-4 Safe working loads of new braided synthetic fiber rope-pounds (rope used six months will have a SWL of 50%) *155*
Fig. 16-4 Standard hoisting signals *156*

17 FORMULA 157

Fig. 17-2 Formulas for circles *158*
Fig. 17-3 Formulas for triangles *159*
Fig. 17-4 Formulas for polygons (four-sided) *160*
Fig. 17-5 Formulas for regular polygons (six-, eight-, and multisided) *161*
Fig. 17-6 Formulas for cubes, prisms, cones, and pyramids *162*
Fig. 17-7 Formulas for cylinders, spheres, torus, and ellipsoids *164*
Table 17-1 Circumference, area, volume of circles and cylinders *166*
Table 17-2 Volume and surface area of spheres *169*

18 TRIANGLES 170

Fig. 18-2 Formulas for right triangles (Pythagorean theorem and Hero's formula) *171*
Fig. 18-3 Right triangle setup for trigonometry *171*
Fig. 18-4 Trigonometry functions *171*
Fig. 18-5 Solving for unknowns with right angle trigonometry *172*
Fig. 18-6 Formulas for non-right angle triangles (oblique-angled) *172*
Table 18-1 Natural trigonometric functions *173*
Table 18-2 Conversion of minutes into decimals of a degree *174*

19 CONVERSIONS 177

Table 19-1 Decimal and metric equivalents of common fractions of an inch *178*
Table 19-2 Unit conversions *179*
Table 19-3 Hardness conversion numbers *179*
Table 19-4 Power required for pumping *180*
Table 19-5 Conversion factors *181*

20 ABBREVIATIONS 187

21 GLOSSARY 195

22 STANDARDS 210

INDEX 217

PREFACE

Pipe Fitting and Piping Handbook is meant to be used as a reference companion for those piping drafters/designers/engineers, and pipe fitters (and students in these areas) who continually need a quick and comprehensive guide on the job or in the classroom. The type and content of reference material differ between the in-house design stage and the out-in-the-field applications of the pipe fitter. Therefore, some of the chapters will be used exclusively by one or the other group. In general, both the pipe fitter and the pipe drafter/designer/engineer will refer to the same chapters, though different sections of each. The author would greatly appreciate suggestions from those people in the piping field. Criticisms, new material, and new ideas on future revisions are welcome and encouraged.

Since other publications in this field cover both the common area of general plumbing or the advanced piping theory area, they are not included here. This is not to say that the plumber or the industrial piping engineer will not find use for the Handbook, but only the more common and useful aspects of piping are included in order to satisfy a greater range of possible users while keeping the price and length of the book within reason.

The author would like to thank the following for contributions to the text: Pat J. Scheetz, for the illustrations; my wife, Margaret K. Lamit, for preliminary work on the glossary and abbreviation chapters; Sue Menkhaus, for inputting (typing) the glossary, abbreviations, and standards chapters. Sincere appreciation is also conveyed to all the companies and organizations that allowed their material to be reprinted. Last, I would like to thank George K. Bachmann for the use of portions of his *Pipefitter's and Plumber's Vest Pocket Reference Book* (Prentice-Hall, Inc.)

Louis Gary Lamit

PIPE FITTING AND PIPING HANDBOOK

1

PIPE DATA

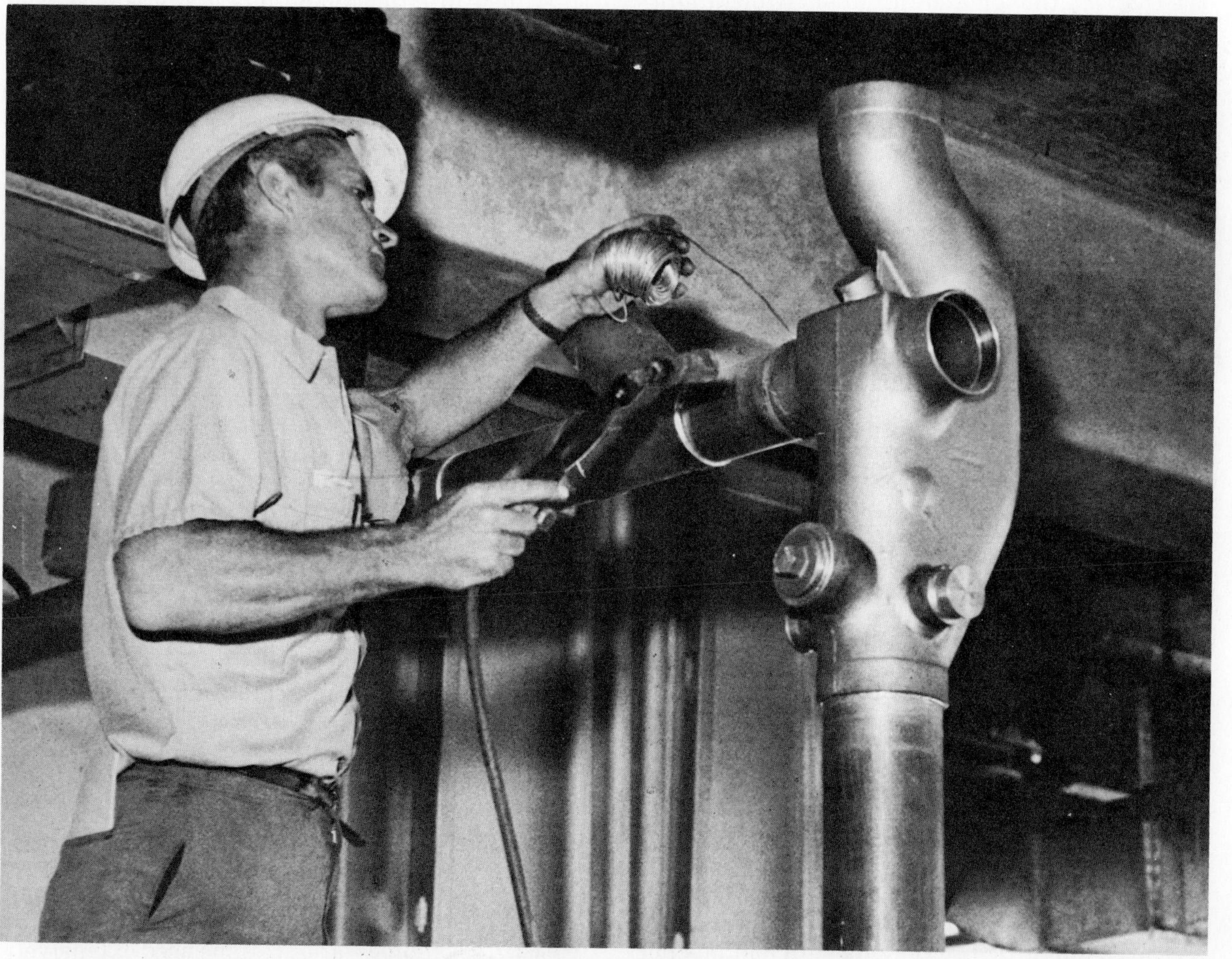

Fig. 1-1 (*Copper Development Association, Inc.*)

PIPING MATERIALS

The service for which a pipe is to be used determines the material and the manufacturing and assembling operation. Pipe can be made from a variety of materials, including glass, rubber, plastic, wood, aluminum, clay, concrete, combinations of metals, copper, lead, brass, wrought iron, and steel. Special pipes are also manufactured with exotic materials such as titanium, with aluminum and steel alloys such as stainless steel, and with plastic. These materials are used for special situations, and their physical properties and temperature and pressure limitations must be taken into account along with their cost. Because service conditions vary so widely regarding heat, strain, pressure, and chemical reactivity, the selection of pipe becomes an important factor in the design of the total system. In general, cast iron, wrought iron, and steel are the most common materials still used today although there are about 40 metallurgical materials on the market.

The chief variables that must be taken into consideration in the selection of pipe are corrosion, cost, pressure, temperature, and safety. With corrosion, the problem cannot be solved by manufacturer's recommendations alone because the conditions under which the system will be operating can affect its corrosion properties even if the same fluid is being transported. It is usually up to the engineers and designers to choose materials for the project. Though cost is always a consideration, it must never take precedence over the safety conditions necessary for high-temperature and high-pressure service.

Metallic Pipe and Tubing

Wrought Iron and Steel The vast majority of pipes are manufactured from carbon steel; chemical composition and manufacturing methods vary throughout the industry. Carbon steel pipe is available in grade A or B and, in a few cases, grade C. Because of the higher carbon content, grades B and C have higher tensile strength than grade A but are less ductile—that is, more brittle. Although the cost usually runs the same for the various grades, care must be taken not to substitute grade A for applications requiring grade B steel (although grade B may usually substitute for grade A). Other common compositions are wrought iron, chrome, moly steel, stainless steel, and nickel steel along with a number of nonferrous metals that are manufactured in iron pipe sizes. The American Society of Testing Materials (ASTM) publishes most of the specifications for piping materials. A great majority of grades recognized in ASTM specifications are approved also by the American Standards Association (ASA) and the American National Standards Institute (ANSI) codes. Moreover, the American Petroleum Institute has established its own material specifications which cover pipelines for the oil and gas and chemical industries.

In highly corrosive situations, wrought iron is preferable to steel although it is not as strong and is somewhat costlier. Both steel and wrought iron are commonly used for water, steam, oil, and gas services. Until the early twentieth century wrought iron was available only as standard, extra strong, and double extra strong, but with the advent of steam services of increasing pressures and temperatures, more diversified wall thicknesses and sizes became necessary. Steel pipe is obtained in either black pipe form or, less commonly, the galvanized type. Both of these materials are available in lengths up to 40 ft (12.2 m) in smaller sizes. Available lengths decrease with increasing wall thickness and size. A majority of pipe comes in random (± 20 ft) or double random (± 40 ft) lengths.

Steel pipe which has been treated with molten zinc to prevent rust is called *galvanized pipe*. This pipe is suitable for drinking water services. Steel pipe is used throughout the industry because of its resistance to high pressure and high temperature, especially chromium or nickel alloy pipe because of the added strength. Various fittings and valves are also obtainable in steel. Standard steel pipe is specified by the nominal diameter. This diameter is always less than the actual outer diameter of the pipe. Nominal wall thickness equals the average wall thickness. In order to use common fittings with different weights of pipe, the outer diameter of each of the different pipes remains the same and the inner diameter is varied to provide for various wall thicknesses. Extra strong and double extra strong steel pipe is produced in this manner. Minimum wall thickness is found by taking the nominal wall thickness and subtracting the mill tolerance—$12\frac{1}{2}$ percent for seamless and 0.01 percent for pipe made from plate stock.

Cast Iron Cast iron pipe is used mainly for the transportation of water or gas or, in some cases, for soil pipe—applications used extensively in underground piping systems. For these applications the push-on joint is used, which makes installation in a trench easier than screwed, flanged, or welded pipe. Generally, water or gas pipes are connected with bell and spigot joints or flanged joints. Soil pipe is available from 2 to 15 in. (5.0 to 38.1 cm) in diameter, in lengths of 5 ft (1.5 m), and in various weights. Soil pipe is almost exclusively manufactured with bell and spigot joints, although threaded joints are sometimes available. Special external loading conditions for buried installations (including ground settlement), along with internal pressure, must be considered by the designer when using cast iron pipe because such loads can fracture the pipeline if it is not sufficiently flexible.

Cast iron has been shown to have an excellent resistance to destruction by external loads when compared to wrought iron or steel, although it cannot be used under conditions of high temperature, high pressure, or vibration. For cast iron, steam temperatures should not exceed 450°F (230°C) and oil not over 300°F (150°C). Although the first years of installation show progressive corrosion caused by rust, the rate of penetration decreases considerably after a number of years of operation.

The nominal size of cast iron pipe, unlike steel pipe, always indicates the inside diameter (ID) regardless of size. When ordering cast iron pipe, one must indicate wall thickness and outside diameter along with nominal size because terms such as strong and extra strong are not used for cast iron pipe. Cast iron piping is available in a variety of standards and weights when manufactured with bell and spigot joints and in sizes 1¼ to 12 in. (3.17 to 30.48 cm) with threaded ends.

Brass and Copper Brass and copper pipes are manufactured in two weights: extra strong and regular. They are available in sizes 1/8 in. to 12 in. (0.31 to 30.4 cm) and are similar in wall thickness and inside diameter to steel pipe classified as strong and extra strong and standard. The outside diameters of brass and copper pipe are exactly the same as the outside diameters of the corresponding nominal sizes of steel pipe. Because of their cost, brass and copper pipes are used only for services where longer life expectancy will compensate for higher price.

Pipe made from these compositions resists many of the chemical solutions in the pulp and paper and chemical industries. Red brass pipe has a composition of 85 percent copper and 15 percent zinc alloy and has a greater allowable stress value than copper at temperatures up to 300°F (150°C). At 400°F (204°C) these stress values become equal at 3000 psi. Standard piping codes do not permit the use of brass or copper pipes at temperatures over 400°F. Brass and copper pipes are available in straight lengths of up to 12 ft (3.6 m) and should be joined with fittings of copper-base alloys. Both copper ASTM B42 and brass ASTM B43 pipe are available in only two grades: regular and extra strong. They are very similar to schedule 40 and 80 steel pipes, respectively. This material, when used in piping runs, must be supported at more frequent intervals than other piping (which is usually stronger). When soft copper tubing is arranged in parallel runs, a trough construction is used to ensure continuous support of the piping.

Copper Water Tubes Copper water tubes are used for waste and vent lines, as well as for heating services and general plumbing. They are available in three weights only: type K, type L, and type M. One should consult the manufacturer's catalog for weights and sizes. Types K and L are available as hard or soft tubes; type M is available as a hard tube only. Hard-tempered tubing is much stiffer than soft tubing and is used where extra rigidity is necessary. Soft tubing is used in situations that require frequent bending without the use of fittings. Copper tubing is available in lengths up to 20 ft (6.09 m) or, in soft tubing, in coils up to 60 ft (18.28 m) long. The ASTM B88 and B251 specifications give the dimensions for copper tubing. The actual diameter is consistently 0.1258 in. (0.3195 cm) greater than the normal size in copper water tubing, which is available from 1/8 in. to 12 in. (0.31 to 30.4 cm) in diameter. The wall thickness is considerably less than that of other piping materials. When specifying the size of tubing, one must specify wall thickness and outside diameter.

Flexible or soft tubing is used in many situations where equipment vibration is present. Tubing whose thickness corresponds to standard steel pipe is referred to as *pipe* instead of tubing and should be avoided in systems that experience severe vibration, high temperatures, and excessive stress.

Seamless brass pipe is used for water condenser tubes and heating lines, but it is expensive. Copper tubing is used in plumbing and heating where vibration and misalignment are factors and in automotive, hydraulic, and pneumatic design. Copper and brass are also excellent for handling liquids containing salts, although because of the extra expense they are used only in special situations. If a copper water system is to be connected to a steel system, special insulating fittings must be used; otherwise, electrolytic corrosion will quickly destroy the fittings.

Lead Lead pipe or lead-lined pipe resists the chemical action of many acids. Therefore this type of pipe is found in chemical work and in systems that transport acids.

Aluminum Aluminum, with a weight one-third that of steel, is used throughout the piping field for industrial applications. Since temperature affects its strength, this material's design capabilities must be thoroughly examined before it is used in high-pressure and high-temperature situations. Many alloys of aluminum are used in piping and for pipe fittings. The manufacturer's catalog of specifications, weights, and alloy composition should be consulted if aluminum piping is considered for use—there are vast differences in manufacturing details throughout the industry.

Alloys In highly corrosive service involving high fluid velocity, shock (thermal and mechanical), and high temperatures and pressures, the use of alloy piping has increased in recent years. Certain alloys are able to resist corrosion even at high temperatures, which generally increase a fluid's chemical activity. It is always important to consider the solution, its chemical concentration, and its velocity when deciding on piping material. In many cases the alloy's resistance to chemical compositions stems from the formation of a surface film which, when the velocity of the fluid is high, may prove unsatisfactory because the fluid will scour the protective film. Thermal shock and rapid fluctuation of temperature also affect this protective coat.

Stainless Steel A common alloy used throughout the nuclear field and in other special situations is stainless steel—a class of alloy highly resistant to corrosion primarily because of the 10 percent or more of chromium added to the steel. This pipe material is extremely resistant to oxidizing solutions. The ability of stainless steel to resist corrosion is directly proportional to the chromium content. The American Iron and Steel Institute codes divide stainless steels into chromium nickel types (the AISI 300 series) and straight chromium types (the AISI 400 series).

Titanium One of the more exotic materials used in special services is titanium, which is found to be of great help in the chemical and pulp industries. There are a number of alloys containing aluminum and also chromium magnesium, which, when added to titanium, change the material's properties for specific applications.

Nonmetallic Pipe and Tubing

Nonmetallic pipe and tubing are used predominantly for low-pressure and low-temperature services. The major types include plastic, glass, clay, and wood pipe.

Plastic Plastic pipe is one of the most commonly used nonmetallic materials. Plastics offer considerable savings in initial expense and cost of support systems and have proved to be resistant to many corrosive chemicals. It is important to consult the manufacturer's catalog when utilizing plastic piping because of the varying strengths and resistances of the many different types on the market today. Fluorocarbon plastics are considered the most resistant to attack by acids, alkalies, and organic compounds. Many organic compounds, such as petroleum products and chlorinated hydrocarbons, are readily handled by some, but not all, of the plastic materials. Plastics can also be used as liners in high-pressure and low-temperature service in steel pipe. Synthetic rubbers are also used to line steel pipe. Major drawbacks of plastic piping are that it deteriorates in direct sunlight and cannot be used for high-temperature service. It is important to weigh the initial cost and corrosion resistance of a piping system against the eventual cost of replacement. Because of the industry's new exotic materials, many plastics and other molten products provide excellent strength and rigidity compared with earlier materials. Polyvinyl chloride (PVC) piping was first produced in Germany and since then has become one of the most widely used forms of plastic piping. It is used predominantly for intermediate-strength and high-flexibility applications, for water services, or for waste piping where temperatures and pressures are not extreme. This type of piping is considered very tough and exceptionally resistant to chemical attack. PVC pipe is manufactured by extrusion; the fittings and flanges are manufactured by injection molding. Plastic pipe, in general, is available in iron pipe sizes. The manufacturer's catalog should be consulted when ordering.

Glass Glass resists many acidic materials and can be found throughout the chemical industry, although it has a 400°F temperature limitation in most instances and is vulnerable to pressure and vibration. This type of piping is used predominantly in the chemical, beverage, pharmaceutical, and food industries where corrosive contamination of the line contents is undesirable. Glass piping can also be found in manufacturing plants such as paper mills and textile plants.

Clay Clay piping has been used for sewage, industrial waste, and water storage in diameters from 4 to 36 in. (10.1 to 91.4 cm). Clay pipe is manufactured predominantly from clays and shales in combination. High-temperature firing of the piping clay produces a strong, chemical-proof composition that has been used with exceptional success throughout the industrial waste and sewage field. It can carry away every known chemical waste with the exception of hydrochloric acid, which is not discharged through normal sewage channels.

Wood Wood pipe has been used throughout history for the transportation of water. It is manufactured by boring holes through solid logs and is used predominantly for transportation of municipal water supplies, sewers, mining, irrigation, and hydroelectric power developments. It is found mainly in the western states where timber is plentiful. Douglas fir and redwood are the most common types of wood pipe, but excessive logging has made redwood extremely rare and expensive.

Continuous-stave wood piping is made in diameters as large as 20 ft (6.09 m) and is shipped unassembled and then erected at the site. The wood is chemically prepared with creosote to ensure against chemical attack and fungus. Continuous-stave wood piping is generally made of Douglas fir bound by steel or iron bands. Certainly wood piping has its positive aspects: It acts as its own insulation; it is not affected by corrosion; and it is affected only slightly by vibration. Wire-wound wood pipe can be purchased in sizes up to 24 in. (60.9 cm). This factory-made product is wrapped with wire instead of steel bands to retain the staves.

Wood pipe is often used in underground installations after it has been treated to withstand rotting. It is also possible to obtain wood-lined steel pipe for service where high pressures make wood pipe unsatisfactory but chemical resistance is needed. This type of pipe is used throughout the paper and pulp mill industry.

Pipe Sizes, Dimensions, and Schedule Numbers

Any tubular product that has sizes governed by the American Petroleum Institute Standards (APIS) is referred to as *pipe*. The pipe's dimensions are set by the American National Standards Institute. Wall thickness varies with schedule number, but the outside diameter remains constant for 14 in. (35.5 cm) diameter and over (Fig. 1-2). As the thickness changes, the inside diameter is altered. To call out a pipe, it is necessary to give both the schedule number and the weight/strength designation for both pipes and fittings. All pipes under 14 in. (35.5 cm) in diameter are designated by the nominal inside diameter and schedule number. Those over 14 in. (35.5 cm) are designated by the actual outside diameter and thickness of walls. In Fig. 1-2, pipe sizes from 1/8 in. to 12 in. (0.31 to 30.4 cm) are shown by the nominal inside diameter (which is different from the actual inside diameter). The inside diameter and the outside diameter will vary with the sched-

Fig. 1-2 Relationship of nominal size to inside diameter (ID), outside diameter (OD), and schedule number.

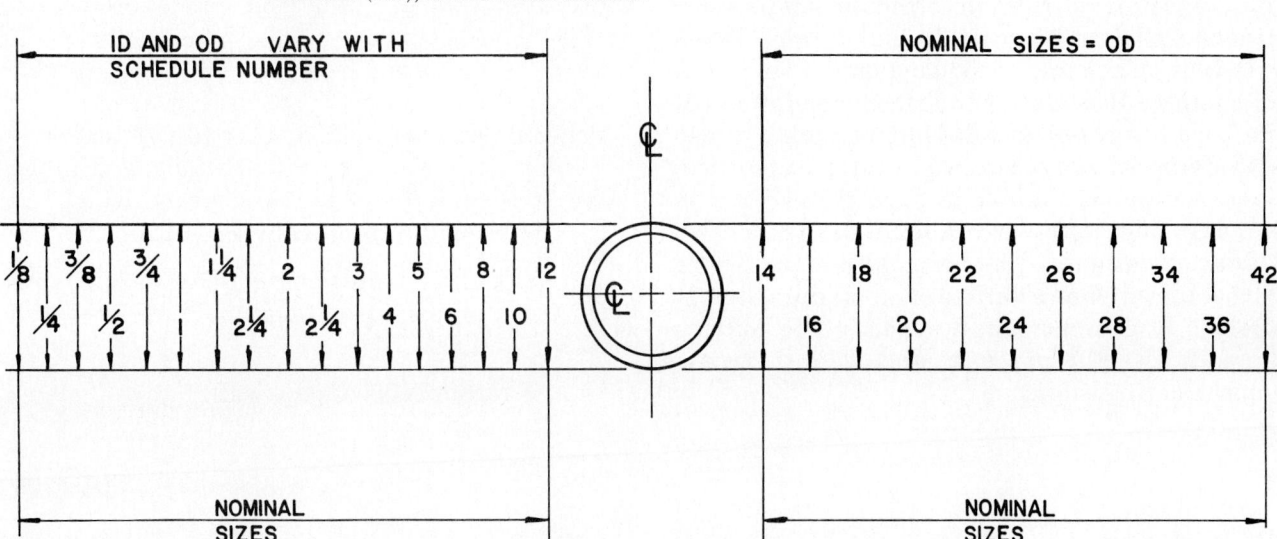

ule number from 1/8 in. to 12 in. (0.31 to 30.4 cm) diameters and the nominal size will equal the outside diameter from 14 to 42 in. (35.5 to 106.6 cm).

During the early years of pipe manufacturing, the walls of the smaller sizes were too thick and manufacturers, in correcting this error, took the excess from inside diameter, thus avoiding changes in fittings. To distinguish pipe sizes from actual measured diameters, the terms IPS (iron pipe size) or NPS (nominal pipe size) are usually used. The outside diameter will remain constant or relatively constant for all pipe sizes. Variations in wall thickness usually affect only the inside diameter. To distinguish weights for pipe, three traditional designations are used—standard wall, extra strong wall, and double extra strong wall (SW, XS, XXS). In some cases extra heavy wall (XH) and double extra heavy wall (XXH) are used instead of extra strong wall and double extra strong wall, respectively.

The American National Standards Institute, sponsored by the ASTM and the ASME, published ANSI B36.10 to standardize pipe dimensions throughout the industry. This publication also contains the broadened range of wall thicknesses available. Schedule numbers from 10 through 160 were adopted for steel pipe. These schedule numbers are indicative of approximate values of 1000 times the pressure/stress ratio. Stainless steel schedule numbers from schedule 5S through 80S are published in ANSI B36.19 for sizes up to 12 in. (30.4 cm). (The additional S pertains to stainless steel.) Refer to Table 1-1 for schedule number and pipe sizes for steel and stainless steel pipes.

The ASTM tolerances on regular pipe products specify that wall thickness should not vary more than 12½ percent under the nominal wall thickness that is specified for regular mill-rolled pipe. Pipe is usually supplied at random length from 16 to 22 ft (4.8 to 6.7 m), although it is possible to get lengths cut to order at a higher price. When specifying pipes and schedule numbers, the drafter should follow closely the recommendations of the design engineer. Besides piping, tubing is also manufactured and controlled by certain specifications. Any round tubular products that are not in standard pipe sizes are called *tubes* or *tubing*. Tubing is designated by an outside diameter for the basic size when a variety of inside diameters is offered. It is important to specify the outside diameter, weight per foot, and wall thickness when ordering tubing.

Calculation of Schedule Numbers

The ASA standard schedule numbers which specify wall thickness can be calculated by using the formula $1000 \times p/s$. Here p is equal to the internal pressure of the pipe (pressure in pounds per square inch) and s is the allowable fiber stress (in pounds per square inch). (Consult manufacturer's table or Table 1-1 for the s value.) As pressure increases, so does the pipe thickness requirement. The temperature of the line medium, besides putting thermal stress on the system, will also affect the pipe thickness. The s value takes into account temperature, pressure, and material.

At any given point along a pipe, the pressure tends to expand the pipe's cross section. This expansion also puts a lengthwise or longitudinal tension in the pipe. This stress can be visualized easily as two taut cords strung between fixed walls. If the two cords are wedged apart by means of a stick, the tension in the cords is increased. Here the cords are like the diametrically opposite walls of a pipe, which are wedged apart by the pressure in the pipe. This tension occurs at every point along the length of the pipe. A small piece of pipe wall is thus being stretched in all directions, but the pressure tends to produce a longitudinal burst in a pipe.

A single formula, with several algebraic variations, can be used to find the maximum pressure for a given pressure, wall thickness, and pipe diameter or to find the required wall thickness if pressure demand is known along with the material and pipe diameter. Usually the diameter is determined first, from flow requirements, but the formula could also be used to select the pipe material if all other factors were determined and the only quantity to vary was the tensile strength of the material. Here are the calculations:

$$t = \frac{p \times D}{2 \times s} = \frac{pD}{2s}$$

$$p = \frac{2 \times s \times t}{D} = \frac{2st}{D}$$

Schedule number = $1000 \times \frac{p}{s} = 1000 \frac{p}{s}$, where

p = pressure in pipe (psi)

D = inside diameter of pipe (in.)

s = unit tensile stress (allowable fiber stress in psi) (from manufacturer's table or Table 1-1)

t = thickness of pipe wall

TABLE 1-1 Allowable (S) values for seamless alloy, carbon steel, and wrought iron pipe (in 100 psi units)

	SEAMLESS ALLOY PIPE				CARBON STEEL & WROUGHT IRON PIPE												CARBON STEEL PIPE							
PIPE MATERIAL	FERRITIC			AUSTENITIC	WROUGHT IRON		CARBON STEEL		CARBON STEEL							CARBON STEEL						BLACK OR GALVANIZED CARBON STEEL		
					BUTT WELD	LAP WELD	ERW (RESISTANCE)	ERW	ELECTRIC FUSION			WELDED				SEAMLESS		SEAMLESS		ELECTRIC RESISTANCE WELDED		BUTT WELD	LAP WELD	
SEAMLESS/WELDED	S	S	S	S																					
ASTM Spec.	A335 A369	A335 A369	A335 A369	A312 A376	A72	A72	A135	A135	A155	A155	A155	A139	A139	A139		A106		A53		A53		A53	A53	A120	A120	A120
GRADE/TYPE	TYPE P5 & FP5	TYPE P2 FP2	TYPE P1 FP1	TYPE TP321			B	A	C55	C50	C45	B	A	A283D	A283C	A283B A283A	A134 A245C A245B A245A	B	A	B	A					
MIN. ULT. STRENGTH	60	55	55	75	40	40	60.0	60.0	55.0	50.0	45.0	60.0	48.0	60.0	55.0	50.0 45.0	55.0 52.0 48.0	60.0	48.0	60.0	48.0	45.0	45.0			
-20 to 100	15	13.75	13.75	18.75	6.0	8.0	12.75	10.02	12.4	11.25	10.1	12.0	9.6	10.1	10.1	9.2 8.3	10.1 9.6 8.8	15.0	12.0	15.0	12.0	6.75	9.0	10.8	6.5	8.8
200	15			18.75																				10.6	6.35	8.6
300	15			17.0																				10.2	6.1	8.2
400	15			15.8																				9.8	5.85	7.8
450	14.7			15.4																				9.6	5.7	7.6
500	14.5			15.2																						
600	14			14.4																						
650	13.7			14.85			12.75	10.02	12.4	11.25	10.1	12.0	9.6	10.1	10.1	9.2 8.3	10.1 9.6 8.8	15.0	12.0	12.75	10.2	6.75	9.0			
700	13.4	13.75	13.75	14.8			12.12	9.9	11.9	10.9	9.8	11.35	9.25					14.33	11.65	12.2	9.9					
750	13.1	13.75	13.75	14.7			11.0	9.1	10.85	9.9	8.9	9.95	8.3					12.95	10.7	11.0	9.1					
800	12.8	13.45	13.45	14.55					9.2	8.45	7.5							10.9	9.0							
850	12.4	13.15	13.15	14.3					7.0	6.55	5.95							7.8	7.1							
900	11.5	12.5	12.5	14.1														5.0	5.0							
950	10	10		13.85																						
1000	7.3	6.25		13.5																						
1050	5.2			13.1																						
1100	3.3			10.3																						
1150	2.2			7.6																						
1200	1.5			5																						

MAXIMUM TEMPERATURES

To find the thickness of the wall we take the pressure per square inch p times the diameter (ID) divided by twice the unit's tensile stress (s). Use the next greater t and schedule number after calculating. This formula enables us to find the required thickness of pipe for the design and therefore the schedule number for ordering.

Sizing Steam Piping

The following review of methods for sizing steam pipelines will give the reader experience in calculating pipe sizes and velocities for steam uses. (They are not to be substituted for manufacturers' charts for suggested product uses for particular services.) It is important to be able to deliver the required quantities of steam at a specified pressure to specific parts of the system. The size of the piping must be adequate for the condition but not oversized. Large pipes are expensive and require more installation, whereas a pipe too small produces excessive pressure drop and high velocities and does not deliver the required quantity of steam at the required pressure. It is impossible to set down definite rules for sizing steam piping; experience is the best guide. Both variation of inside diameter and condition of pipe influence the calculations for pipe sizing. The demands of the installation must be taken into account along with its performance and economy when determining pipe size for steam piping.

The following formulas will enable the designer and drafter to find the necessary pipe size for a service. The following variables are involved:

d = inside pipe diameter

S = superheat (in degrees Fahrenheit)

p = absolute pressure (in psi)

W = steam flow (in pounds per minute)

V = velocity

Pipe sizes for saturated steam:

$$d = \sqrt{\frac{80,000 W}{pV}}$$

Thomas formula for sizing well-insulated steam lines:

$$d = 4.5(1 + 0.00025S)\frac{W}{p}$$

Determining Pipe Sizes by Velocity

Before laying out piping systems, it is important to determine all pipe sizes that will allow for reasonable velocity, taking into account friction losses. As far as economics is concerned, keep the velocity high without exceeding the maximum to avoid excessive operating losses through pressure drops. Maximum allowable velocity corresponds to the permissible pressure drop from the point of consumption to the point of supply. The chart on steam design velocity for the flow of fluids in pipes is used for the determination of reasonable velocity drops in saturated and superheated steam lines. The higher limits are used for larger pipes and the lower limits for smaller ones. In extreme services, use calculated velocities by computing the actual pressure drop. Steam with a high moisture content has a greater erosive action on seats and valves and exposed parts and will also reduce the velocity. Another general rule is to use the velocity of 1000 to 1200 feet per minute (fpm) per inch of inside pipe diameter—1200 fpm for inside diameters over 12 in. and 1000 fpm for inside diameters below 12 in.

SATURATED STEAM DESIGN VELOCITIES FOR FLOW OF FLUIDS IN PIPE

Pressure (psig)	Use	Velocity (fpm)
0–15	Low pressures (heating)	4000–6000
50–150	Medium pressures	6000–10,000
200+	High pressures (turbines, boilers for power plants)	10,000–20,000

SUMMARY OF MISCELLANEOUS PIPE SPECIFICATIONS

API 5LX Covers high-test line pipe with higher tensile strengths than pipe made to API 5L. Seamless or welded, sizes 2⅜" OD through 48" OD, grades X-42 through X-65. Numerical designation in grade indicates minimum yield, thus X-42 = 42000 # min. yield steel, etc. OD, weight and wall tolerances, tensile and flattening tests, and marking requirements are similar to API 5L.

ASTM A135 Covers two grades (A and B) of electric-resistance-welded pipe, 30" and under, intended for conveying liquid, gas, or vapor. Only grade A is adapted for flanging and bending. Purpose for which pipe is intended should be stated in the order.

ASTM A252 Covers three grades (physical properties) of welded and seamless steel pipe piles of cylindrical shape either as permanent load-carrying members or shells to form cast-in-place concrete piles. Sizes 6″ OD through 24″ OD.

ASTM A312 Covers 15 grades (designated as TP 304H, TP 316H, TP 347H, etc.) of seamless and welded austenitic steel pipe intended for high-temperature and general corrosive service.

ASTM A333 Covers nominal seamless and welded carbon and alloy steel pipe for low-temperature service.

ASTM A335 Covers 12 grades of alloy seamless steel pipe intended for high-temperature service and suitable for bending, flanging (vanstoning), other forming operations, and fusion welding.

ASTM A501 Covers hot formed welded and seamless carbon steel square, round, rectangular, or special shape structural tubing for welded, riveted, or bolted construction of bridges and buildings, and for general structural purposes.

ASTM A589 Covers four specific types of threaded and coupled carbon steel pipe for use in water wells.

- Type I—Drive Pipe
- Type II—Water-Well Reamed and Drifted Pipe
- Type III—Driven Well Pipe
- Type IV—Water-Well Casing Pipe

ASTM A618 Covers Grade 1 equivalent to USS COR-TEN, Grade 2 equivalent to USS TRI-TEN, and Grade 3 equivalent to USS EX-TEN 50 hot formed welded and seamless high-strength low-alloy square, round, rectangular, or special shape structural tubing for welded, riveted, or bolted construction of bridges and buildings and for general structural purposes. Inquire for available sizes, ranges, and properties.

TABLE 1-2 ASTM carbon steel pipe and flange specifications (*Courtesy ITT Grinnell Corporation.*)

	Description and Applications	ASTM Spec No.	Grade or Type	Tensile Strength PSI	Yield Pt or Strength PSI	Elongation (% in 2″) Std Round	Rectangular t ⅛″	Rectangular ³⁄₁₆″	C	Mn	P	S
PIPE AND TUBING	Seamless milled steel PIPE for high-temperature service, suitable for bending, flanging and similar forming operations.	(1) A106	A	48,000	30,000	28 long. or (4) 20 trans.	17.5+56t or 12.5+40t	35 25	.25 max	.27 to .93	.048 max	.058 max
	As above, except use Grade A for close coiling, cold bending or forge welding.	(1) A106	B	60,000	35,000	22 long or (4) 12 trans.	15.0+48t or 6.5+32t	30 16.5	.30 max	.29 to 1.06	.048 max	.058 max
	Black or hot-dip galv. seamless or res.-welded steel PIPE suitable for coiling, bending, flanging, and other special purposes, suitable for welding.	A 53	A	48,000	30,000	28	17.5+56t	35	(2)	—	(3)	
	As above, except use Grade A for close coiling, cold bending or forge welding.	A 53	B	60,000	35,000	22	15.0+48t	30	(2)	—	(3)	
	Black or hot-dip galv. seamless or res.-welded steel PIPE for ordinary uses. (When tension, flattening or bend test required, order to A-53).	A120	—									
	Resistance welded steel PIPE for liquid, gas or vapor.	A135	A	48,000	30,000	—	17.5+56t	35			.050 max	.060 max
	As above, except use Grade A for flanging and bending.	A135	B	60,000	35,000		15.0+48t	30			.050 max	.060 max
	Electric-fusion-welded straight- or spiral-seam PIPE for liquid, gas or vapor from mill grades of plate.	A139	A	48,000	30,000	—	17.5+56t	35	—	.30 to 1.00	.040 max	.050 max
	As above.	A139	B	60,000	35,000	—	15.0+48t	30	.30 max	.30 to 1.00	.040 max	.050 max
FORGED PIPE, FLANGES,	Forged or rolled steel pipe flanges, fittings (6) valves and parts for high temperature service. Heat treatment required; may be annealed or normalized.	A105	I	60,000	30,000	25	—	—	.35 (5) max	.90 max	.05 max	.05 max
	As above.	A105	II	70,000	36,000	22	—	—	.35 (5) max	.90 max	.05 max	.05 max
	As above except for general service. Heat treatment is not required.	A181	I	60,000	30,000	22	—	—	.35 (5) max	.90 max	.05 max	.05 max
	As above.	A181	II	70,000	36,000	18	—	—	.35 (5) max	.90 max	.05 max	.05 max

(1) .10% silicon minimum.
(2) Open hearth, .13 max for ⅛″ and ¼″ size resistance welded pipe only.
(3) Seamless: open hearth .048 max, acid bessemer .11 max; Res.-welded: open hearth .050 max.
(4) Longitudinal or transverse direction of test specimen with respect to pipe axis.
(5) When flanges will be subject to fusion welding, the carbon content shall not exceed .35%. When carbon is restricted to .35% max, it may be necessary to add silicon to meet required tensile properties. The silicon content shall not exceed .35%.
(6) Factory-made Wrought Carbon Steel and Ferritic Alloy Steel Welding Fitting Specifications are covered under ASTM A234.

PIPE DATA

TABLE 1-3 Schedule numbers and pipe sizes for steel and stainless steel pipe
(Courtesy ITT Grinnell Corporation.)

The following formulas are used in the computation of the values shown in the table:

† weight of pipe per foot (pounds) $= 10.6802t(D-t)$
weight of water per foot (pounds) $= 0.3405d^2$
square feet outside surface per foot $= 0.2618D$
square feet inside surface per foot $= 0.2618d$
inside area (square inches) $= 0.785d^2$
area of metal (square inches) $= 0.785(D^2-d^2)$
moment of inertia (inches^4) $= 0.0491(D^4-d^4)$
$= A_m R_g^2$
section modulus (inches^3) $= \dfrac{0.0982(D^4-d^4)}{D}$
radius of gyration (inches) $= 0.25\sqrt{D^2+d^2}$

$A_m =$ area of metal (square inches)
$d =$ inside diameter (inches)
$D =$ outside diameter (inches)
$R_g =$ radius of gyration (inches)
$t =$ pipe wall thickness (inches)

† The ferritic steels may be about 5% less, and the austenitic stainless steels about 2% greater than the values shown in this table which are based on weights for carbon steel.

* schedule numbers

Standard weight pipe and schedule 40 are the same in all sizes through 10-inch; from 12-inch through 24-inch, standard weight pipe has a wall thickness of ⅜-inch.

Extra strong weight pipe and schedule 80 are the same in all sizes through 8-inch; from 8-inch through 24-inch, extra strong weight pipe has a wall thickness of ½-inch.

Double extra strong weight pipe has no corresponding schedule number.

a: ANSI B36.10 steel pipe schedule numbers
b: ANSI B36.10 steel pipe nominal wall thickness designation
c: ANSI B36.19 stainless steel pipe schedule numbers

nominal pipe size *outside diameter*, in.	schedule number* a	b	c	wall thickness, in.	inside diameter, in.	inside area, sq. in.	metal area, sq. in.	sq ft outside surface, per ft	sq ft inside surface, per ft	weight per ft, lb†	weight of water per ft, lb	moment of inertia, in.⁴	section modulus, in.³	radius gyration, in.
⅛ 0.405			10S	0.049	0.307	0.0740	0.0548	0.106	0.0804	0.186	0.0321	0.00088	0.00437	0.1271
	40	Std	40S	0.068	0.269	0.0568	0.0720	0.106	0.0705	0.245	0.0246	0.00106	0.00525	0.1215
	80	XS	80S	0.095	0.215	0.0364	0.0925	0.106	0.0563	0.315	0.0157	0.00122	0.00600	0.1146
¼ 0.540			10S	0.065	0.410	0.1320	0.0970	0.141	0.1073	0.330	0.0572	0.00279	0.01032	0.1694
	40	Std	40S	0.088	0.364	0.1041	0.1250	0.141	0.0955	0.425	0.0451	0.00331	0.01230	0.1628
	80	XS	80S	0.119	0.302	0.0716	0.1574	0.141	0.0794	0.535	0.0310	0.00378	0.01395	0.1547
⅜ 0.675			5S	0.065	0.710	0.396	0.1582	0.220	0.1859	0.538	0.1716	0.01197	0.0285	0.2750
			10S	0.065	0.545	0.2333	0.1246	0.177	0.1427	0.423	0.1011	0.00586	0.01737	0.2169
	40	Std	40S	0.091	0.493	0.1910	0.1670	0.177	0.1295	0.568	0.0827	0.00730	0.02160	0.2090
	80	XS	80S	0.126	0.423	0.1405	0.2173	0.177	0.1106	0.739	0.0609	0.00862	0.02554	0.1991
½ 0.840			5S	0.065	0.710	0.3959	0.1583	0.220	0.1859	0.538	0.171	0.0120	0.0285	0.2750
			10S	0.083	0.674	0.357	0.1974	0.220	0.1765	0.671	0.1547	0.01431	0.0341	0.2692
	40	Std	40S	0.109	0.622	0.304	0.2503	0.220	0.1628	0.851	0.1316	0.01710	0.0407	0.2613
	80	XS	80S	0.147	0.546	0.2340	0.320	0.220	0.1433	1.088	0.1013	0.02010	0.0478	0.2505
	160			0.187	0.466	0.1706	0.383	0.220	0.1220	1.304	0.0740	0.02213	0.0527	0.2402
		XXS		0.294	0.252	0.0499	0.504	0.220	0.0660	1.714	0.0216	0.02425	0.0577	0.2192
¾ 1.050			5S	0.065	0.920	0.665	0.2011	0.275	0.2409	0.684	0.2882	0.02451	0.0467	0.349
			10S	0.083	0.884	0.614	0.2521	0.275	0.2314	0.857	0.2661	0.02970	0.0566	0.343
	40	Std	40S	0.113	0.824	0.533	0.333	0.275	0.2157	1.131	0.2301	0.0370	0.0706	0.334
	80	XS	80S	0.154	0.742	0.432	0.435	0.275	0.1943	1.474	0.1875	0.0448	0.0853	0.321
	160			0.218	0.614	0.2961	0.570	0.275	0.1607	1.937	0.1284	0.0527	0.1004	0.304
		XXS		0.308	0.434	0.1479	0.718	0.275	0.1137	2.441	0.0641	0.0579	0.1104	0.2840
1 1.315			5S	0.065	1.185	1.103	0.2553	0.344	0.310	0.868	0.478	0.0500	0.0760	0.443
			10S	0.109	1.097	0.945	0.413	0.344	0.2872	1.404	0.409	0.0757	0.1151	0.428
	40	Std	40S	0.133	1.049	0.864	0.494	0.344	0.2746	1.679	0.374	0.0874	0.1329	0.421
	80	XS	80S	0.179	0.957	0.719	0.639	0.344	0.2520	2.172	0.311	0.1056	0.1606	0.407
	160			0.250	0.815	0.522	0.836	0.344	0.2134	2.844	0.2261	0.1252	0.1903	0.387
		XXS		0.358	0.599	0.2818	1.076	0.344	0.1570	3.659	0.1221	0.1405	0.2137	0.361
1¼ 1.660			5S	0.065	1.530	1.839	0.326	0.434	0.401	1.107	0.797	0.1038	0.1250	0.564
			10S	0.109	1.442	1.633	0.531	0.434	0.378	1.805	0.707	0.1605	0.1934	0.550
	40	Std	40S	0.140	1.380	1.496	0.669	0.434	0.361	2.273	0.648	0.1948	0.2346	0.540
	80	XS	80S	0.191	1.278	1.283	0.881	0.434	0.335	2.997	0.555	0.2418	0.2913	0.524
	160			0.250	1.160	1.057	1.107	0.434	0.304	3.765	0.458	0.2839	0.342	0.506
		XXS		0.382	0.896	0.631	1.534	0.434	0.2346	5.214	0.2732	0.341	0.411	0.472
1½ 1.900			5S	0.065	1.770	2.461	0.375	0.497	0.463	1.274	1.067	0.1580	0.1663	0.649
			10S	0.109	1.682	2.222	0.613	0.497	0.440	2.085	0.962	0.2469	0.2599	0.634

TABLE 1-3 (contd.)

nominal pipe size outside diameter, in.	schedule number* a	b	c	wall thickness, in.	inside diameter, in.	inside area, sq. in.	metal area, sq. in.	sq ft outside surface, per ft	sq ft inside surface, per ft	weight per ft, lb†	weight of water per ft, lb	moment of inertia, in.^{4}	section modulus, in.^{3}	radius gyration, in.
1½ 1.900	40	Std	40S	0.145	1.610	2.036	0.799	0.497	0.421	2.718	0.882	0.310	0.326	0.623
	80	XS	80S	0.200	1.500	1.767	1.068	0.497	0.393	3.631	0.765	0.391	0.412	0.605
	160	......	......	0.281	1.338	1.406	1.429	0.497	0.350	4.859	0.608	0.483	0.508	0.581
	......	XXS	......	0.400	1.100	0.950	1.885	0.497	0.288	6.408	0.412	0.568	0.598	0.549
	......	......	......	0.525	0.850	0.567	2.267	0.497	0.223	7.710	0.246	0.6140	0.6470	0.5200
	......	......	......	0.650	0.600	0.283	2.551	0.497	0.157	8.678	0.123	0.6340	0.6670	0.4980
2 2.375	......	......	5S	0.065	2.245	3.96	0.472	0.622	0.588	1.604	1.716	0.315	0.2652	0.817
	......	......	10S	0.109	2.157	3.65	0.776	0.622	0.565	2.638	1.582	0.499	0.420	0.802
	40	Std	40S	0.154	2.067	3.36	1.075	0.622	0.541	3.653	1.455	0.666	0.561	0.787
	80	XS	80S	0.218	1.939	2.953	1.477	0.622	0.508	5.022	1.280	0.868	0.731	0.766
	160	......	......	0.343	1.689	2.240	2.190	0.622	0.442	7.444	0.971	1.163	0.979	0.729
	......	XXS	......	0.436	1.503	1.774	2.656	0.622	0.393	9.029	0.769	1.312	1.104	0.703
	......	......	......	0.562	1.251	1.229	3.199	0.622	0.328	10.882	0.533	1.442	1.2140	0.6710
	......	......	......	0.687	1.001	0.787	3.641	0.622	0.262	12.385	0.341	1.5130	1.2740	0.6440
2½ 2.875	......	......	5S	0.083	2.709	5.76	0.728	0.753	0.709	2.475	2.499	0.710	0.494	0.988
	......	......	10S	0.120	2.635	5.45	1.039	0.753	0.690	3.531	2.361	0.988	0.687	0.975
	40	Std	40S	0.203	2.469	4.79	1.704	0.753	0.646	5.793	2.076	1.530	1.064	0.947
	80	XS	80S	0.276	2.323	4.24	2.254	0.753	0.608	7.661	1.837	1.925	1.339	0.924
	160	......	......	0.375	2.125	3.55	2.945	0.753	0.556	10.01	1.535	2.353	1.637	0.894
	......	XXS	......	0.552	1.771	2.464	4.03	0.753	0.464	13.70	1.067	2.872	1.998	0.844
	......	......	......	0.675	1.525	1.826	4.663	0.753	0.399	15.860	0.792	3.0890	2.1490	0.8140
	......	......	......	0.800	1.275	1.276	5.212	0.753	0.334	17.729	0.554	3.2250	2.2430	0.7860
3 3.500	......	......	5S	0.083	3.334	8.73	0.891	0.916	0.873	3.03	3.78	1.301	0.744	1.208
	......	......	10S	0.120	3.260	8.35	1.274	0.916	0.853	4.33	3.61	1.822	1.041	1.196
	40	Std	40S	0.216	3.068	7.39	2.228	0.916	0.803	7.58	3.20	3.02	1.724	1.164
	80	XS	80S	0.300	2.900	6.61	3.02	0.916	0.759	10.25	2.864	3.90	2.226	1.136
	160	......	......	0.437	2.626	5.42	4.21	0.916	0.687	14.32	2.348	5.03	2.876	1.094
	......	XXS	......	0.600	2.300	4.15	5.47	0.916	0.602	18.58	1.801	5.99	3.43	1.047
	......	......	......	0.725	2.050	3.299	6.317	0.916	0.537	21.487	1.431	6.5010	3.7150	1.0140
	......	......	......	0.850	1.800	2.543	7.073	0.916	0.471	24.057	1.103	6.8530	3.9160	0.9840
3½ 4.000	......	......	5S	0.083	3.834	11.55	1.021	1.047	1.004	3.47	5.01	1.960	0.980	1.385
	......	......	10S	0.120	3.760	11.10	1.463	1.047	0.984	4.97	4.81	2.756	1.378	1.372
	40	Std	40S	0.226	3.548	9.89	2.680	1.047	0.929	9.11	4.28	4.79	2.394	1.337
	80	XS	80S	0.318	3.364	8.89	3.68	1.047	0.881	12.51	3.85	6.28	3.14	1.307
	......	XXS	......	0.636	2.728	5.845	6.721	1.047	0.716	22.850	2.530	9.8480	4.9240	1.2100
4 4.500	......	......	5S	0.083	4.334	14.75	1.152	1.178	1.135	3.92	6.40	2.811	1.249	1.562
	......	......	10S	0.120	4.260	14.25	1.651	1.178	1.115	5.61	6.17	3.96	1.762	1.549
	......	......	......	0.188	4.124	13.357	2.547	1.178	1.082	8.560	5.800	5.8500	2.6000	1.5250
	40	Std	40S	0.237	4.026	12.73	3.17	1.178	1.054	10.79	5.51	7.23	3.21	1.510
	80	XS	80S	0.337	3.826	11.50	4.41	1.178	1.002	14.98	4.98	9.61	4.27	1.477
	120	......	......	0.437	3.626	10.33	5.58	1.178	0.949	18.96	4.48	11.65	5.18	1.445
	......	......	......	0.500	3.500	9.621	6.283	1.178	0.916	21.360	4.160	12.7710	5.6760	1.4250
	160	......	......	0.531	3.438	9.28	6.62	1.178	0.900	22.51	4.02	13.27	5.90	1.416
	......	XXS	......	0.674	3.152	7.80	8.10	1.178	0.825	27.54	3.38	15.29	6.79	1.374
	......	......	......	0.800	2.900	6.602	9.294	1.178	0.759	31.613	2.864	16.6610	7.4050	1.3380
	......	......	......	0.925	2.650	5.513	10.384	1.178	0.694	35.318	2.391	17.7130	7.8720	1.3060
5 5.563	......	......	5S	0.109	5.345	22.44	1.868	1.456	1.399	6.35	9.73	6.95	2.498	1.929
	......	......	10S	0.134	5.295	22.02	2.285	1.456	1.386	7.77	9.53	8.43	3.03	1.920
	40	Std	40S	0.258	5.047	20.01	4.30	1.456	1.321	14.62	8.66	15.17	5.45	1.878
	80	XS	80S	0.375	4.813	18.19	6.11	1.456	1.260	20.78	7.89	20.68	7.43	1.839
	120	......	......	0.500	4.563	16.35	7.95	1.456	1.195	27.04	7.09	25.74	9.25	1.799
	160	......	......	0.625	4.313	14.61	9.70	1.456	1.129	32.96	6.33	30.0	10.80	1.760
	......	XXS	......	0.750	4.063	12.97	11.34	1.456	1.064	38.55	5.62	33.6	12.10	1.722
	......	......	......	0.875	3.813	11.413	12.880	1.456	0.998	43.810	4.951	36.6450	13.1750	1.6860
	......	......	......	1.000	3.563	9.966	14.328	1.456	0.933	47.734	4.232	39.1110	14.0610	1.6520

PIPE DATA

TABLE 1-3 (contd.)

nominal pipe size outside diameter, in.	schedule number* a	b	c	wall thickness, in.	inside diameter, in.	inside area, sq. in.	metal area, sq. in.	sq ft outside surface, per ft	sq ft inside surface, per ft	weight per ft, lb†	weight of water per ft, lb	moment of inertia, in.⁴	section modulus, in.³	radius gyration, in.
6 6.625			5S	0.109	6.407	32.2	2.231	1.734	1.677	5.37	13.98	11.85	3.58	2.304
			10S	0.134	6.357	31.7	2.733	1.734	1.664	9.29	13.74	14.40	4.35	2.295
				0.219	6.187	30.100	4.410	1.734	1.620	15.020	13.100	22.6600	6.8400	2.2700
	40	Std	40S	0.280	6.065	28.89	5.58	1.734	1.588	18.97	12.51	28.14	8.50	2.245
	80	XS	80S	0.432	5.761	26.07	8.40	1.734	1.508	28.57	11.29	40.5	12.23	2.195
	120			0.562	5.501	23.77	10.70	1.734	1.440	36.39	10.30	49.6	14.98	2.153
	160			0.718	5.189	21.15	13.33	1.734	1.358	45.30	9.16	59.0	17.81	2.104
		XXS		0.864	4.897	18.83	15.64	1.734	1.282	53.16	8.17	66.3	20.03	2.060
				1.000	4.625	16.792	17.662	1.734	1.211	60.076	7.284	72.1190	21.7720	2.0200
				1.125	4.375	15.025	19.429	1.734	1.145	66.084	6.517	76.5970	23.1240	1.9850
8 8.625			5S	0.109	8.407	55.5	2.916	2.258	2.201	9.91	24.07	26.45	6.13	3.01
			10S	0.148	8.329	54.5	3.94	2.258	2.180	13.40	23.59	35.4	8.21	3.00
				0.219	8.187	52.630	5.800	2.258	2.150	19.640	22.900	51.3200	11.9000	2.9700
	20			0.250	8.125	51.8	6.58	2.258	2.127	22.36	22.48	57.7	13.39	2.962
	30			0.277	8.071	51.2	7.26	2.258	2.113	24.70	22.18	63.4	14.69	2.953
	40	Std	40S	0.322	7.981	50.0	8.40	2.258	2.089	28.55	21.69	72.5	16.81	2.938
	60			0.406	7.813	47.9	10.48	2.258	2.045	35.64	20.79	88.8	20.58	2.909
	80	XS	80S	0.500	7.625	45.7	12.76	2.258	1.996	43.39	19.80	105.7	24.52	2.878
8 8.625	100			0.593	7.439	43.5	14.96	2.258	1.948	50.87	18.84	121.4	28.14	2.847
	120			0.718	7.189	40.6	17.84	2.258	1.882	60.63	17.60	140.6	32.6	2.807
	140			0.812	7.001	38.5	19.93	2.258	1.833	67.76	16.69	153.8	35.7	2.777
	160			0.906	6.813	36.5	21.97	2.258	1.784	74.69	15.80	165.9	38.5	2.748
				1.000	6.625	34.454	23.942	2.258	1.734	81.437	14.945	177.1320	41.0740	2.7190
				1.125	6.375	31.903	26.494	2.258	1.669	90.114	13.838	190.6210	44.2020	2.6810
10 10.750			5S	0.134	10.482	86.3	4.52	2.815	2.744	15.15	37.4	63.7	11.85	3.75
			10S	0.165	10.420	85.3	5.49	2.815	2.728	18.70	36.9	76.9	14.30	3.74
				0.219	10.312	83.52	7.24	2.815	2.70	24.63	36.2	100.46	18.69	3.72
	20			0.250	10.250	82.5	8.26	2.815	2.683	28.04	35.8	113.7	21.16	3.71
	30			0.307	10.136	80.7	10.07	2.815	2.654	34.24	35.0	137.5	25.57	3.69
	40	Std	40S	0.365	10.020	78.9	11.91	2.815	2.623	40.48	34.1	160.8	29.90	3.67
	60	XS	80S	0.500	9.750	74.7	16.10	2.815	2.553	54.74	32.3	212.0	39.4	3.63
	80			0.593	9.564	71.8	18.92	2.815	2.504	64.33	31.1	244.9	45.6	3.60
	100			0.718	9.314	68.1	22.63	2.815	2.438	76.93	29.5	286.2	53.2	3.56
	120			0.843	9.064	64.5	26.24	2.815	2.373	89.20	28.0	324	60.3	3.52
				0.875	9.000	63.62	27.14	2.815	2.36	92.28	27.6	333.46	62.04	3.50
	140			1.000	8.750	60.1	30.6	2.815	2.291	104.13	26.1	368	68.4	3.47
	160			1.125	8.500	56.7	34.0	2.815	2.225	115.65	24.6	399	74.3	3.43
				1.250	8.250	53.45	37.31	2.815	2.16	126.82	23.2	428.17	79.66	3.39
				1.500	7.750	47.15	43.57	2.815	2.03	148.19	20.5	478.59	89.04	3.31
12 12.750			5S	0.156	12.438	121.4	6.17	3.34	3.26	20.99	52.7	122.2	19.20	4.45
			10S	0.180	12.390	120.6	7.11	3.34	3.24	24.20	52.2	140.5	22.03	4.44
	20			0.250	12.250	117.9	9.84	3.34	3.21	33.38	51.1	191.9	30.1	4.42
	30			0.330	12.090	114.8	12.88	3.34	3.17	43.77	49.7	248.5	39.0	4.39
		Std	40S	0.375	12.000	113.1	14.58	3.34	3.14	49.56	49.0	279.3	43.8	4.38
	40			0.406	11.938	111.9	15.74	3.34	3.13	53.53	48.5	300	47.1	4.37
		XS	80S	0.500	11.750	108.4	19.24	3.34	3.08	65.42	47.0	362	56.7	4.33
	60			0.562	11.626	106.2	21.52	3.34	3.04	73.16	46.0	401	62.8	4.31
	80			0.687	11.376	101.6	26.04	3.34	2.978	88.51	44.0	475	74.5	4.27
				0.750	11.250	99.40	28.27	3.34	2.94	96.2	43.1	510.7	80.1	4.25
	100			0.843	11.064	96.1	31.5	3.34	2.897	107.20	41.6	562	88.1	4.22
				0.875	11.000	95.00	32.64	3.34	2.88	110.9	41.1	578.5	90.7	4.21
	120			1.000	10.750	90.8	36.9	3.34	2.814	125.49	39.3	642	100.7	4.17
	140			1.125	10.500	86.6	41.1	3.34	2.749	139.68	37.5	701	109.9	4.13
				1.250	10.250	82.50	45.16	3.34	2.68	153.6	35.8	755.5	118.5	4.09
	160			1.312	10.126	80.5	47.1	3.34	2.651	160.27	34.9	781	122.6	4.07

TABLE 1-3 (contd.)

nominal pipe size outside diameter, in.	schedule number* a	schedule number* b	schedule number* c	wall thickness, in.	inside diameter, in.	inside area, sq. in.	metal area, sq. in.	sq ft outside surface, per ft	sq ft inside surface, per ft	weight per ft, lb†	weight of water per ft, lb	moment of inertia, in.⁴	section modulus, in.³	radius gyration, in.
14 14.000			5S	0.156	13.688	147.20	6.78	3.67	3.58	23.0	63.7	162.6	23.2	4.90
			10S	0.188	13.624	145.80	8.16	3.67	3.57	27.7	63.1	194.6	27.8	4.88
				0.210	13.580	144.80	9.10	3.67	3.55	30.9	62.8	216.2	30.9	4.87
				0.219	13.562	144.50	9.48	3.67	3.55	32.2	62.6	225.1	32.2	4.87
	10			0.250	13.500	143.1	10.80	3.67	3.53	36.71	62.1	255.4	36.5	4.86
				0.281	13.438	141.80	12.11	3.67	3.52	41.2	61.5	285.2	40.7	4.85
	20			0.312	13.376	140.5	13.42	3.67	3.50	45.68	60.9	314	44.9	4.84
				0.344	13.312	139.20	14.76	3.67	3.48	50.2	60.3	344.3	49.2	4.83
	30	Std		0.375	13.250	137.9	16.05	3.67	3.47	54.57	59.7	373	53.3	4.82
	40			0.437	13.126	135.3	18.62	3.67	3.44	63.37	58.7	429	61.2	4.80
				0.469	13.062	134.00	19.94	3.67	3.42	67.8	58.0	456.8	65.3	4.79
		XS		0.500	13.000	132.7	21.21	3.67	3.40	72.09	57.5	484	69.1	4.78
	60			0.593	12.814	129.0	24.98	3.67	3.35	84.91	55.9	562	80.3	4.74
				0.625	12.750	127.7	26.26	3.67	3.34	89.28	55.3	589	84.1	4.73
	80			0.750	12.500	122.7	31.2	3.67	3.27	106.13	53.2	687	98.2	4.69
	100			0.937	12.126	115.5	38.5	3.67	3.17	130.73	50.0	825	117.8	4.63
	120			1.093	11.814	109.6	44.3	3.67	3.09	150.67	47.5	930	132.8	4.58
	140			1.250	11.500	103.9	50.1	3.67	3.01	170.22	45.0	1127	146.8	4.53
	160			1.406	11.188	98.3	55.6	3.67	2.929	189.12	42.6	1017	159.6	4.48
16 16.000			5S	0.165	15.670	192.90	8.21	4.19	4.10	28	83.5	257	32.2	5.60
			10S	0.188	15.624	191.70	9.34	4.19	4.09	32	83.0	292	36.5	5.59
	10			0.250	15.500	188.7	12.37	4.19	4.06	42.05	81.8	384	48.0	5.57
	20			0.312	15.376	185.7	15.38	4.19	4.03	52.36	80.5	473	59.2	5.55
	30	Std		0.375	15.250	182.6	18.41	4.19	3.99	62.58	79.1	562	70.3	5.53
	40	XS		0.500	15.000	176.7	24.35	4.19	3.93	82.77	76.5	732	91.5	5.48
	60			0.656	14.688	169.4	31.6	4.19	3.85	107.50	73.4	933	116.6	5.43
	80			0.843	14.314	160.9	40.1	4.19	3.75	136.46	69.7	1157	144.6	5.37
	100			1.031	13.938	152.6	48.5	4.19	3.65	164.83	66.1	1365	170.6	5.30
	120			1.218	13.564	144.5	56.6	4.19	3.55	192.29	62.6	1556	194.5	5.24
	140			1.437	13.126	135.3	65.7	4.19	3.44	223.64	58.6	1760	220.0	5.17
	160			1.593	12.814	129.0	72.1	4.19	3.35	245.11	55.9	1894	236.7	5.12
18 18.000			5S	0.165	17.670	245.20	9.24	4.71	4.63	31	106.2	368	40.8	6.31
			10S	0.188	17.624	243.90	10.52	4.71	4.61	36	105.7	417	46.4	6.30
	10			0.250	17.500	240.5	13.94	4.71	4.58	47.39	104.3	549	61.0	6.28
	20			0.312	17.376	237.1	17.34	4.71	4.55	59.03	102.8	678	75.5	6.25
		Std		0.375	17.250	233.7	20.76	4.71	4.52	70.59	101.2	807	89.6	6.23
	30			0.437	17.126	230.4	24.11	4.71	4.48	82.06	99.9	931	103.4	6.21
		XS		0.500	17.00	227.0	27.49	4.71	4.45	93.45	98.4	1053	117.0	6.19
	40			0.562	16.876	223.7	30.8	4.71	4.42	104.75	97.0	1172	130.2	6.17
	60			0.750	16.500	213.8	40.6	4.71	4.32	138.17	92.7	1515	168.3	6.10
	80			0.937	16.126	204.2	50.2	4.71	4.22	170.75	88.5	1834	203.8	6.04
	100			1.156	15.688	193.3	61.2	4.71	4.11	207.96	83.7	2180	242.2	5.97
	120			1.375	15.250	182.6	71.8	4.71	3.99	244.14	79.2	2499	277.6	5.90
	140			1.562	14.876	173.8	80.7	4.71	3.89	274.23	75.3	2750	306	5.84
	160			1.781	14.438	163.7	90.7	4.71	3.78	308.51	71.0	3020	336	5.77
20 20.000			5S	0.188	19.634	302.40	11.70	5.24	5.14	40	131.0	574	57.4	7.00
			10S	0.218	19.564	300.60	13.55	5.24	5.12	46	130.2	663	66.3	6.99
	10			0.250	19.500	298.6	15.51	5.24	5.11	52.73	129.5	757	75.7	6.98
	20	Std		0.375	19.250	291.0	23.12	5.24	5.04	78.60	126.0	1114	111.4	6.94
	30	XS		0.500	19.000	283.5	30.6	5.24	4.97	104.13	122.8	1457	145.7	6.90
	40			0.593	18.814	278.0	36.2	5.24	4.93	122.91	120.4	1704	170.4	6.86
	60			0.812	18.376	265.2	48.9	5.24	4.81	166.40	115.0	2257	225.7	6.79
				0.875	18.250	261.6	52.6	5.24	4.78	178.73	113.4	2409	240.9	6.77
	80			1.031	17.938	252.7	61.4	5.24	4.70	208.87	109.4	2772	277.2	6.72
	100			1.281	17.438	238.8	75.3	5.24	4.57	256.10	103.4	3320	332	6.63

PIPE DATA

TABLE 1-3 (contd.)

nominal pipe size outside diameter, in.	schedule number* a	b	c	wall thickness, in.	inside diameter, in.	inside area, sq in.	metal area, sq in.	sq ft outside surface, per ft	sq ft inside surface, per ft	weight per ft, lb†	weight of water per ft, lb	moment of inertia, in.⁴	section modulus, in.³	radius gyration, in.
20 20.000	120			1.500	17.000	227.0	87.2	5.24	4.45	296.37	98.3	3760	376	6.56
	140			1.750	16.500	213.8	100.3	5.24	4.32	341.10	92.6	4220	422	6.48
	160			1.968	16.064	202.7	111.5	5.24	4.21	379.01	87.9	4590	459	6.41
22 22.000			5S	0.188	21.624	367.3	12.88	5.76	5.66	44	159.1	766	69.7	7.71
			10S	0.218	21.564	365.2	14.92	5.76	5.65	51	158.2	885	80.4	7.70
	10			0.250	21.500	363.1	17.18	5.76	5.63	58	157.4	1010	91.8	7.69
	20	Std		0.375	21.250	354.7	25.48	5.76	5.56	87	153.7	1490	135.4	7.65
	30	XS		0.500	21.000	346.4	33.77	5.76	5.50	115	150.2	1953	177.5	7.61
				0.625	20.750	338.2	41.97	5.76	5.43	143	146.6	2400	218.2	7.56
				0.750	20.500	330.1	50.07	5.76	5.37	170	143.1	2829	257.2	7.52
	60			0.875	20.250	322.1	58.07	5.76	5.30	197	139.6	3245	295.0	7.47
	80			1.125	19.750	306.4	73.78	5.76	5.17	251	132.8	4029	366.3	7.39
	100			1.375	19.250	291.0	89.09	5.76	5.04	303	126.2	4758	432.6	7.31
	120			1.625	18.750	276.1	104.02	5.76	4.91	354	119.6	5432	493.8	7.23
	140			1.875	18.250	261.6	118.55	5.76	4.78	403	113.3	6054	550.3	7.15
	160			2.125	17.750	247.4	132.68	5.76	4.65	451	107.2	6626	602.4	7.07
24 24.000	10			0.250	23.500	434	18.65	6.28	6.15	63.41	188.0	1316	109.6	8.40
	20	Std		0.375	23.250	425	27.83	6.28	6.09	94.62	183.8	1943	161.9	8.35
		XS		0.500	23.000	415	36.9	6.28	6.02	125.49	180.1	2550	212.5	8.31
	30			0.562	22.876	411	41.4	6.28	5.99	140.80	178.1	2840	237.0	8.29
				0.625	22.750	406	45.9	6.28	5.96	156.03	176.2	3140	261.4	8.27
	40			0.687	22.626	402	50.3	6.28	5.92	171.17	174.3	3420	285.2	8.25
				0.750	22.500	398	54.8	6.28	5.89	186.24	172.4	3710	309	8.22
			5S	0.218	23.564	436.1	16.29	6.28	6.17	55	188.9	1152	96.0	8.41
				0.875	22.250	388.6	63.54	6.28	5.83	216	168.6	4256	354.7	8.18
	60			0.968	22.064	382	70.0	6.28	5.78	238.11	165.8	4650	388	8.15
	80			1.218	21.564	365	87.2	6.28	5.65	296.36	158.3	5670	473	8.07
	100			1.531	20.938	344	108.1	6.28	5.48	367.40	149.3	6850	571	7.96
	120			1.812	20.376	326	126.3	6.28	5.33	429.39	141.4	7830	652	7.87
	140			2.062	19.876	310	142.1	6.28	5.20	483.13	134.5	8630	719	7.79
	160			2.343	19.314	293	159.4	6.28	5.06	541.94	127.0	9460	788	7.70
26 26.000				0.250	25.500	510.7	19.85	6.81	6.68	67	221.4	1646	126.6	9.10
	10			0.312	25.376	505.8	25.18	6.81	6.64	86	219.2	2076	159.7	9.08
		Std		0.375	25.250	500.7	30.19	6.81	6.61	103	217.1	2478	190.6	9.06
	20	XS		0.500	25.000	490.9	40.06	6.81	6.54	136	212.8	3259	250.7	9.02
				0.625	24.750	481.1	49.82	6.81	6.48	169	208.6	4013	308.7	8.98
				0.750	24.500	471.4	59.49	6.81	6.41	202	204.4	4744	364.9	8.93
				0.875	24.250	461.9	69.07	6.81	6.35	235	200.2	5458	419.9	8.89
				1.000	24.000	452.4	78.54	6.81	6.28	267	196.1	6149	473.0	8.85
				1.125	23.750	443.0	87.91	6.81	6.22	299	192.1	6813	524.1	8.80
28 28.000				0.250	27.500	594.0	21.80	7.33	7.20	74	257.3	2098	149.8	9.81
	10			0.312	27.376	588.6	27.14	7.33	7.17	92	255.0	2601	185.8	9.79
		Std		0.375	27.250	583.2	32.54	7.33	7.13	111	252.6	3105	221.8	9.77
	20	XS		0.500	27.000	572.6	43.20	7.33	7.07	147	248.0	4085	291.8	9.72
	30			0.625	26.750	562.0	53.75	7.33	7.00	183	243.4	5038	359.8	9.68
				0.750	26.500	551.6	64.21	7.33	6.94	218	238.9	5964	426.0	9.64
				0.875	26.250	541.2	74.56	7.33	6.87	253	234.4	6865	490.3	9.60
				1.000	26.000	530.9	84.82	7.33	6.81	288	230.0	7740	552.8	9.55
				1.125	25.750	520.8	94.98	7.33	6.74	323	225.6	8590	613.6	9.51
30 30.000			5S	0.250	29.500	683.4	23.37	7.85	7.72	79	296.3	2585	172.3	10.52
	10		10S	0.312	29.376	677.8	29.19	7.85	7.69	99	293.7	3201	213.4	10.50
		Std		0.375	29.250	672.0	34.90	7.85	7.66	119	291.2	3823	254.8	10.48
	20	XS		0.500	29.000	660.5	46.34	7.85	7.59	158	286.2	5033	335.5	10.43
	30			0.625	28.750	649.2	57.68	7.85	7.53	196	281.3	6213	414.2	10.39

TABLE 1-3 (contd.)

nominal pipe size outside diameter, in.	schedule number* a	schedule number* b	schedule number* c	wall thickness, in.	inside diameter, in.	inside area, sq. in.	metal area, sq. in.	sq ft outside surface, per ft	sq ft inside surface, per ft	weight per ft, lb†	weight of water per ft lb	moment of inertia, in.⁴	section modulus, in.³	radius gyration, in.
30 30.000	40			0.750	28.500	637.9	68.92	7.85	7.46	234	276.6	7371	491.4	10.34
				0.875	28.250	620.7	80.06	7.85	7.39	272	271.8	8494	566.2	10.30
				1.000	28.000	615.7	91.11	7.85	7.33	310	267.0	9591	639.4	10.26
				1.125	27.750	604.7	102.05	7.85	7.26	347	262.2	10653	710.2	10.22
32 32.000				0.250	31.500	779.2	24.93	8.38	8.25	85	337.8	3141	196.3	11.22
	10			0.312	31.376	773.2	31.02	8.38	8.21	106	335.2	3891	243.2	11.20
		Std		0.375	31.250	766.9	37.25	8.38	8.18	127	332.5	4656	291.0	11.18
	20	XS		0.500	31.000	754.7	49.48	8.38	8.11	168	327.2	6140	383.8	11.14
	30			0.625	30.750	742.5	61.59	8.38	8.05	209	321.9	7578	473.6	11.09
	40			0.688	30.624	736.6	67.68	8.38	8.02	230	319.0	8298	518.6	11.07
				0.750	30.500	730.5	73.63	8.38	7.98	250	316.7	8990	561.9	11.05
				0.875	30.250	718.3	85.52	8.38	7.92	291	311.6	10372	648.2	11.01
				1.000	30.000	706.8	97.38	8.38	7.85	331	306.4	11680	730.0	10.95
				1.125	29.750	694.7	109.0	8.38	7.79	371	301.3	13023	814.0	10.92
34 34.000				0.250	33.500	881.2	26.50	8.90	8.77	90	382.0	3773	221.9	11.93
	10			0.312	33.376	874.9	32.99	8.90	8.74	112	379.3	4680	275.3	11.91
		Std		0.375	33.250	867.8	39.61	8.90	8.70	135	376.2	5597	329.2	11.89
	20	XS		0.500	33.000	855.3	52.62	8.90	8.64	179	370.8	7385	434.4	11.85
	30			0.625	32.750	841.9	65.53	8.90	8.57	223	365.0	9124	536.7	11.80
	40			0.688	32.624	835.9	72.00	8.90	8.54	245	362.1	9992	587.8	11.78
				0.750	32.500	829.3	78.34	8.90	8.51	266	359.5	10829	637.0	11.76
				0.875	32.250	816.4	91.01	8.90	8.44	310	354.1	12501	735.4	11.72
				1.000	32.000	804.2	103.67	8.90	8.38	353	348.6	14114	830.2	11.67
				1.125	31.750	791.3	116.13	8.90	8.31	395	343.2	15719	924.7	11.63
36 36.000				0.250	35.500	989.7	28.11	9.42	9.29	96	429.1	4491	249.5	12.64
	10			0.312	35.376	982.9	34.95	9.42	9.26	119	426.1	5565	309.1	12.62
		Std		0.375	35.250	975.8	42.01	9.42	9.23	143	423.1	6664	370.2	12.59
	20	XS		0.500	35.000	962.1	55.76	9.42	9.16	190	417.1	8785	488.1	12.55
	30			0.625	34.750	948.3	69.50	9.42	9.10	236	411.1	10872	604.0	12.51
	40			0.750	34.500	934.7	83.01	9.42	9.03	282	405.3	12898	716.5	12.46
				0.875	34.250	920.6	96.50	9.42	8.97	328	399.4	14903	827.9	12.42
				1.000	34.000	907.9	109.96	9.42	8.90	374	393.6	16851	936.2	12.38
				1.125	33.750	894.2	123.19	9.42	8.89	419	387.9	18763	1042.4	12.34
42 42.000				0.250	41.500	1352.6	32.82	10.99	10.86	112	586.4	7126	339.3	14.73
		Std		0.375	41.250	1336.3	49.08	10.99	10.80	167	579.3	10627	506.1	14.71
	20	XS		0.500	41.000	1320.2	65.18	10.99	10.73	222	572.3	14037	668.4	14.67
	30			0.625	40.750	1304.1	81.28	10.99	10.67	276	565.4	17373	827.3	14.62
	40			0.750	40.500	1288.2	97.23	10.99	10.60	330	558.4	20689	985.2	14.59
				1.000	40.000	1256.6	128.81	10.99	10.47	438	544.8	27080	1289.5	14.50
				1.250	39.500	1225.3	160.03	10.99	10.34	544	531.2	33233	1582.5	14.41
				1.500	39.000	1194.5	190.85	10.99	10.21	649	517.9	39181	1865.7	14.33

PIPE DATA

TABLE 1-5 Cast iron flanged pipe for water. ASA B16.1 class 125 or ASA B16.2 class 250 *(Courtesy United States Steel Corporation.)*

Size	Actual O.D., in.	Diam of flange, in.	Diam of raised face, in.	Diam of bolt circle, in.	No. of bolts	Diam of bolts, in.	Thickness of flange, in.	Max working pressure, psi	Wall thickness, in.	Wt per ft plain end, lb	Wt of one flange, lb	Wt 16 ft. length with 2 flanges, lb
3	3.96	7.50		6.00	4	5/8	0.75	150	0.38	13.3	7	225
3	3.96	7.50		6.00	4	5/8	0.75	250	0.38	13.3	7	225
3	3.96	8.25	5.69	6.62	8	3/4	1.12	250	0.38	13.3	12	235
4	4.80	9.00		7.50	8	5/8	0.94	150	0.38	16.5	13	290
4	4.80	9.00		7.50	8	5/8	0.94	250	0.38	16.5	13	290
4	4.80	10.00	6.94	7.88	8	3/4	1.25	250	0.38	16.5	20	305
6	6.90	11.00		9.50	8	3/4	1.00	150	0.38	24.3	17	425
6	6.90	11.00		9.50	8	3/4	1.00	250	0.38	24.3	17	425
6	6.90	12.50	9.69	10.62	12	3/4	1.44	250	0.38	24.3	34	455
8	9.05	13.50		11.75	8	3/4	1.12	150	0.41	34.7	27	610
8	9.05	13.50		11.75	8	3/4	1.12	250	0.41	34.7	27	610
8	9.05	15.00	11.94	13.00	12	7/8	1.62	250	0.44	34.7	50	655
10	11.10	16.00		14.25	12	7/8	1.19	150	0.44	46.0	38	810
10	11.10	16.00		14.25	12	7/8	1.19	250	0.44	46.0	38	810
10	11.10	17.50	14.06	15.25	16	1	1.88	250	0.44	46.0	70	875
12	13.20	19.00		17.00	12	7/8	1.25	150	0.48	59.8	58	1,075
12	13.20	19.00		17.00	12	7/8	1.25	250	0.52	64.6	58	1,150
12	13.20	20.50	16.44	17.75	16	1	2.00	250	0.52	64.6	102	1,240
14	15.30	21.00		18.75	12	1	1.38	150	0.51	73.9	72	1,325
14	15.30	21.00		18.75	12	1	1.38	250	0.59	85.1	72	1,505
14	15.30	23.00	18.94	20.25	20	1 1/8	2.12	250	0.59	85.1	130	1,620
16	17.40	23.50		21.25	16	1	1.44	150	0.54	89.2	90	1,605
16	17.40	23.50		21.25	16	1	1.44	250	0.63	103.6	90	1,840
16	17.40	25.50	21.06	22.50	20	1 1/8	2.25	250	0.63	103.6	162	1,980
18	19.50	25.00		22.75	16	1 1/8	1.56	150	0.58	107.6	90	1,900
18	19.50	25.00		22.75	16	1 1/8	1.56	250	0.68	125.4	90	2,185
18	19.50	28.00	23.31	24.75	24	1 1/4	2.38	250	0.68	125.4	200	2,405
20	21.60	27.50		25.00	20	1 1/8	1.69	150	0.62	127.5	115	2,270
20	21.60	27.50		25.00	20	1 1/8	1.69	250	0.72	147.4	115	2,590
20	21.60	30.50	25.56	27.00	24	1 1/4	2.50	250	0.72	147.4	245	2,850
24	25.80	32.00		29.50	20	1 1/4	1.88	150	0.73	179.4	160	3,190
24	25.80	32.00		29.50	20	1 1/4	1.88	250	0.79	193.7	160	3,420
24	25.80	36.00	30.25	32.00	24	1 1/2	2.75	250	0.79	193.7	370	3,840
30	32.00	38.75		36.00	28	1 1/4	2.12	150	0.85	259.5	240	4,630
30	32.00	38.75		36.00	28	1 1/4	2.12	250	0.99	300.9	240	5,295
30	32.00	43.0	37.19	39.25	28	1 3/8	3.00	250	0.99	300.9	530	5,875
36	38.30	46.00		42.75	32	1 1/2	2.38	150	0.94	344.2	350	6,205
36	38.30	46.00		42.75	32	1 1/2	2.38	250	1.10	401.1	350	7,120
36	38.30	50.0	43.69	46.0	32	2	3.38	250	1.10	401.1	710	7,840
42	44.50	53.00		49.5	36	1 1/2	2.62	150	1.05	447.2	500	8,155
42	44.50	53.00		49.5	36	1 1/2	2.62	250	1.22	517.6	500	9,280
42	44.50	57.0	50.44	52.75	36	2	3.69	250	1.22	517.6	900	10,080
48	50.80	59.50		56.0	44	1 1/2	2.75	150	1.14	554.9	625	10,130
48	50.80	59.50		56.0	44	1 1/2	2.75	250	1.33	644.9	625	11,570
48	50.80	65.0	58.44	60.75	40	2	4.00	250	1.33	644.9	1350	13,020

TABLE 1-4 API standard line pipe, threaded. Weights and dimensions are nominal *(Courtesy United States Steel Corporation.)*

Size, in.	Weight per ft, lb Threads and coupling	Weight per ft, lb Plain end	Thickness, in.	Diameter, in. O.D.	Diameter, in. I.D.	Threads per in.	Couplings Length, in.	Couplings O.D., in.	Couplings Weight, lb	Test pressure,* psi Lap-welded or Grade A seamless	Test pressure,* psi Grade B seamless	Test pressure,* psi Grade C seamless
1/8	0.25	0.24	0.068	0.405	0.269	27	1 1/16	0.563	0.04	700	700	700
1/4	0.43	0.42	0.088	0.540	0.364	18	1 5/8	0.719	0.09	700	700	700
3/8	0.57	0.57	0.091	0.675	0.493	18	1 5/8	0.875	0.13	700	700	700
1/2	0.86	0.85	0.109	0.840	0.622	14	2 3/8	1.063	0.24	700	700	700
3/4	1.14	1.13	0.113	1.050	0.824	14	2 5/8	1.313	0.34	700	700	700
1	1.70	1.68	0.133	1.315	1.049	11 1/2	2 5/8	1.576	0.54	700	700	700
1 1/4	2.30	2.27	0.140	1.660	1.380	11 1/2	2 3/4	2.054	1.03	1,000	1,100	1,300
1 1/2	2.75	2.72	0.145	1.900	1.610	11 1/2	2 3/4	2.200	1.17	1,000	1,100	1,300
2	3.75	3.65	0.154	2.375	2.067	11 1/2	3 1/4	2.875	2.13	1,000	1,100	1,300
2 1/2	5.90	5.79	0.203	2.875	2.469	8	4 1/8	3.375	3.27	1,000	1,100	1,300
3	7.70	7.58	0.216	3.500	3.068	8	4 1/4	4.000	4.09	1,000	1,300	1,600
3 1/2	9.25	9.11	0.226	4.000	3.548	8	4 3/8	4.625	5.92	1,200	1,300	1,600
4	11.00	10.79	0.237	4.500	4.026	8	4 1/2	5.200	7.59	1,200	1,300	1,600
5	15.00	14.62	0.258	5.563	5.047	8	4 5/8	6.296	9.98	1,200	1,300	1,600
6	19.45	18.97	0.280	6.625	6.065	8	4 7/8	7.390	12.92	1,200	1,300	1,600
8	25.55	24.70	0.277	8.625	8.071	8	5 1/4	9.625	23.18	1,000	1,200	1,400
8	29.35	28.55	0.322	8.625	7.981	8	5 1/4	9.625	23.18	1,200	1,300	1,600
10	32.75	31.20	0.279	10.750	10.192	8	5 3/4	11.750	31.55	1,000	1,200	1,400
10	35.75	34.24	0.307	10.750	10.136	8	5 3/4	11.750	31.55	1,000	1,200	1,400
10	41.85	40.48	0.365	10.750	10.020	8	5 3/4	11.750	31.55	1,200	1,300	1,600
12	45.45	43.77	0.330	12.750	12.090	8	6 3/8	14.000	49.27	1,000	1,200	1,400
12	51.15	49.56	0.375	12.750	12.000	8	6 3/8	14.000	49.27	1,000	1,200	1,400
14 O.D.	57.00	54.57	0.375	14.000	13.250	8	6 3/8	15.000	45.83	950	1,100	1,400
15 O.D.	61.15	58.57	0.375	15.000	14.250	8	6 5/8	16.000	51.26	900	1,000	1,400
16 O.D.	65.30	62.58	0.375	16.000	15.250	8	6 3/4	17.000	55.83	850	1,000	1,300
17 O.D.	73.20	69.70	0.393	17.000	16.214	8	7	18.000	61.67	850	950	1,200
18 O.D.	81.20	76.84	0.409	18.000	17.182	8	7 1/8	19.000	66.53	800	950	1,200
20 O.D.	90.00	85.58	0.409	20.000	19.182	8	7 7/8	21.000	79.37	750	850	1,100

The permissible variation in weight for any length of pipe is 10 percent above and 3 1/2 percent below; but the carload weight shall not be more than 1 3/4 percent under the calculated weight.

Furnished with threads and couplings and in random lengths, unless otherwise ordered.

The weight per foot of pipe with threads and couplings is based on a length of 20 ft, including the coupling.

* Test pressure butt-welded pipe 1/8 to 1 in. = 700 psi; 1 1/4 to 3 in. = 800 psi.

TABLE 1-8 Weights and dimensions of lead pipe (*Courtesy ITT Grinnell Corporation.*)

I.D., in.	Classification East	Classification West	O.D., in.	Weight per ft, lb
3/8	E	AQ	0.520	0.50
	D	XL	0.549	0.63
	C	L	0.577	0.75
	B	M	0.631	1.00
	A	S	0.725	1.50
	AA	XS	0.811	2.00
	AAA	XXS	0.888	2.50
1/2	E	AQ	0.628	0.56
	D	XL	0.666	0.75
	C	L	0.712	1.00
	B	M	0.756	1.25
	A	S	0.798	1.50
	AA	XS	0.876	2.00
	AAA	XXS	1.012	3.00
5/8	E	AQ	0.765	0.75
	D	XL	0.803	1.00
	C	L	0.881	1.50
	B	M	0.953	2.00
	A	S	1.019	2.50
	AA	XS	1.082	3.00
	AAA	XXS	1.137	3.50
3/4	E	AQ	0.906	1.00
	D	XL	0.940	1.25
	C	L	1.006	1.75
	B	M	1.068	2.25
	A	S	1.156	3.00
	AA	XS	1.212	3.50
	AAA	XXS	1.336	4.75
1	E	AQ	1.192	1.63
	D	XL	1.232	2.00
	C	L	1.284	2.50
	B	M	1.356	3.25
	A	S	1.428	4.00
	AA	XS	1.492	4.75
	AAA	XXS	1.596	6.00
1 1/4	E	AQ	1.442	2.00
	D	XL	1.486	2.50
	C	L	1.528	3.00
	B	M	1.592	3.75
	A	S	1.670	4.75
	AA	XS	1.765	6.00
	AAA	XXS	1.889	7.75
1 1/2	E	AQ	1.740	3.00
	D	XL	1.776	3.50
	C	L	1.830	4.25
	B	M	1.882	5.00
	A	S	1.984	6.50
	AA	XS	2.076	8.00
	AAA	XXS	2.272	11.25
1 3/4	D	XL	2.024	4.00
	C	L	2.086	5.00
	B	M	2.146	6.00
	A	S	2.193	6.75
	AA	XS	2.404	10.50
	AAA	XXS	2.624	14.75

Additional standard sizes are 2, 2 1/2, 3, 4, 5, and 6 in.

TABLE 1-6 Approx. weight of round seamless cold-finished carbon-steel mechanical tubing (pound per foot). Carbon, 0.25% maximum standard sizes for warehouse stocks random lengths (*Courtesy United States Steel Corporation.*)

Wall thickness In.	G or in.	3/8	1/2	5/8	3/4	7/8	1	1 1/8	1 1/4	1 3/8	1 1/2	1 5/8	1 3/4	1 7/8	2	2 1/8	2 1/4	2 3/8	2 1/2
0.035	20 G	0.127	0.174	0.221	0.267	0.314	0.361	0.407	0.454		0.548								
0.049	18 G	0.171	0.236	0.301	0.367	0.432	0.498	0.563	0.629		0.759	0.825	0.890		1.02				
0.058	17 G		0.274	0.351	0.429		0.584			0.694									
0.065	16 G	0.215	0.302	0.389	0.476	0.562	0.649	0.736	0.823	0.909	0.996	1.08	1.17	1.26	1.34	1.43	1.52	1.60	1.69
0.083	14 G		0.370	0.480	0.591	0.702	0.813	0.924	1.03	1.15	1.26		1.68	1.81	1.70				
0.095	13 G		0.411	0.538	0.665	0.791	0.918	1.05	1.17	1.30	1.43	1.55			1.93	2.06	2.19	2.31	2.44
0.109	12 G		0.455	0.601	0.746	0.892	1.04	1.18	1.33		1.62								
0.120	11 G		0.487	0.647	0.807	0.968	1.13	1.29	1.45	1.61	1.77	1.93	2.09	2.25	2.41	2.57	2.73	2.89	3.05
0.134	10 G			0.703	0.882		1.24	1.42	1.60	1.78	1.96	2.13	2.31		2.67				
0.156	9/32			0.781	0.990	1.20	1.41	1.61	1.82	2.03	2.24	2.45	2.66	2.86	3.07	3.28	3.49	3.70	3.91
0.188	3/16				1.13	1.38	1.63	1.88	2.13	2.38	2.63	2.89	3.14	3.39	3.64	3.89	4.14	4.39	4.64
0.219	7/32					1.53	1.83	2.12	2.41	2.70	3.00	3.29	3.58	3.87	4.17	4.46	4.75	5.04	5.34
0.250	1/4				1.34	1.67	2.00	2.34	2.67	3.00	3.34	3.67	4.01	4.34	4.67	5.01	5.34	5.67	6.01
0.281	9/32								2.91		3.66	4.03	4.41	4.78	5.16		5.91	6.28	6.66
0.313	5/16								3.13	3.55	3.97	4.39	4.80	5.22	5.64	6.06	6.48	6.89	7.31
0.375	3/8								3.50	4.01	4.51	5.01	5.51	6.01	6.51	7.01	7.51	8.01	8.51
0.438	7/16										4.97	5.53	6.14	6.72	7.31	7.89	8.48	9.06	9.65
0.500	1/2										5.34	6.01	6.68	7.34	8.01	8.68	9.35	10.0	10.7
0.625	5/8														9.18		10.8	11.7	12.5

* Other standard sizes, in certain standard wall thicknesses, vary by 1/8 in. increments from 2 1/2 to 3 1/2 in.; by 1/4 in. increments from 3 1/2 to 7 1/2 in.; 1/2 in. increments from 7 1/2 to 10 1/2 in. O.D. There are also standard sizes for every 1/16 in. from 3/8 to 1 5/8 in. O.D.

TABLE 1-7 Sizes and weights of SPS copper and 85 red brass pipe* (*Courtesy ITT Grinnell Corporation.*)

Standard pipe size (S. P. S.), in.	O.D., in.	I.D., in.	Wall thickness, in.	Weight, lb per ft 85 Red brass	Weight, lb per ft Copper
1/8	0.405	0.281	0.0620	0.2533	0.2590
1/4	0.540	0.375	0.0825	0.4496	0.4596
3/8	0.675	0.494	0.0905	0.6302	0.6441
1/2	0.840	0.625	0.1075	0.9381	0.9588
3/4	1.050	0.822	0.1140	1.271	1.299
1	1.315	1.062	0.1265	1.791	1.831
1 1/4	1.660	1.368	0.1460	2.633	2.692
1 1/2	1.900	1.600	0.1500	3.127	3.196
2	2.375	2.062	0.1565	4.136	4.228
2 1/2	2.875	2.500	0.1875	6.003	6.136
3	3.500	3.062	0.2190	8.560	8.750
3 1/2	4.000	3.500	0.2500	11.17	11.42
4	4.500	4.000	0.2500	12.66	12.94
5	5.563	5.062	0.2505	15.85	16.20
6	6.625	6.125	0.2500	18.99	19.41
8	8.625	8.000	0.3125	30.95	31.63
10	10.750	10.019	0.3655	45.22	46.22
11	11.750	11.000	0.3750	50.82	51.94
12	12.750	12.000	0.3750	55.28	56.51

*85% copper, 15% zinc.

17

TABLE 1-9 Sizes and weights for aluminum pipe *(Courtesy ITT Grinnell Corporation.)*

Nominal pipe size, in.	Schedule number*	O.D., in.	I.D., in.	Wall thickness, in.	Weight per linear foot, lb, plain ends†	Cross-sectional wall area, sq in.	Inside cross-sectional area, sq in.	Moment of inertia, in.⁴	Section modulus, in.³	Radius of gyration, in.
⅛	40‡	0.405	0.269	0.068	0.085	0.0720	0.0568	0.0011	0.0053	0.1215
	80§	0.405	0.215	0.095	0.109	0.0925	0.0363	0.0012	0.0060	0.1146
¼	40‡	0.540	0.364	0.088	0.147	0.1250	0.1041	0.0033	0.0123	0.1628
	80§	0.540	0.302	0.119	0.185	0.1574	0.0716	0.0038	0.0139	0.1547
⅜	40‡	0.675	0.493	0.091	0.196	0.1670	0.1909	0.0073	0.0216	0.2090
	80§	0.675	0.423	0.126	0.256	0.2173	0.1405	0.0086	0.0255	0.1991
½	40‡	0.840	0.622	0.109	0.294	0.2503	0.3039	0.0171	0.0407	0.2613
	80§	0.840	0.546	0.147	0.376	0.3200	0.2341	0.0201	0.0478	0.2505
¾	10	1.050	0.884	0.083	0.297	0.2521	0.6138	0.0297	0.0566	0.3432
	40‡	1.050	0.824	0.113	0.391	0.3326	0.5333	0.0370	0.0705	0.3337
	80§	1.050	0.742	0.154	0.510	0.4335	0.4324	0.0448	0.0853	0.3214
1	5	1.315	1.185	0.065	0.300	0.2553	1.103	0.0500	0.0760	0.4425
	10	1.315	1.097	0.109	0.486	0.4130	0.9452	0.0757	0.1151	0.4382
	40‡	1.315	1.049	0.133	0.581	0.4939	0.8643	0.0873	0.1328	0.4205
	80§	1.315	0.957	0.179	0.751	0.6388	0.7193	0.1056	0.1606	0.4066
1¼	5	1.660	1.530	0.065	0.383	0.3257	1.839	0.1037	0.1250	0.5644
	10	1.660	1.442	0.109	0.625	0.5311	1.633	0.1605	0.1934	0.5497
	40‡	1.660	1.380	0.140	0.786	0.6685	1.496	0.1947	0.2346	0.5397
	80§	1.660	1.278	0.191	1.037	0.8815	1.283	0.2418	0.2913	0.5238
1½	5	1.900	1.770	0.065	0.441	0.3747	2.461	0.1579	0.1662	0.6492
	10	1.900	1.682	0.109	0.721	0.6133	2.222	0.2468	0.2598	0.6344
	40‡	1.900	1.610	0.145	0.940	0.7995	2.036	0.3099	0.3262	0.6226
	80§	1.900	1.500	0.200	1.256	1.068	1.767	0.3912	0.4118	0.6052
2	5	2.375	2.245	0.065	0.555	0.4717	3.958	0.3149	0.2652	0.8170
	10	2.375	2.157	0.109	0.913	0.7760	3.654	0.4992	0.4204	0.8021
	40‡	2.375	2.067	0.154	1.264	1.074	3.356	0.6657	0.5606	0.7871
	80§	2.375	1.939	0.218	1.737	1.477	2.953	0.8679	0.7309	0.7665
2½	5	2.875	2.709	0.083	0.856	0.7280	5.764	0.7100	0.4939	0.9876
	10	2.875	2.635	0.120	1.221	1.039	5.453	0.9873	0.6868	0.9750
	40‡	2.875	2.469	0.203	2.004	1.704	4.788	1.530	1.064	0.9474
	80§	2.875	2.323	0.276	2.650	2.254	4.238	1.924	1.339	0.9241
3	5	3.500	3.334	0.083	1.048	0.8910	8.730	1.301	0.7435	1.208
	10	3.500	3.260	0.120	1.498	1.274	8.346	1.822	1.041	1.196
	40‡	3.500	3.068	0.216	2.621	2.228	7.393	3.017	1.724	1.164
	80§	3.500	2.900	0.300	3.547	3.016	6.605	3.894	2.225	1.136
3½	5	4.000	3.834	0.083	1.201	1.021	11.55	1.960	0.9799	1.385
	10	4.000	3.760	0.120	1.720	1.463	11.10	2.755	1.378	1.372
	40‡	4.000	3.548	0.226	3.151	2.680	9.887	4.788	2.394	1.337
	80§	4.000	3.364	0.318	4.326	3.678	8.888	6.281	3.140	1.307
4	5	4.500	4.334	0.083	1.354	1.152	14.75	2.810	1.249	1.562
	10	4.500	4.260	0.120	1.942	1.651	14.25	3.963	1.761	1.549
	40‡	4.500	4.026	0.237	3.733	3.174	12.73	7.232	3.214	1.510
	80§	4.500	3.826	0.337	5.183	4.407	11.50	9.611	4.272	1.477
5	40‡	5.563	5.047	0.258	5.057	4.300	20.01	15.16	5.451	1.878
	80§	5.563	4.813	0.375	7.188	6.112	18.19	20.67	7.432	1.839
6	40‡	6.625	6.065	0.280	6.564	5.581	28.89	28.14	8.496	2.246
	80§	6.625	5.761	0.432	9.884	8.405	26.07	40.49	12.22	2.195
8	30	8.625	8.071	0.277	8.543	7.265	51.16	63.35	14.69	2.953
	40‡	8.625	7.981	0.322	9.878	8.399	50.03	72.49	16.81	2.938
	80§	8.625	7.625	0.500	15.01	12.76	45.66	105.7	24.51	2.878
10	...	10.750	10.192	0.279	10.79	9.178	81.59	125.9	23.42	3.704
	30	10.750	10.136	0.307	11.84	10.07	80.69	137.4	25.57	3.694
	40‡	10.750	10.020	0.365	14.00	11.91	78.85	160.7	29.90	3.674
	80§	10.750	9.750	0.500	18.93	16.10	74.66	211.9	39.43	3.628
12	30	12.750	12.090	0.330	15.14	12.88	114.8	248.5	38.97	4.393
	..‡	12.750	12.000	0.375	17.14	14.58	113.1	279.3	43.81	4.377
	..§	12.750	11.750	0.500	22.63	19.24	108.4	361.5	56.71	4.335

* Schedule numbers conform to American Standard for Wrought-iron and Wrought-steel Pipe, ASA B36.10.
† Weights calculated for 6061 and 6063. For 3003 multiply by 1.010.
‡ Also designated as standard pipe.
§ Also designated as extra-heavy or extra-strong pipe. All calculations based on nominal dimensions.

TABLE 1-11 Corrosion-resistance data, polyethylene pipe (*Courtesy United States Steel Corporation.*)

Reagent	Performance at 75 F	Performance at 120 F
Acetic acid, glacial*	F	NG
Acetic acid, 10 percent*	E	E
Acetone	NG	NG
Ammonia, dry gas	E	E
Ammonium hydroxide, 10 percent	E	E
Ammonium hydroxide, 28 percent	E	E
Amyl acetate	NG	NG
Aniline	NG	F
Benzene	NG	NG
Bromine	NG	NG
Butyraldehyde	E	G
Calcium chloride, saturated	E	E
Calcium hydroxide	E	E
Calcium hypochlorite	E	E
Carbon disulphide	NG	NG
Carbon tetrachloride	NG	NG
Carbonic acid	E	E
Chlorine, dry gas	F	NG
Chlorine, liquid	NG	NG
Chlorosulphonic acid	NG	NG
Citric acid, saturated	E	E
Copper sulphate	E	E
Cyclohexanone	NG	NG
Diethylene glycol	E	E
Dioxane	E	G
Ethyl acetate	F	NG
Ethyl alcohol, 35 percent	NG	NG
Ethyl butyrate	F	NG
Ethylene dichloride	NG	NG
Ferric chloride	E	E
Ferrous sulphate, 15 percent aq	E	E
Fluorine	E	F
Fluosilicic acid, concentrated	E	E
Formaldehyde, 40 percent	E	E
Formic acid, 50 percent	E	E
Furfuryl alcohol	NG	NG
Gasoline	NG	NG
Hydrobromic acid	E	E
Hydrochloric acid, 10 percent	E	E
Hydrochloric acid, 37 percent	E	E
Hydrofluoric acid, 48 percent	E	E
Hydrofluoric acid, 75 percent	E	F
Hydrogen peroxide, 30 percent	E	G
Hydrogen peroxide, 90 percent	G	NG

Reagent	Performance at 75 F	Performance at 120 F
Lactic acid, 90 percent	E	E
Linseed oil	E	E
Lubricating oil	NG	NG
Magnesium chloride	E	E
Magnesium sulphate	E	E
Methyl bromide	NG	E
Methyl isobutyl ketone	F	NG
Nitric acid, 10 percent	E	E
Nitric acid, 30–50 percent	E	E
Nitric acid, 70 percent	E	E
Oleic acid	F	NG
Phosphoric acid, 30 percent	E	E
Phosphoric acid, 90 percent	E	NG
Photographic developer	E	E
Potassium borate	E	E
Potassium carbonate	E	E
Potassium chloride, saturated	E	E
Potassium dichromate	E	E
Potassium hydroxide	E	E
Potassium nitrate	E	E
Potassium permanganate	E	E
Silicic acid	E	E
Silver nitrate	E	E
Sodium benzoate	E	E
Sodium bisulphite	E	E
Sodium carbonate, concentrated	E	E
Sodium chloride, saturated solution	E	E
Sodium hydroxide, 10 percent	E	E
Sodium hydroxide, 50 percent	E	E
Sodium sulphate	E	E
Stannic chloride, saturated	E	E
Stearic acid, 100 percent	E	E
Sulphuric acid, 10 percent	E	E
Sulphuric acid, 30 percent	E	E
Sulphuric acid, 60 percent	E	E
Sulphuric acid, 98 percent	F	E
Tannic acid	E	E
Toluene	NG	NG
Trichlorobenzene	F	NG
Trichloroethylene	NG	NG
Vinegar	E	E
Xylene	NG	NG
Zinc chloride	E	E
Zinc sulphate	E	E

Corrosion resistance data given in this table are based on laboratory tests conducted by the manufacturers of the materials covered, and are indicative only of the conditions under which the tests were made. This information may be considered as a basis for recommendation, but not as a guarantee. Materials should be tested under actual service to determine suitability for a particular purpose.
E = excellent, G = good, F = fair, NG = not good.
* Polyethylene is permeable to acetic acid.

TABLE 1-10 Polyethylene plastic pipe dimensions (*Courtesy United States Steel Corporation.*)

Nominal size, in.	O.D., in.	I.D., in.	Wall thickness, in.	Max operating press at 75 F, psi	Wt per ft, lb	Shipping length, ft
½	0.782	0.622	0.080	75	0.0704	100 and 400 (coiled)
	0.842	0.622	0.110	100	0.101	100 and 400 (coiled)
¾	1.024	0.824	0.100	75	0.1157	100 and 400 (coiled)
	1.114	0.824	0.145	100	0.176	100 and 400 (coiled)
1	1.300	1.050	0.125	75	0.1838	100 and 400 (coiled)
	1.410	1.050	0.180	100	0.277	100 and 400 (coiled)
1¼	1.660	1.380	0.140	70	0.266	100 and 400 (coiled)
	1.710	1.380	0.165	75	0.3190	100 and 400 (coiled)
	1.860	1.380	0.240	100	0.487	100 and 400 (coiled)
1½	1.900	1.610	0.145	60	0.318	100 and 400 (coiled)
	2.000	1.610	0.195	75	0.441	100 and 400 (coiled)
	2.170	1.610	0.280	100	0.662	100 and 400 (coiled)
2	2.375	2.067	0.154	50	0.428	100 and 400 (coiled)
	2.567	2.067	0.250	75	0.725	100 and 400 (coiled)
	2.777	2.067	0.355	100	1.077	100 and 400 (coiled)
3	3.500	3.068	0.216	50	0.888	100 (coiled)
	3.670	3.068	0.301	75	1.280	100 (coiled)
	4.068	3.068	0.500	100	2.230	100 (coiled)
4	4.500	4.026	0.237	40	1.265	25 (straight)
	4.820	4.026	0.397	75	2.200	25 (straight)
	5.386	4.026	0.680	100	4.000	25 (straight)
6	6.625	6.065	0.280	35	2.200	25 (straight)

TABLE 1-12 Cast iron soil pipe (*Courtesy Cast Iron Soil Pipe Institute.*)

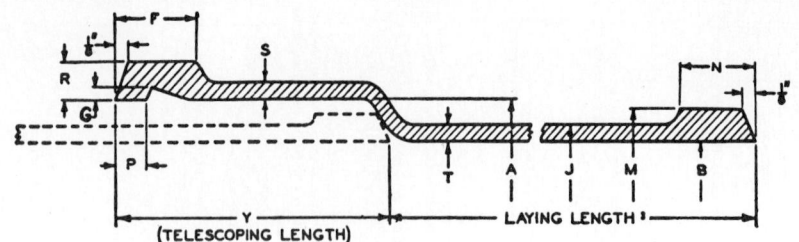

DIMENSIONS OF HUBS, SPIGOTS, AND BARRELS FOR EXTRA-HEAVY CAST IRON SOIL PIPE AND FITTINGS

SIZE	INSIDE DIAMETER OF HUB	OUTSIDE DIAMETER OF SPIGOT BEAD	OUTSIDE DIAMETER OF BARREL	TELESCOPING LENGTH	INSIDE DIAMETER OF BARREL	THICKNESS OF BARREL
	A	M	J	Y	B	T
INCHES	INCHES	INCHES	INCHES	INCHES	INCHES	INCHES
2	3.06	2.75	2.38	2.50	2.00	.19
3	4.19	3.88	3.50	2.75	3.00	.25
4	5.19	4.88	4.50	3.00	4.00	.25
5	6.19	5.88	5.50	3.00	5.00	.25
6	7.19	6.88	6.50	3.00	6.00	.25
8	9.50	9.00	8.62	3.50	8.00	.31
10	11.62	11.13	10.75	3.50	10.00	.37
12	13.75	13.13	12.75	4.25	12.00	.37
15	17.00	16.25	15.88	4.25	15.00	.44

SIZE	THICKNESS OF HUB - HUB BODY	THICKNESS OF HUB - OVER BEAD	WIDTH OF HUB BEAD	WIDTH OF SPIGOT BEAD	DISTANCE FROM LEAD GROOVE TO END, PIPE AND FITTINGS	DEPTH OF LEAD GROOVE G (MIN.)	DEPTH OF LEAD GROOVE G (MAX.)
	S (MIN.)	R (MIN.)	F	N	P	G (MIN.)	G (MAX.)
INCHES	INCHES	INCHES	INCHES	INCHES	INCHES	INCHES	INCHES
2	0.18	0.37	0.75	0.69	0.28	0.10	0.13
3	.25	.43	.81	.75	.28	.10	.13
4	.25	.43	.88	.81	.28	.10	.13
5	.25	.43	.88	.81	.28	.10	.13
6	.25	.43	.88	.81	.28	.10	.13
8	.34	.59	1.19	1.12	.38	.15	.19
10	.40	.65	1.19	1.12	.38	.15	.19
12	.40	.65	1.44	1.38	.47	.15	.19
15	.46	.71	1.44	1.38	.47	.15	.19

TABLE 1-13 Heat loss from bare pipe and fittings (*Courtesy ITT Grinnell Corporation.*)

HEAT LOSS FROM BARE SURFACES
Flat, Curved, and Cylindrical Surfaces
Heat losses given in B.t.u. per hour, per square foot of bare surface, at various temperature differences.

Temp. difference degrees F.	Total heat loss at temp. indicated	Increments per degree for total losses	Loss per degree temp. difference
50	97.5		1.950
100	215.2	2.35	2.152
150	360.0	2.90	2.400
200	533.0	3.46	2.665
250	737.8	4.10	2.951
300	978.0	4.80	3.260
350	1269.5	5.83	3.627
400	1614.0	6.89	4.035
450	2007.5	7.87	4.461
500	2460.0	9.05	4.920
550	2953.5	9.87	5.370
600	3510.0	11.10	5.850
650	4108.0	11.96	6.320
700	4760.0	13.04	6.800
750	5456.3	13.93	7.275
800	6200.0	14.87	7.750
850	6991.3	15.83	8.225
900	7830.0	16.77	8.700
950	8716.3	17.73	9.175
1000	9650.0	18.67	9.650

Total Radiation Areas of Fittings
Standard flanged fittings

Nominal pipe size	Equivalent pipe length in linear feet				
	Flanged coupling	90-degree ell	Long-radius ell	Tee	Cross
1″	0.93	2.31	2.59	3.59	4.72
1 1/4″	0.88	2.20	2.49	3.40	4.47
1 1/2″	0.95	2.35	2.68	3.64	4.78
2″	1.08	2.65	2.96	4.08	5.34
2 1/2″	1.17	2.78	3.08	4.26	5.56
3″	1.03	2.60	2.93	3.99	5.70
3 1/2″	1.07	2.85	3.13	4.28	5.56
4″	1.14	2.90	3.36	4.59	5.97
4 1/2″	1.13	3.01	3.38	4.63	6.01
5″	1.11	3.049	3.43	4.67	6.06
6″	1.049	2.950	3.45	4.53	5.81
7″	1.097	3.090	3.697	4.69	6.01
8″	1.067	3.090	3.790	4.67	5.96
9″	1.190	3.457	4.200	5.23	6.66
10″	1.220	3.610	4.380	4.47	6.95
12″	1.320	3.920	4.900	5.89	7.45
14″ O.D.	1.465	4.470	5.470	6.78	8.60
15″ O.D.	1.572	4.720	5.830	7.10	9.04
16″ O.D.	1.600	4.820	6.070	7.23	9.15

*Figures indicate number of linear feet of pipe having an area equivalent to that of the fittings. Area calculated includes accompanying flanges.

TABLE 1-14 Pipe size metric conversion (*Courtesy United States Steel Corporation.*)

Diameter

Nominal Size		Outside Diameter		Nominal Size		Outside Diameter	
Millimetres	Inches	Millimetres	Inches	Millimetres	Inches	Millimetres	Inches
6	⅛	10.3	0.405	350	14	355.6	14.000
8	¼	13.7	0.540	400	16	406.4	16.000
10	⅜	17.1	0.675	450	18	457.2	18.000
15	½	21.3	0.840	500	20	508.0	20.000
20	¾	26.7	1.050	550	22	558.8	22.000
25	1	33.4	1.315	600	24	609.6	24.000
32	1¼	42.2	1.660	650	26	660.4	26.000
40	1½	48.3	1.900	700	28	711.2	28.000
50	2	60.3	2.375	750	30	762.0	30.000
65	2½	73.0	2.875	800	32	812.8	32.000
80	3	88.9	3.500	850	34	863.6	34.000
90	3½	101.6	4.000	900	36	914.4	36.000
100	4	114.3	4.500	950	38	965.2	38.000
125	5	141.3	5.563	1000	40	1016.0	40.000
150	6	168.3	6.625	1050	42	1066.8	42.000
200	8	219.1	8.625	1100	44	1117.6	44.000
250	10	273.0	10.750	1150	46	1143.0	46.000
300	12	323.8	12.750	1200	48	1219.2	48.000

Wall Thickness

Millimetres	Inches	Millimetres	Inches	Millimetres	Inches	Millimetres	Inches
1.73	.068	5.16	.203	8.74	.344	21.95	.864
2.24	.088	5.49	.216	9.27	.365	22.23	.875
2.31	.091	5.54	.218	9.53	.375	23.01	.906
2.41	.095	5.56	.219	10.31	.406	23.83	.938
2.77	.109	5.74	.226	10.97	.432	24.61	.969
2.79	.110	6.02	.237	11.07	.436	25.40	1.000
2.87	.113	6.35	.250	11.13	.438	26.19	1.031
2.90	.114	6.55	.258	11.91	.469	27.79	1.094
3.02	.119	7.01	.276	12.70	.500	28.58	1.125
3.07	.121	7.04	.277	14.02	.552	29.36	1.156
3.20	.126	7.09	.279	14.27	.562	30.96	1.219
3.38	.133	7.11	.280	15.09	.594	32.54	1.281
3.56	.140	7.14	.281	15.88	.625	33.32	1.312
3.68	.145	7.62	.300	16.66	.656	35.71	1.406
3.91	.154	7.80	.307	17.12	.674	38.10	1.500
3.96	.156	7.92	.312	17.48	.688	44.45	1.750
4.55	.179	8.08	.318	18.26	.719	50.80	2.000
4.78	.188	8.18	.322	19.05	.750		
4.85	.191	8.38	.330	20.62	.812		
5.08	.200	8.56	.337	21.44	.844		

Bold face indicates preferred designation—
Under 2⅜" OD—nominal. 2⅜" OD and larger—OD.

PIPE DATA

TABLE 1-15 Pipe and water weight per line foot (*Courtesy ITT Grinnel Corporation.*)

NOM. PIPE SIZE	WEIGHT OF: STD. PIPE	WEIGHT OF: WATER	WEIGHT OF: XS PIPE	WEIGHT OF: WATER
½	.851	.132	1.088	.101
¾	1.131	.231	1.474	.187
1	1.679	.374	2.172	.311
1¼	2.273	.648	2.997	.555
1½	2.718	.882	3.632	.765
2	3.653	1.453	5.022	1.278
2½	5.794	2.073	7.662	1.835
3	7.580	3.200	10.250	2.860
3½	9.110	4.280	12.510	3.850
4	10.790	5.510	14.990	4.980
5	14.620	8.660	20.780	7.880
6	18.980	12.510	28.580	11.290
8	28.560	21.680	43.400	19.800
10	40.500	34.100	54.700	32.300
12	49.600	49.000	65.400	47.000
14	54.600	59.700	72.100	57.500
16	62.600	79.100	82.800	76.500
18	70.600	101.200	93.500	98.300
20	78.600	126.000	104.100	122.800
24	94.600	183.800	125.500	179.900
30	118.700	291.000	157.600	286.000

TABLE 1-17 Heat loss from horizontal bare steel pipe (BTU per hour per linear foot at 70° room temperature) (*Courtesy ITT Grinnell Corporation.*)

NOM. PIPE SIZE	HOT WATER (180°F)	STEAM 5 PSIG (20 PSIA)
½	60	96
¾	73	118
1	90	144
1¼	112	179
1½	126	202
2	155	248
2½	185	296
3	221	355
3½	244	401
4	279	448

TABLE 1-16 Weights of metals (*Courtesy ITT Grinnell Corporation.*)

MATERIAL	CHEMICAL SYMBOL	WEIGHT, IN POUNDS PER CUBIC INCH	WEIGHT, IN POUNDS PER CUBIC FOOT
Aluminum	Al	.093	160
Antimony	Sb	.2422	418
Brass	—	.303	524
Bronze	—	.320	552
Chromium	Cr	.2348	406
Copper	Cu	.323	558
Gold	Au	.6975	1205
Iron (cast)	Fe	.260	450
Iron (wrought)	Fe	.2834	490
Lead	Pb	.4105	710
Manganese	Mn	.2679	463
Mercury	Hg	.491	849
Molybdenum	Mo	.309	534
Monel	—	.318	550
Platinum	Pt	.818	1413
Steel (mild)	—	.2816	490
Steel (stainless)	—	.277	484
Tin	Sn	.265	459
Titanium	Ti	.1278	221
Zinc	Zn	.258	446

TABLE 1-18 Total thermal expansion of piping materials (inches per 100 ft above 32° F) (*Courtesy ITT Grinnel Corporation.*)

TEMPERATURE °F	CARBON AND CARBON MOLY STEEL	CAST IRON	COPPER	BRASS AND BRONZE	WROUGHT IRON
32	0	0	0	0	0
100	0.5	0.5	0.8	0.8	0.5
150	0.8	0.8	1.4	1.4	0.9
200	1.2	1.2	2.0	2.0	1.3
250	1.7	1.5	2.7	2.6	1.7
300	2.0	1.9	3.3	3.2	2.2
350	2.5	2.3	4.0	3.9	2.6
400	2.9	2.7	4.7	4.6	3.1
450	3.4	3.1	5.3	5.2	3.6
500	3.8	3.5	6.0	5.9	4.1
550	4.3	3.9	6.7	6.5	4.6
600	4.8	4.4	7.4	7.2	5.2
650	5.3	4.8	8.2	7.9	5.6
700	5.9	5.3	9.0	8.5	6.1
750	6.4	5.8	—	—	6.7
800	7.0	6.3	—	—	7.2
850	7.4	—	—	—	—
900	8.0	—	—	—	—
950	8.5	—	—	—	—
1000	9.1	—	—	—	—

2

FITTINGS

Fig. 2-1

FITTINGS

Pipe fittings are the components which tie together pipelines, valves, and other parts of a piping system. They are used in "making up" a pipeline. Fittings may come in screwed, welded, soldered, or flanged varieties and are used to change the size of the line or its direction and to join together the various parts that make up a piping system. Standard fittings are available in cast iron, malleable iron, forged steel, cast steel, wrought steel, brass, and copper. It is also possible to use other materials such as plastic and glass.

The majority of pipe fittings are specified by the nominal pipe size, type, material, and the name of the fitting. Besides the end connections mentioned above (screwed, welded, soldered, flanged), it is also possible to order bell and spigot fittings, which are usually cast iron and used for low-pressure service.

One of the major functions of fittings in a piping system is to change direction. There are three procedures for changing direction in a piping system: bending, the single-miter bend, and the double-miter bend. Bending a pipe is common practice, but in some cases it stretches and thins the pipe on the outer radius while the inner wall tends to bunch or thicken. Because of this tendency, flow resistance and erosion increase at the bend. Moreover, the pipe's ability to withstand pressure is reduced because of the thin outer wall.

To change direction with a single- or double-miter joint it is only necessary to cut straight pieces to the proper angle and then weld them together. The main drawback with this type of bend is the great increase in friction and resistance caused by the drastic change in direction, which also increases the turbulence of the flow at the turn. A full 90-degree elbow fitting offers six times less resistance than a single 90-degree miter bend. Because of these considerations, fittings that change direction are used more often than bending or mitered joints.

In general, then, a fitting is any component in a piping system that changes its direction, alters its function, or simply makes end connections. A fitting is joined to the system by bolting, welding, or screwing, depending on many variables in the system. Fittings, therefore, are the most economical and efficient way of redirecting or constructing a piping system, especially where flow rate, pressure, and other considerations are important.

One of the most commonly used means of joining fittings to pipe is the butt weld, which employs fittings made just for that purpose. In this type of fitting, the ends are machine-grooved to form a bevel which fits together with a piece of pipe which is also beveled at the end. They are fused together by welding. Welded piping systems have many advantages over the screwed or flanged variety. They require less maintenance and provide a permanent leak-proof bond. Moreover, because of their high-pressure/high-temperature safety record, they cut down on the total maintenance cost of the system. Welded construction also weighs much less than the flanged type and does not require as much clearance, since flanged or screwed fittings are larger and heavier. Therefore welded construction requires a minimum of space, which enables the designer to place the system closer to ceilings, walls, or other equipment. Another advantage of welded construction is that it is easier to insulate.

Flanged fittings are bolted to the piping system by means of welded or screwed flanges. This type of connection is used in petrochemical work at places where it is necessary to dismantle lines frequently. Flanges are also used where welding would be impossible because of fire hazard. One drawback of an all-flanged system is that a flanged fitting may weigh three to four times as much as its welded or screwed counterpart, making the whole system heavier and more difficult to support.

Threaded or screwed fittings are usually limited to smaller-sized piping. They come machined and threaded with standard size pipe threads. Forged steel screwed fittings are used primarily for high-temperature/high-pressure service.

Cast iron screwed fittings are more common for heating systems and low-pressure/low-temperature requirements. One disadvantage of the threaded system is the loss of pipe thickness—cutting the threads on a pipe can remove up to 50 percent of the pipe wall thickness. Exposure of threads can also lead to faster corrosion. All fittings are manufactured to standard dimensions. Threaded fittings are manufactured to national pipe size standards.

Butt-Welded Fittings

Because of simplifications in welding equipment it is now possible to use welding on piping systems that in the past could only accept screwed or flanged fittings. Today, even in small jobs such as buildings, small industrial plants, and schools, a completely welded system is within reach, thus providing a permanent, self-contained system.

Welded fittings are used primarily in systems meant to be permanent. They have the same wall thickness as the mating pipe. Among the many advantages of butt-welded systems are the following:

1. They have a smooth inner surface and offer gradual direction change with minimum turbulence.
2. They require much less space for constructing and hanging the pipe system.
3. They form leakproof connections.
4. They are almost maintenance free.
5. They have a higher temperature and pressure limit.
6. They form a self-contained system.
7. They are easy to insulate.
8. They offer a uniform wall thickness throughout the system.

One of the major disadvantages of butt-welded systems is that they are not easy to dismantle. Therefore it is often advisable to provide the system with enough flanged joints so that it can be broken down at intervals. (One of the main uses of the butt-welded system is for steam lines, which are usually in high-temperature/high-pressure service.)

Socket-Welded Fittings

Socket-welded fittings have certain advantages over butt-welded fittings. They are easier to use on small-size pipelines and the ends of the pipes need not be beveled since the pipe end slips into the socket of the joint (Fig. 2-2). With socket-welded fittings there is no danger of the weld protruding into the pipeline and restricting flow or creating turbulence. Thus the advantages of the socket-welded system are:

1. The pipe does not need to be beveled.
2. No tack welding is necessary for alignment since the joint and the pipe are self-aligning.
3. Weld material cannot extend into the pipeline.
4. It can be used in place of threaded fittings, therefore reducing the likelihood of leaks, which usually accompany the use of threaded fittings.
5. It is less expensive and easier to construct than other welded systems.

One of the major disadvantages of this type of fitting is the possibility of a mismatch inside the fitting where improperly aligned or mated parts may create a recess where erosion or corrosion could start.

Socket-welded fittings have the same inside diameter as standard (schedule 40), extra strong (schedule 80), and double extra strong (schedule 160) pipe, depending on the weight of the fitting and mating pipe. Socket-welded fittings are covered in ASA B16.11. They are drilled to match the internal diameter of schedule 40 or schedule 80 pipe.

Flanged Fittings

Flanged connections are found on piping systems throughout the petrochemical and power generation fields on pipelines that are a minimum of 2 in. (5.08 cm) in diameter. The majority of flanged fittings are made of cast steel or cast iron.

Flanged steel fittings are used in place of cast iron where the system is subject to shock or in high-temperature/high-pressure situations where the danger of fire is prevalent, because cast iron has a tendency to crack or rupture under certain stresses. A flange may be cast or forged onto the ends of the fitting or valve and bolted to a connecting flange which is screwed or welded onto the pipeline, thereby providing a tight joint. An assortment of facings, ring joint grooves, and connections are available in flange variations.

One advantage of flanged systems is that they are easily dismantled and assembled. One of the disadvantages is that they are considerably heavier than an equally rated butt-welded system, because of the large amounts of metal that go into making up the joints and flanges. Moreover, flanged fittings occupy far more space than the butt-welded or screwed equivalents. Because of this higher weight load, a flanged system becomes far more expensive to support or hang from the existing structure.

Cast steel and forged steel flanged fittings are available in ASA ratings from 150 to 2500 lb. The cast steel flange fittings comply with the specifications of ASA B16.5.

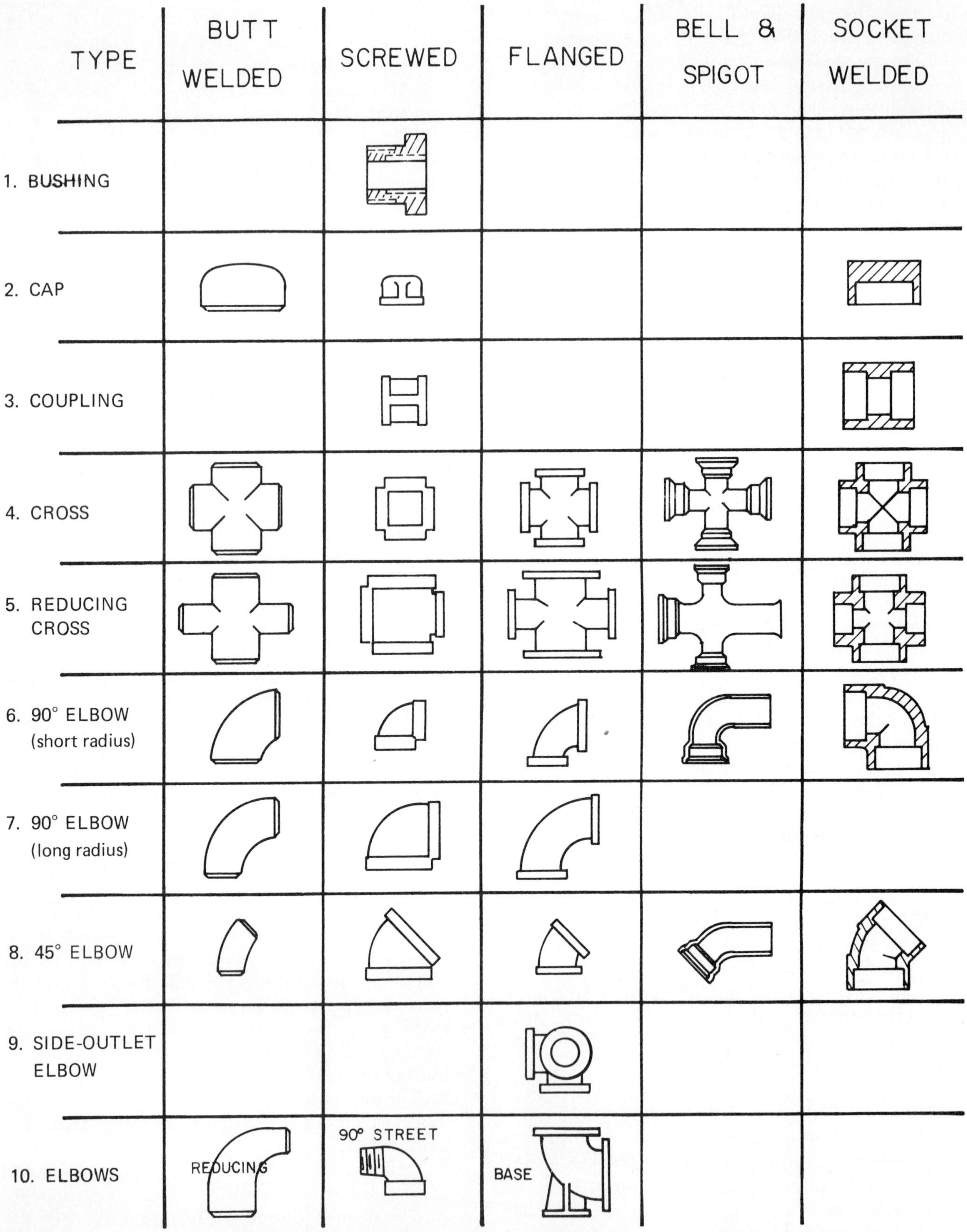

TYPE	BUTT WELDED	SCREWED	FLANGED	BELL & SPIGOT	SOCKET WELDED
1. BUSHING		▨			
2. CAP	▨	▨			▨
3. COUPLING		▨			▨
4. CROSS	▨	▨	▨	▨	▨
5. REDUCING CROSS	▨	▨	▨	▨	▨
6. 90° ELBOW (short radius)	▨	▨	▨	▨	▨
7. 90° ELBOW (long radius)	▨	▨	▨		
8. 45° ELBOW	▨	▨	▨	▨	▨
9. SIDE-OUTLET ELBOW			▨		
10. ELBOWS	REDUCING	90° STREET	BASE		

Fig. 2-2 Common pipe fittings.

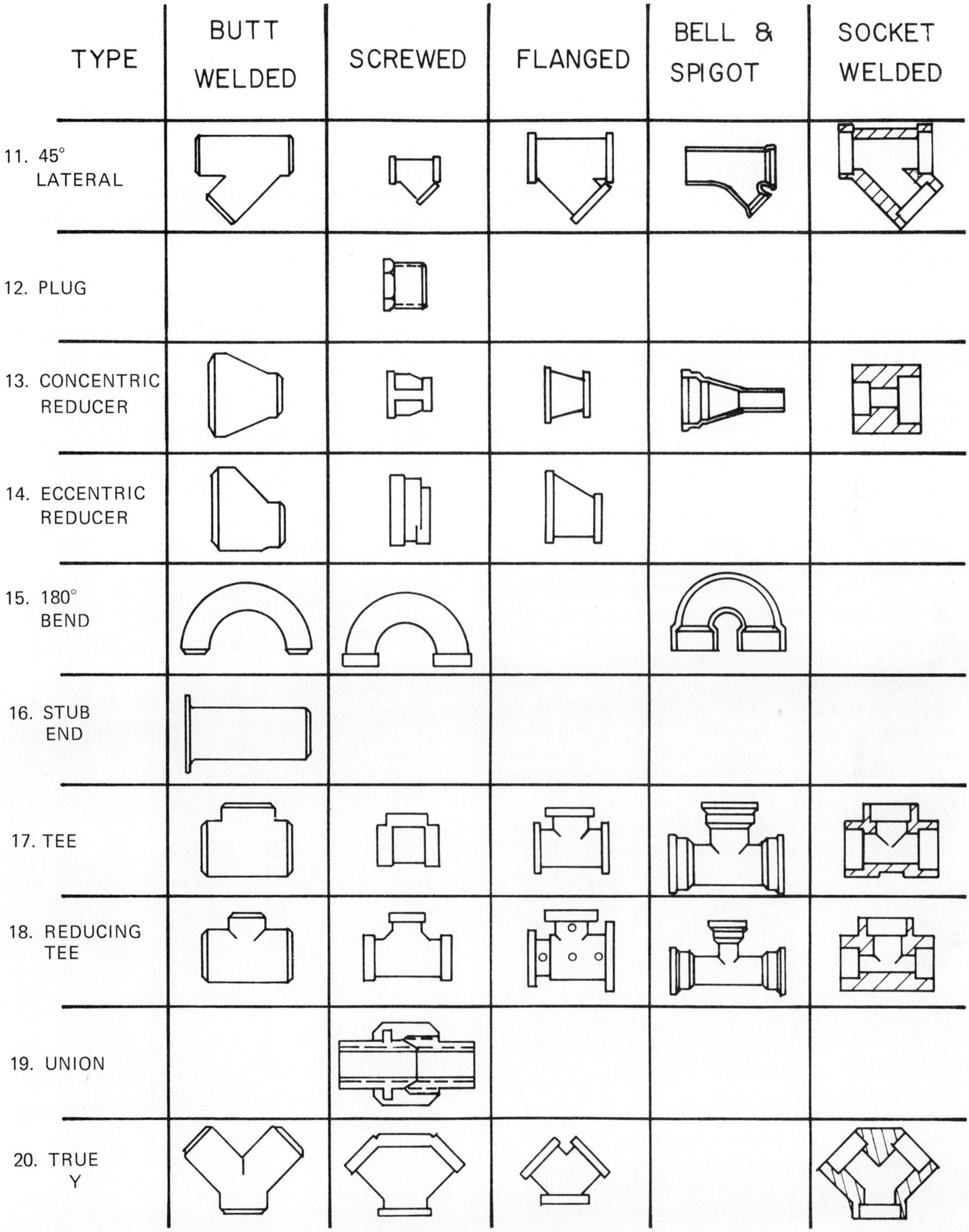

TYPE	BUTT WELDED	SCREWED	FLANGED	BELL & SPIGOT	SOCKET WELDED
11. 45° LATERAL					
12. PLUG					
13. CONCENTRIC REDUCER					
14. ECCENTRIC REDUCER					
15. 180° BEND					
16. STUB END					
17. TEE					
18. REDUCING TEE					
19. UNION					
20. TRUE Y					

Fig. 2-2 (*contd.*)

FITTINGS

Screwed or Threaded Fittings

Screwed fittings are used primarily on low-temperature/low-pressure installations and can be found throughout the typical house or apartment building (in water systems, for example). They are used for small-size pipes usually $2\frac{1}{2}$ in. (6.35 cm) or less in diameter. The threads on a screwed fitting correspond to the American Standard pipe threads.

Fittings can come with a tapered thread, which is more common, or with a straight thread for special applications. In practice the pipe fitter will often use a fitting with a straight thread and pipe with a tapered thread, thereby providing a tighter joint. To form a tight joint with a threaded fitting, dope or pipe compounds are employed. Threaded or screwed connections are considered the least leakproof joint possible in a piping system. In general, pipes 2 in. (5.08 cm) or larger are not fitted with screwed connections. Because of the advantages of all-welded systems, they are being employed in place of screwed fittings because of ease in makeup and leak resistance. Threaded joints should be used with caution in systems that have a high-temperature/high-pressure requirement, since cutting the threads reduces the wall thickness of the affected pipe and thereby increases the chance of failure.

Screwed fittings are available in steel, malleable iron, bronze, brass, and cast iron. Cast brass screwed fittings are covered by ASA B16.15 and used in systems with a maximum working steam pressure of 250 psig. Cast iron screwed fittings are covered by ASA B16.4 and rated at a maximum saturated steam pressure of 250 psig. Malleable iron screwed fittings are covered under ASA B16.3 and are used at a maximum of 300 psig. All screwed fittings have been standardized by the ASA.

Soldered Fittings and Flared Fittings

Soldered joint fittings are used primarily for brass or copper water tubes. These fittings are covered by ASA H23.1 and manufactured in accordance with ASA B16.22 and B16.18. Soldered joints are used primarily for thin-wall tubing sizes 1/4 in. to 4 in. (0.64 to 10.16 cm) OD. This type of joint has a low-temperature/low-pressure rating and therefore will not be encountered often in petrochemical or power generation facilities. Soldered joint fittings come in wrought metal, cast brass, or bronze and are used mainly for fittings on water systems for domestic use.

Flared fittings or joints are used on soft copper tubing for low-pressure/low-temperature situations.

Bell and Spigot Fittings

The bell and spigot fitting is used for cast iron, low-pressure/low-temperature situations and is not found very often in petrochemical, power generation, or heating systems. It is used primarily in sewage, water, and gas lines for underground service. There are also other methods of connecting cast iron pipe, including welding, sleeve coupling, universal joints, and ball and socket types. For steel and wrought iron pipes the welded, threaded or flanged fittings are the most common.

TABLE 2-1 DIMENSIONS FOR WELDED FITTINGS
(*COURTESY ITT GRINNELL CORPORATION.*)

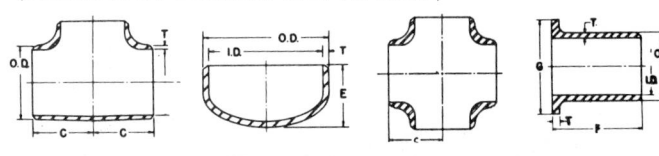

180° RETURNS LONG R K	180° RETURNS SHORT R K	45° ELBOWS B	TEES C	CAPS E	CROSSES C	STUB ENDS F	STUB ENDS G
1⅞	—	⅝	1	1	—	3	1⅜
1¹¹⁄₁₆	—	⁷⁄₁₆	1⅛	1	—	3	1¹¹⁄₁₆
2³⁄₁₆	1⅝	⅞	1½	1½	—	4	2
2¾	2¹⁄₁₆	1	1⅞	1½	1⅞	4	2½
3¼	2⁷⁄₁₆	1⅛	2¼	1½	2¼	4	2⅞
4³⁄₁₆	3³⁄₁₆	1⅜	2½	1½*	2½	6	3⅝
5³⁄₁₆	3¹⁵⁄₁₆	1¾	3	1½*	3	6	4⅛
6¼	4¾	2	3⅜	2*	3⅜	6	5
7¼	5½	2¼	3¾	2½*	3¾	6	5½
8¼	6¼	2½	4⅛	2½*	4⅛	6	6³⁄₁₆
10⁵⁄₁₆	7¾	3⅛	4⅞	3*	4⅞	8	7⁵⁄₁₆
12⁵⁄₁₆	9⁵⁄₁₆	3¾	5⅝	3½*	5⅝	8	8½
16⁵⁄₁₆	12⁵⁄₁₆	5	7	4*	7	8	10⅝
20⅜	15⅜	6¼	8½	5*	8½	10	12¾
24⅜	18⅜	7½	10	6*	10	10	15
28	21	8¾	11	6½*	11	12	16¼
32	24	10	12	7*	12	12	18½
36	27	11¼	13½	8*	13½	12	21
40	30	12½	15	9*	15	12	23
44	—	13½	16½	10	16½	—	—
48	36	15	17	10½	17	12	27¼
52	—	16	19½	10½	—	—	—
60	45	18½	22	10½	—	—	—
—	—	21	25	10½	—	—	—
—	54	22¼	26½	10½	—	—	—
—	—	26	—	12	—	—	—

*Dimensions apply to STD and XS only.

TABLE 2-1 *(CONTD.)*

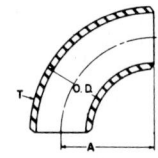

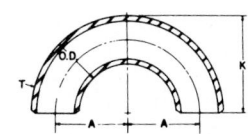

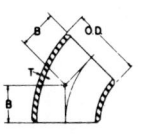

NOM PIPE SIZE	OD	WALL THICKNESS T				90° ELBOWS	
		STD.	XS	160	XX	LONG R A	SHORT R A
½	.840	.109	.147	—	—	1½	—
¾	1.050	.113	.154	—	.308	1⅛	—
1	1.315	.133	.179	.250	.358	1½	1
1¼	1.660	.140	.191	.250	.382	1⅞	1¼
1½	1.900	.145	.200	.281	.400	2¼	1½
2	2.375	.154	.218	.344	.436	3	2
2½	2.875	.203	.276	.375	.552	3¾	2½
3	3.500	.216	.300	.438	.600	4½	3
3½	4.000	.226	.318	—	.636	5¼	3½
4	4.500	.237	.337	.531	.674	6	4
5	5.563	.258	.375	.625	.750	7½	5
6	6.625	.280	.432	.719	.864	9	6
8	8.625	.322	.500	.906	.875	12	8
10	10.750	.365	.500	1.125	1.000	15	10
12	12.750	.375	.500	1.312	1.000	18	12
14	14.000	.375	.500	—	—	21	14
16	16.000	.375	.500	—	—	24	16
18	18.000	.375	.500	—	—	27	18
20	20.000	.375	.500	—	—	30	20
22	22.000	.375	.500	—	—	33	—
24	24.000	.375	.500	—	—	36	24
26	26.000	.375	.500	—	—	39	—
30	30.000	.375	.500	—	—	45	30
34	34.000	.375	.500	—	—	51	—
36	36.000	.375	.500	—	—	54	36
42	42.000	.375	.500	—	—	63	48

TABLE 2-1 *(CONTD.)*

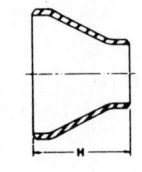

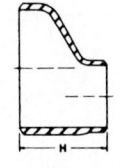

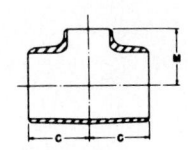

NOM PIPE SIZE		CONCENTRIC AND ECCENTRIC REDUCERS	REDUCING OUTLET TEES	
		H	C	M
½×	¼ ⅜	—	1	1 1
¾×	⅜ ½	1½	1⅛	1⅛ 1⅛
1×	⅜ ½ ¾	2	1½	1½ 1½ 1½
1¼×	½ ¾ 1	2	1⅞	1⅞ 1⅞ 1⅞
1½×	½ ¾ 1 1¼	2½	2¼	2¼ 2¼ 2¼ 2¼
2×	¾ 1 1¼ 1½	3	2½	1¾ 2 2¼ 2⅜
2½×	1 1¼ 1½ 2	3½	3	2¼ 2½ 2⅝ 2¾
3×	1 1¼ 1½ 2 2½	— 3½	3⅜	2⅝ 2¾ 2⅞ 3 3¼
3½×	1¼ 1½ 2 2½ 3	4	— 3¾	— 3⅛ 3¼ 3½ 3⅝
4×	1½ 2 2½ 3 3½	4	4⅛	3⅜ 3½ 3¾ 3⅞ 4
5×	2 2½ 3 3½ 4	5	4⅞	4⅛ 4¼ 4⅜ 4½ 4⅝
6×	2½ 3 3½ 4 5	5½	5⅝	4¾ 4⅞ 5 5⅛ 5⅜
8×	3 3½ 4 5 6	— 6	7	6 6 6⅛ 6⅜ 6⅝
10×	4 5 6 8	7	8½	7¼ 7½ 7⅝ 8
12×	5 6 8 10	8	10	8½ 8⅝ 9 9½
14×	6 8 10 12	13	11	9⅜ 9¼ 10⅛ 10⅝

TABLE 2-2 FORGED STEEL SOCKET-WELDING FITTINGS (DIMENSIONS IN INCHES) (COURTESY CRANE COMPANY.)

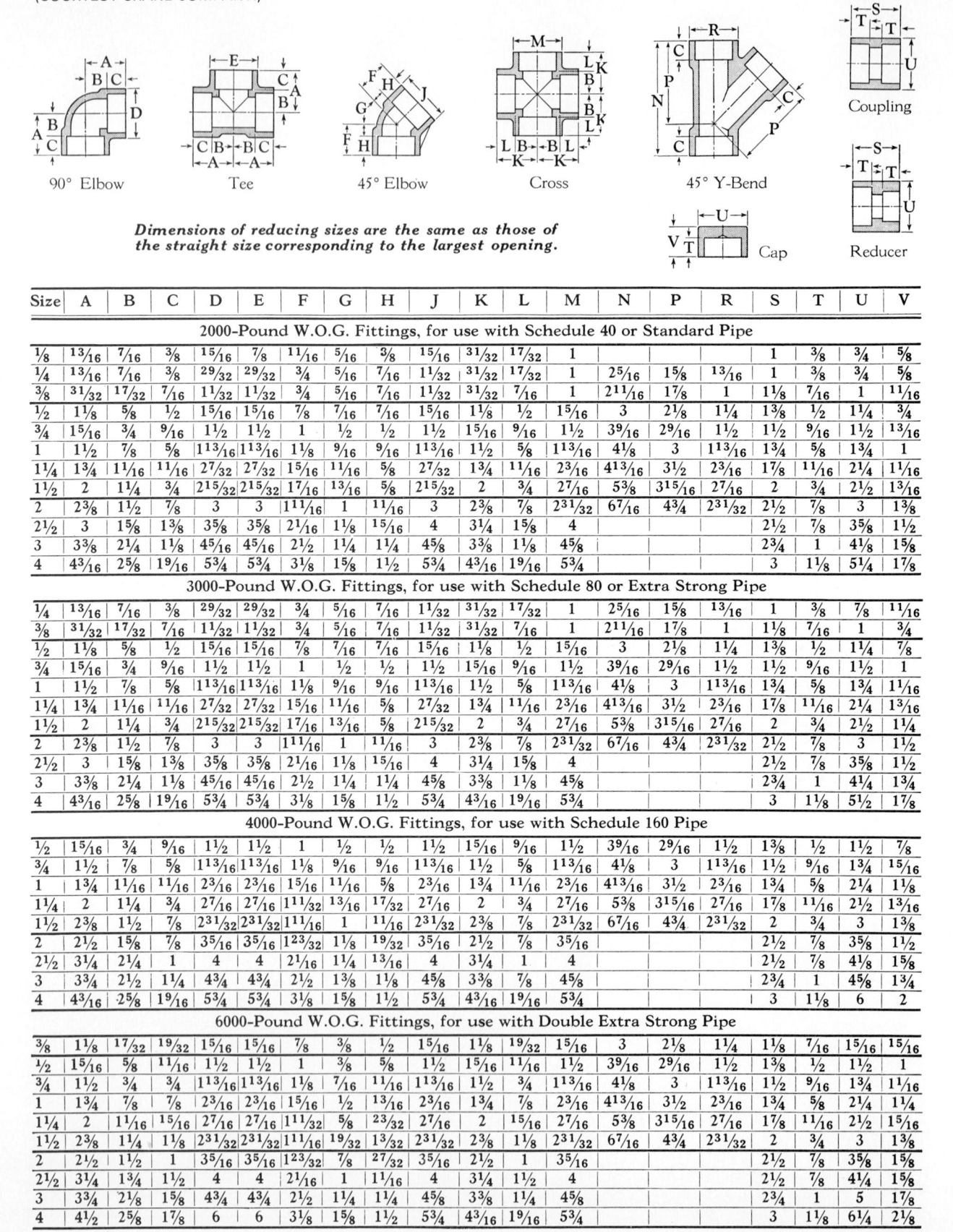

90° Elbow   *Tee*   *45° Elbow*   *Cross*   *45° Y-Bend*   *Coupling*   *Cap*   *Reducer*

Dimensions of reducing sizes are the same as those of the straight size corresponding to the largest opening.

Size	A	B	C	D	E	F	G	H	J	K	L	M	N	P	R	S	T	U	V
\multicolumn{20}{c}{2000-Pound W.O.G. Fittings, for use with Schedule 40 or Standard Pipe}
1/8	13/16	7/16	3/8	15/16	7/8	11/16	5/16	3/8	15/16	31/32	17/32	1			1	3/8	3/4	5/8
1/4	13/16	7/16	3/8	29/32	29/32	3/4	5/16	7/16	1 1/32	31/32	17/32	1	2 5/16	1 5/8	13/16	1	3/8	3/4	5/8
3/8	31/32	17/32	7/16	1 1/32	1 1/32	3/4	5/16	7/16	1 1/32	31/32	7/16	1	2 11/16	1 7/8	1	1 1/8	7/16	1	11/16
1/2	1 1/8	5/8	1/2	15/16	15/16	7/8	7/16	7/16	15/16	1 1/8	1/2	15/16	3	2 1/8	1 1/4	1 3/8	1/2	1 1/4	3/4
3/4	15/16	3/4	9/16	1 1/2	1 1/2	1	1/2	1/2	1 1/2	15/16	9/16	1 1/2	3 9/16	2 9/16	1 1/2	1 1/2	9/16	1 1/2	13/16
1	1 1/2	7/8	5/8	1 13/16	1 13/16	1 1/8	9/16	9/16	1 13/16	1 1/2	5/8	1 13/16	4 1/8	3	1 13/16	1 3/4	5/8	1 3/4	1
1 1/4	1 3/4	1 1/16	11/16	2 7/32	2 7/32	15/16	11/16	5/8	2 7/32	1 3/4	11/16	2 3/16	4 13/16	3 1/2	2 3/16	1 7/8	11/16	2 1/4	1 1/16
1 1/2	2	1 1/4	3/4	2 15/32	2 15/32	17/16	13/16	5/8	2 15/32	2	3/4	2 7/16	5 3/8	3 15/16	2 7/16	2	3/4	2 1/2	13/16
2	2 3/8	1 1/2	7/8	3	3	1 11/16	1	11/16	3	2 3/8	7/8	2 31/32	6 7/16	4 3/4	2 31/32	2 1/2	7/8	3	1 3/8
2 1/2	3	1 5/8	1 3/8	3 5/8	3 5/8	2 1/16	1 1/8	15/16	4	3 1/4	1 5/8	4			2 1/2	7/8	3 5/8	1 1/2
3	3 3/8	2 1/4	1 1/8	4 5/16	4 5/16	2 1/2	1 1/4	1 1/4	4 5/8	3 3/8	1 1/8	4 5/8			2 3/4	1	4 1/4	1 5/8
4	4 3/16	2 5/8	1 9/16	5 3/4	5 3/4	3 1/8	1 5/8	1 1/2	5 3/4	4 3/16	1 9/16	5 3/4			3	1 1/8	5 1/4	1 7/8
\multicolumn{20}{c}{3000-Pound W.O.G. Fittings, for use with Schedule 80 or Extra Strong Pipe}
1/4	13/16	7/16	3/8	29/32	29/32	3/4	5/16	7/16	1 1/32	31/32	17/32	1	2 5/16	1 5/8	13/16	1	3/8	7/8	11/16
3/8	31/32	17/32	7/16	1 1/32	1 1/32	3/4	5/16	7/16	1 1/32	31/32	7/16	1	2 11/16	1 7/8	1	1 1/8	7/16	1	3/4
1/2	1 1/8	5/8	1/2	15/16	15/16	7/8	7/16	7/16	15/16	1 1/8	1/2	15/16	3	2 1/8	1 1/4	1 3/8	1/2	1 1/4	7/8
3/4	15/16	3/4	9/16	1 1/2	1 1/2	1	1/2	1/2	1 1/2	15/16	9/16	1 1/2	3 9/16	2 9/16	1 1/2	1 1/2	9/16	1 1/2	1
1	1 1/2	7/8	5/8	1 13/16	1 13/16	1 1/8	9/16	9/16	1 13/16	1 1/2	5/8	1 13/16	4 1/8	3	1 13/16	1 3/4	5/8	1 3/4	1 1/16
1 1/4	1 3/4	1 1/16	11/16	2 7/32	2 7/32	15/16	11/16	5/8	2 7/32	1 3/4	11/16	2 3/16	4 13/16	3 1/2	2 3/16	1 7/8	11/16	2 1/4	1 3/16
1 1/2	2	1 1/4	3/4	2 15/32	2 15/32	17/16	13/16	5/8	2 15/32	2	3/4	2 7/16	5 3/8	3 15/16	2 7/16	2	3/4	2 1/2	1 1/4
2	2 3/8	1 1/2	7/8	3	3	1 11/16	1	11/16	3	2 3/8	7/8	2 31/32	6 7/16	4 3/4	2 31/32	2 1/2	7/8	3	1 1/2
2 1/2	3	1 5/8	1 3/8	3 5/8	3 5/8	2 1/16	1 1/8	15/16	4	3 1/4	1 5/8	4			2 1/2	7/8	3 5/8	1 1/2
3	3 3/8	2 1/4	1 1/8	4 5/16	4 5/16	2 1/2	1 1/4	1 1/4	4 5/8	3 3/8	1 1/8	4 5/8			2 3/4	1	4 1/4	1 3/4
4	4 3/16	2 5/8	1 9/16	5 3/4	5 3/4	3 1/8	1 5/8	1 1/2	5 3/4	4 3/16	1 9/16	5 3/4			3	1 1/8	5 1/2	1 7/8
\multicolumn{20}{c}{4000-Pound W.O.G. Fittings, for use with Schedule 160 Pipe}
1/2	15/16	3/4	9/16	1 1/2	1 1/2	1	1/2	1/2	1 1/2	15/16	9/16	1 1/2	3 9/16	2 9/16	1 1/2	1 3/8	1/2	1 1/2	7/8
3/4	1 1/2	7/8	5/8	1 13/16	1 13/16	1 1/8	9/16	9/16	1 13/16	1 1/2	5/8	1 13/16	4 1/8	3	1 13/16	1 1/2	9/16	1 3/4	15/16
1	1 3/4	1 1/16	11/16	2 3/16	2 3/16	15/16	11/16	5/8	2 3/16	1 3/4	11/16	2 3/16	4 13/16	3 1/2	2 3/16	1 3/4	5/8	2 1/4	1 1/8
1 1/4	2	1 1/4	3/4	2 7/16	2 7/16	1 11/32	13/16	17/32	2 7/16	2	3/4	2 7/16	5 3/8	3 15/16	2 7/16	1 7/8	11/16	2 1/2	1 3/16
1 1/2	2 3/8	1 1/2	7/8	2 31/32	2 31/32	1 11/16	1	11/16	2 31/32	2 3/8	7/8	2 31/32	6 7/16	4 3/4	2 31/32	2	3/4	3	1 3/8
2	2 1/2	1 5/8	7/8	3 5/16	3 5/16	1 23/32	1 1/8	19/32	3 5/16	2 1/2	7/8	3 5/16			2 1/2	7/8	3 5/8	1 1/2
2 1/2	3 1/4	2 1/4	1	4	4	2 1/16	1 1/4	13/16	4	3 1/4	1	4			2 1/2	7/8	4 1/8	1 5/8
3	3 3/4	2 1/2	1 1/4	4 3/4	4 3/4	2 1/2	1 3/8	1 1/8	4 5/8	3 3/8	7/8	4 5/8			2 3/4	1	4 5/8	1 3/4
4	4 3/16	2 5/8	1 9/16	5 3/4	5 3/4	3 1/8	1 5/8	1 1/2	5 3/4	4 3/16	1 9/16	5 3/4			3	1 1/8	6	2
\multicolumn{20}{c}{6000-Pound W.O.G. Fittings, for use with Double Extra Strong Pipe}
3/8	1 1/8	17/32	19/32	15/16	15/16	7/8	3/8	1/2	15/16	1 1/8	19/32	15/16	3	2 1/8	1 1/4	1 1/8	7/16	15/16	15/16
1/2	15/16	5/8	11/16	1 1/2	1 1/2	1	3/8	5/8	1 1/2	15/16	11/16	1 1/2	3 9/16	2 9/16	1 1/2	1 3/8	1/2	1 1/2	1
3/4	1 1/2	3/4	3/4	1 13/16	1 13/16	1 1/8	7/16	11/16	1 13/16	1 1/2	3/4	1 13/16	4 1/8	3	1 13/16	1 1/2	9/16	1 3/4	1 1/16
1	1 3/4	7/8	7/8	2 3/16	2 3/16	15/16	1/2	13/16	2 3/16	1 3/4	7/8	2 3/16	4 13/16	3 1/2	2 3/16	1 3/4	5/8	2 1/4	1 1/4
1 1/4	2	1 1/16	15/16	2 7/16	2 7/16	1 11/32	5/8	23/32	2 7/16	2	15/16	2 7/16	5 3/8	3 15/16	2 7/16	1 7/8	11/16	2 1/2	15/16
1 1/2	2 3/8	1 1/4	1 1/8	2 31/32	2 31/32	1 11/16	19/32	13/32	2 31/32	2 3/8	1 1/8	2 31/32	6 7/16	4 3/4	2 31/32	2	3/4	3	1 3/8
2	2 1/2	1 1/2	1	3 5/16	3 5/16	1 23/32	7/8	27/32	3 5/16	2 1/2	1	3 5/16			2 1/2	7/8	3 5/8	1 5/8
2 1/2	3 1/4	1 3/4	1 1/2	4	4	2 1/16	1	1 11/16	4	3 1/4	1 1/2	4			2 1/2	7/8	4 1/4	1 5/8
3	3 3/8	2 1/8	1 5/8	4 3/4	4 3/4	2 1/4	1 1/4	1 1/4	4 5/8	3 3/8	1 1/4	4 5/8			2 3/4	1	5	1 7/8
4	4 1/2	2 5/8	1 7/8	6	6	3 1/8	1 5/8	1 1/2	5 3/4	4 3/16	1 9/16	5 3/4			3	1 1/8	6 1/4	2 1/8

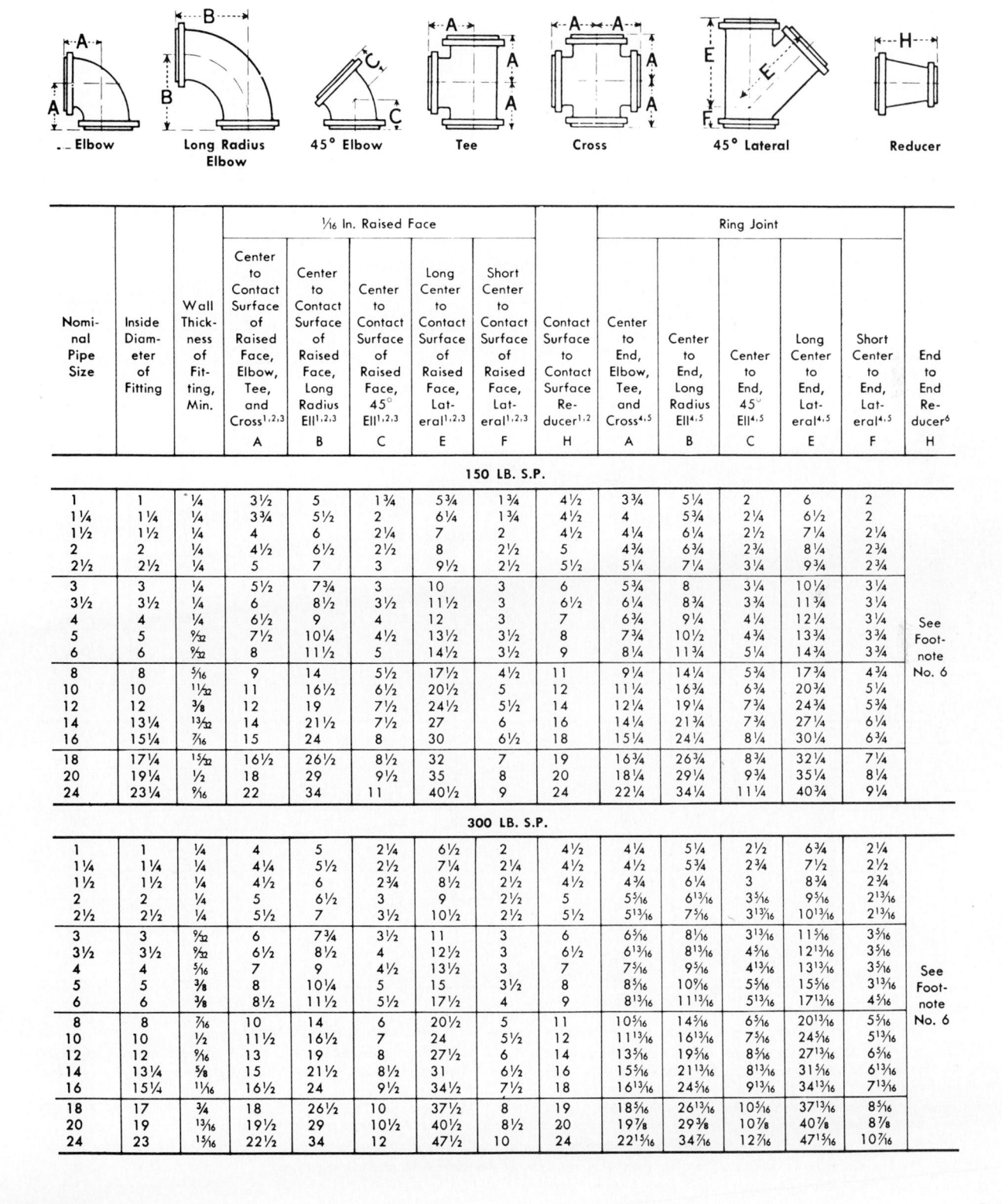

TABLE 2-3 DIMENSIONS OF STEEL FLANGED FITTINGS (*COURTESY LUNKENHEIMER COMPANY, CINCINNATI, OH 45214.*)

150-300 LB. S.P.

American National Standard Institute ANSI B16.5— 1973

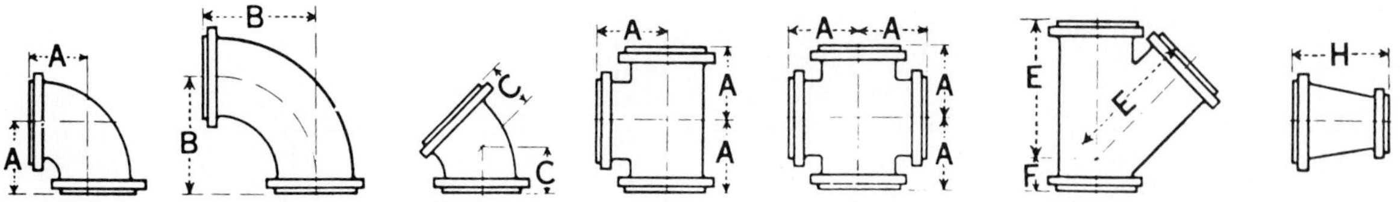

Elbow | Long Radius Elbow | 45° Elbow | Tee | Cross | 45° Lateral | Reducer

Nominal Pipe Size	Inside Diameter of Fitting	Wall Thickness of Fitting, Min.	1/16 In. Raised Face						Ring Joint					
			Center to Contact Surface of Raised Face, Elbow, Tee, and Cross[1,2,3] A	Center to Contact Surface of Raised Face, Long Radius Ell[1,2,3] B	Center to Contact Surface of Raised Face, 45° Ell[1,2,3] C	Long Center to Contact Surface of Raised Face, Lateral[1,2,3] E	Short Center to Contact Surface of Raised Face, Lateral[1,2,3] F	Contact Surface to Contact Surface Reducer[1,2] H	Center to End, Elbow, Tee, and Cross[4,5] A	Center to End, Long Radius Ell[4,5] B	Center to End, 45° Ell[4,5] C	Long Center to End, Lateral[4,5] E	Short Center to End, Lateral[4,5] F	End to End Reducer[6] H
150 LB. S.P.														
1	1	1/4	3 1/2	5	1 3/4	5 3/4	1 3/4	4 1/2	3 3/4	5 1/4	2	6	2	
1 1/4	1 1/4	1/4	3 3/4	5 1/2	2	6 1/4	1 3/4	4 1/2	4	5 3/4	2 1/4	6 1/2	2	
1 1/2	1 1/2	1/4	4	6	2 1/4	7	2	4 1/2	4 1/4	6 1/4	2 1/2	7 1/4	2 1/4	
2	2	1/4	4 1/2	6 1/2	2 1/2	8	2 1/2	5	4 3/4	6 3/4	2 3/4	8 1/4	2 3/4	
2 1/2	2 1/2	1/4	5	7	3	9 1/2	2 1/2	5 1/2	5 1/4	7 1/4	3 1/4	9 3/4	2 3/4	
3	3	1/4	5 1/2	7 3/4	3	10	3	6	5 3/4	8	3 1/4	10 1/4	3 1/4	
3 1/2	3 1/2	1/4	6	8 1/2	3 1/2	11 1/2	3	6 1/2	6 1/4	8 3/4	3 3/4	11 3/4	3 1/4	
4	4	1/4	6 1/2	9	4	12	3	7	6 3/4	9 1/4	4 1/4	12 1/4	3 1/4	
5	5	9/32	7 1/2	10 1/4	4 1/2	13 1/2	3 1/2	8	7 3/4	10 1/2	4 3/4	13 3/4	3 3/4	See Footnote No. 6
6	6	9/32	8	11 1/2	5	14 1/2	3 1/2	9	8 1/4	11 3/4	5 1/4	14 3/4	3 3/4	
8	8	5/16	9	14	5 1/2	17 1/2	4 1/2	11	9 1/4	14 1/4	5 3/4	17 3/4	4 3/4	
10	10	11/32	11	16 1/2	6 1/2	20 1/2	5	12	11 1/4	16 3/4	6 3/4	20 3/4	5 1/4	
12	12	3/8	12	19	7 1/2	24 1/2	5 1/2	14	12 1/4	19 1/4	7 3/4	24 3/4	5 3/4	
14	13 1/4	13/32	14	21 1/2	7 1/2	27	6	16	14 1/4	21 3/4	7 3/4	27 1/4	6 1/4	
16	15 1/4	7/16	15	24	8	30	6 1/2	18	15 1/4	24 1/4	8 1/4	30 1/4	6 3/4	
18	17 1/4	15/32	16 1/2	26 1/2	8 1/2	32	7	19	16 3/4	26 3/4	8 3/4	32 1/4	7 1/4	
20	19 1/4	1/2	18	29	9 1/2	35	8	20	18 1/4	29 1/4	9 3/4	35 1/4	8 1/4	
24	23 1/4	9/16	22	34	11	40 1/2	9	24	22 1/4	34 1/4	11 1/4	40 3/4	9 1/4	
300 LB. S.P.														
1	1	1/4	4	5	2 1/4	6 1/2	2	4 1/2	4 1/4	5 1/4	2 1/2	6 3/4	2 1/4	
1 1/4	1 1/4	1/4	4 1/4	5 1/2	2 1/2	7 1/4	2 1/4	4 1/2	4 1/2	5 3/4	2 3/4	7 1/2	2 1/2	
1 1/2	1 1/2	1/4	4 1/2	6	2 3/4	8 1/2	2 1/2	4 1/2	4 3/4	6 1/4	3	8 3/4	2 3/4	
2	2	1/4	5	6 1/2	3	9	2 1/2	5	5 5/16	6 13/16	3 5/16	9 5/16	2 13/16	
2 1/2	2 1/2	1/4	5 1/2	7	3 1/2	10 1/2	2 1/2	5 1/2	5 13/16	7 5/16	3 13/16	10 13/16	2 13/16	
3	3	9/32	6	7 3/4	3 1/2	11	3	6	6 5/16	8 1/16	3 13/16	11 5/16	3 5/16	
3 1/2	3 1/2	9/32	6 1/2	8 1/2	4	12 1/2	3	6 1/2	6 13/16	8 13/16	4 5/16	12 13/16	3 5/16	
4	4	5/16	7	9	4 1/2	13 1/2	3	7	7 5/16	9 5/16	4 13/16	13 13/16	3 5/16	
5	5	3/8	8	10 1/4	5	15	3 1/2	8	8 5/16	10 9/16	5 5/16	15 5/16	3 13/16	See Footnote No. 6
6	6	3/8	8 1/2	11 1/2	5 1/2	17 1/2	4	9	8 13/16	11 13/16	5 13/16	17 13/16	4 5/16	
8	8	7/16	10	14	6	20 1/2	5	11	10 5/16	14 5/16	6 5/16	20 13/16	5 5/16	
10	10	1/2	11 1/2	16 1/2	7	24	5 1/2	12	11 13/16	16 13/16	7 5/16	24 5/16	5 13/16	
12	12	9/16	13	19	8	27 1/2	6	14	13 5/16	19 5/16	8 5/16	27 13/16	6 5/16	
14	13 1/4	5/8	15	21 1/2	8 1/2	31	6 1/2	16	15 5/16	21 13/16	8 13/16	31 5/16	6 13/16	
16	15 1/4	11/16	16 1/2	24	9 1/2	34 1/2	7 1/2	18	16 13/16	24 5/16	9 13/16	34 13/16	7 13/16	
18	17	3/4	18	26 1/2	10	37 1/2	8	19	18 5/16	26 13/16	10 5/16	37 13/16	8 5/16	
20	19	13/16	19 1/2	29	10 1/2	40 1/2	8 1/2	20	19 7/8	29 3/8	10 7/8	40 7/8	8 7/8	
24	23	15/16	22 1/2	34	12	47 1/2	10	24	22 15/16	34 7/16	12 7/16	47 15/16	10 7/16	

TABLE 2-4 DIMENSIONS OF CAST-IRON FLANGED FITTINGS
(COURTESY LUNKENHEIMER COMPANY, CINCINNATI, OH 45214.)

CLASS 125 LB. S.P. American National Standards Institute ANSI B16.1—1967
CLASS 250 LB. S.P. American National Standards Institute ANSI B16.2—1960

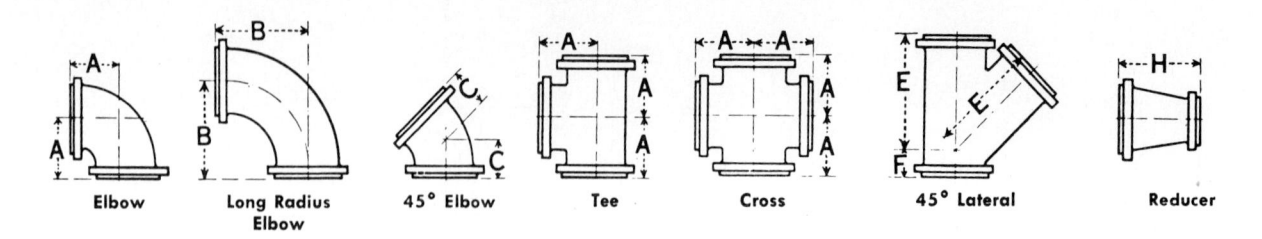

Nominal Pipe Size	Inside Diameter of Fitting	Wall Thickness	Center to Contact Surface, Metal Elbow, Tee and Cross [1,2,3] A	Center to Contact Surface, Long Radius Ell [1,2,3] B	Center to Contact Surface, 45° Ell [1,2,3] C	Long Center to Contact Surface, Lateral [1,2,8] E	Short Center to Contact Surface, Lateral [1,2,8] F	Contact Surface to Contact Surface, Reducer [1,2,8] H
	CLASS 125 LB.
1	1	5/16	3½	5	1¾	5¾	1¾	....
1¼	1¼	5/16	3¾	5½	2	6¼	1¾	....
1½	1½	5/16	4	6	2¼	7	2	....
2	2	5/16	4½	6½	2½	8	2½	5
2½	2½	5/16	5	7	3	9½	2½	5½
3	3	3/8	5½	7¾	3	10	3	6
3½	3½	7/16	6	8½	3½	11½	3	6½
4	4	½	6½	9	4	12	3	7
5	5	½	7½	10¼	4½	13½	3½	8
6	6	9/16	8	11½	5	14½	3½	9
8	8	5/8	9	14	5½	17½	4½	11
10	10	¾	11	16½	6½	20½	5	12
12	12	13/16	12	19	7½	24½	5½	14
14	14	7/8	14	21½	7½	27	6	16
16	16	1	15	24	8	30	6½	18
18	18	1 1/16	16½	26½	8½	32	7	19
20	20	1 1/8	18	29	9½	35	8	20
24	24	1 ¼	22	34	11	40½	9	24
	CLASS 250 LB.
2	2	7/16	5	6½	3	9	2½	5
2½	2½	½	5½	7	3½	10½	2½	5½
3	3	9/16	6	7¾	3½	11	3	6
3½	3½	9/16	6½	8½	4	12½	3	6½
4	4	5/8	7	9	4½	13½	3	7
5	5	11/16	8	10¼	5	15	3½	8
6	6	¾	8½	11½	5½	17½	4	9
8	8	13/16	10	14	6	20½	5	11
10	10	15/16	11½	16½	7	24	5½	12
12	12	1	13	19	8	27½	6	14
14	13¼	1 1/8	15	21½	8½	31	6½	16
16	15¼	1 ¼	16½	24	9½	34½	7½	18
18	17	1 3/8	18	26½	10	37½	8	19
20	19	1 ½	19½	29	10½	40½	8½	20
24	23	1 5/8	22½	34	12	47½	10	24

Notes: All dimensions are given in inches.

[1] All Class 125 cast-iron flanged fittings shall be plain faced; i.e., without projection or raised face.

All Class 250 cast-iron flanges and flanged fittings have a 1/16 in. raised face provided on the flange of each opening of these fittings and is included in (a) "thickness of flange," (b) "center to contact surface," and (c) "contact surface to contact surface dimenison."

[2] Where facings other than the 1/16 in. raised face are used (Class 250 only), the "center to contact surface" dimensions shall remain unchanged, and new "center to contact surface" dimensions shall be established to suit the facing used.

[3] Reducing fittings shall have the same "center to contact surface" dimensions as those of straight size fittings of the largest opening.

[4] Wall thickness at no point shall be less than 87½ per cent of the dimensions given in the table.

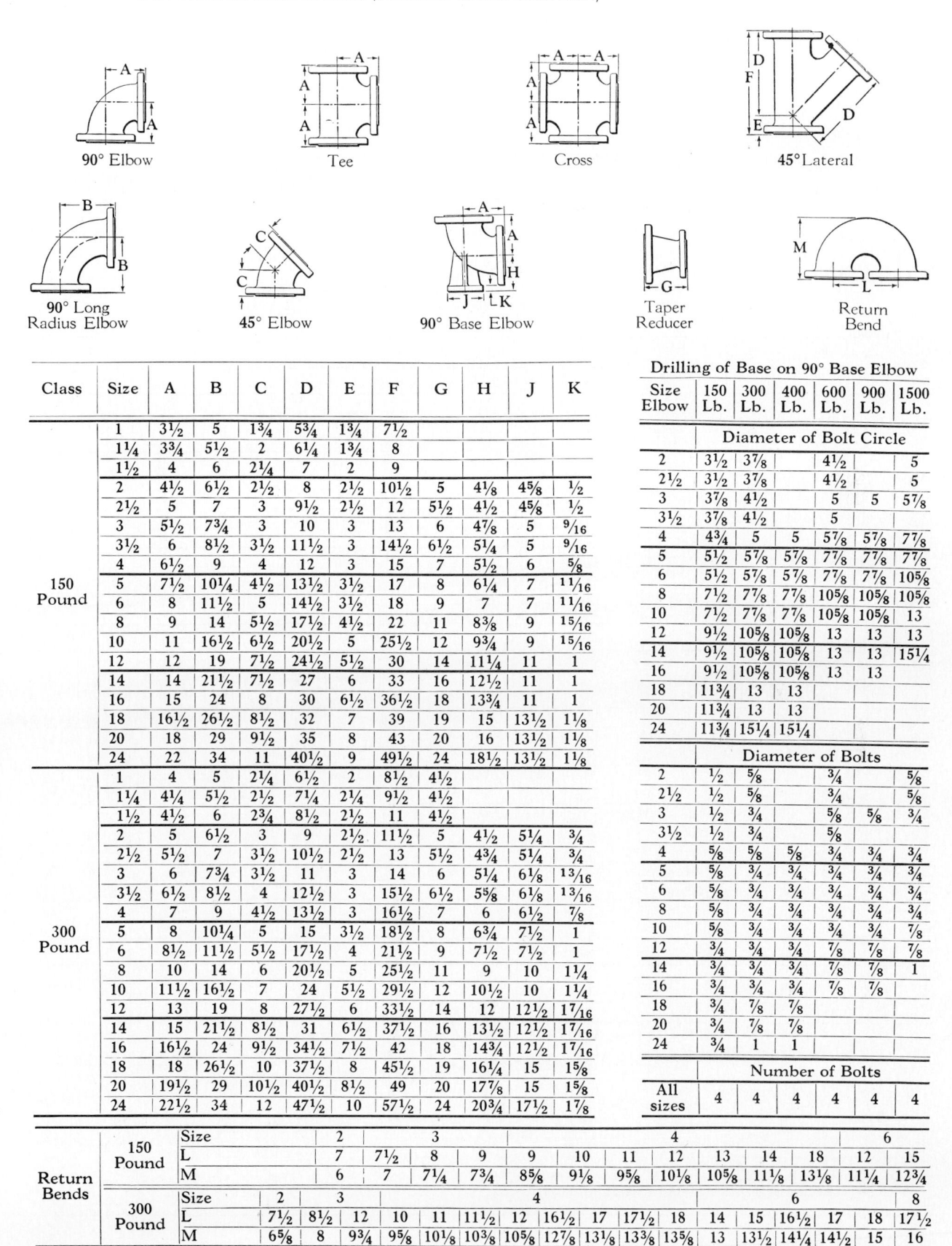

TABLE 2-5   CAST STEEL FLANGED FITTINGS (*COURTESY CRANE COMPANY.*)

90° Elbow | Tee | Cross | 45° Lateral

90° Long Radius Elbow | 45° Elbow | 90° Base Elbow | Taper Reducer | Return Bend

Class	Size	A	B	C	D	E	F	G	H	J	K
150 Pound	1	3½	5	1¾	5¾	1¾	7½				
	1¼	3¾	5½	2	6¼	1¾	8				
	1½	4	6	2¼	7	2	9				
	2	4½	6½	2½	8	2½	10½	5	4⅛	4⅝	½
	2½	5	7	3	9½	2½	12	5½	4¼	4⅝	½
	3	5½	7¾	3	10	3	13	6	4⅞	5	9/16
	3½	6	8½	3½	11½	3	14½	6½	5¼	5	9/16
	4	6½	9	4	12	3	15	7	5½	6	⅝
	5	7½	10¼	4½	13½	3½	17	8	6¼	7	11/16
	6	8	11½	5	14½	3½	18	9	7	7	11/16
	8	9	14	5½	17½	4½	22	11	8⅜	9	15/16
	10	11	16½	6½	20½	5	25½	12	9¾	9	15/16
	12	12	19	7½	24½	5½	30	14	11¼	11	1
	14	14	21½	7½	27	6	33	16	12½	11	1
	16	15	24	8	30	6½	36½	18	13¾	11	1
	18	16½	26½	8½	32	7	39	19	15	13½	1⅛
	20	18	29	9½	35	8	43	20	16	13½	1⅛
	24	22	34	11	40½	9	49½	24	18½	13½	1⅛
300 Pound	1	4	5	2¼	6½	2	8½	4½			
	1¼	4¼	5½	2½	7¼	2¼	9½	4½			
	1½	4½	6	2¾	8½	2½	11	4½			
	2	5	6½	3	9	2½	11½	5	4½	5¼	¾
	2½	5½	7	3½	10½	2½	13	5½	4¾	5¼	¾
	3	6	7¾	3½	11	3	14	6	5¼	6⅛	13/16
	3½	6½	8½	4	12½	3	15½	6½	5⅝	6⅛	13/16
	4	7	9	4½	13½	3	16½	7	6	6½	⅞
	5	8	10¼	5	15	3½	18½	8	6¾	7½	1
	6	8½	11½	5½	17½	4	21½	9	7½	7½	1
	8	10	14	6	20½	5	25½	11	9	10	1¼
	10	11½	16½	7	24	5½	29½	12	10½	10	1¼
	12	13	19	8	27½	6	33½	14	12	12½	1 7/16
	14	15	21½	8½	31	6½	37½	16	13½	12½	1 7/16
	16	16½	24	9½	34½	7½	42	18	14¾	12½	1 7/16
	18	18	26½	10	37½	8	45½	19	16¼	15	1⅝
	20	19½	29	10½	40½	8½	49	20	17⅞	15	1⅝
	24	22½	34	12	47½	10	57½	24	20¾	17½	1⅞

Drilling of Base on 90° Base Elbow

Size Elbow	150 Lb.	300 Lb.	400 Lb.	600 Lb.	900 Lb.	1500 Lb.
Diameter of Bolt Circle							
2	3½	3⅞		4½		5
2½	3½	3⅞		4½		5
3	3⅞	4½		5	5	5⅞
3½	3⅞	4½		5		
4	4¾	5	5	5⅞	5⅞	7⅞
5	5½	5⅞	5⅞	7⅞	7⅞	7⅞
6	5½	5⅞	5⅞	7⅞	7⅞	10⅝
8	7½	7⅞	7⅞	10⅝	10⅝	10⅝
10	7½	7⅞	7⅞	10⅝	10⅝	13
12	9½	10⅝	10⅝	13	13	13
14	9½	10⅝	10⅝	13	13	15¼
16	9½	10⅝	10⅝	13	13	
18	11¾	13	13			
20	11¾	13	13			
24	11¾	15¼	15¼			
Diameter of Bolts							
2	½	⅝		¾		⅝
2½	½	⅝		¾		⅝
3	½	¾		⅝	⅝	¾
3½	½	¾		⅝		
4	⅝	⅝	⅝	¾	¾	¾
5	⅝	¾	¾	¾	¾	¾
6	⅝	¾	¾	¾	¾	¾
8	⅝	¾	¾	¾	¾	¾
10	⅝	¾	¾	¾	¾	⅞
12	¾	¾	¾	⅞	⅞	⅞
14	¾	¾	¾	⅞	⅞	1
16	¾	¾	¾	⅞	⅞	
18	¾	⅞	⅞			
20	¾	⅞	⅞			
24	¾	1	1			
Number of Bolts							
All sizes	4	4	4	4	4	4

Return Bends		Size	2		3		4				6		
	150 Pound	L	7	7½	8	9	9	10	11	12	13	14	18	12	15
		M	6	7	7¼	7¾	8⅝	9⅛	9⅝	10⅛	10⅝	11⅛	13⅛	11¼	12¾
		Size	2		3		4				6		8
	300 Pound	L	7½	8½	12	10	11	11½	12	16½	17	17½	18	14	15	16½	17	18	17½
		M	6⅝	8	9¾	9⅝	10⅛	10⅜	10⅝	12⅞	13⅛	13⅜	13⅝	13	13½	14¼	14½	15	16

FITTINGS

TABLE 2-6 FORGED STEEL SCREWED FITTINGS (*COURTESY CRANE COMPANY.*)

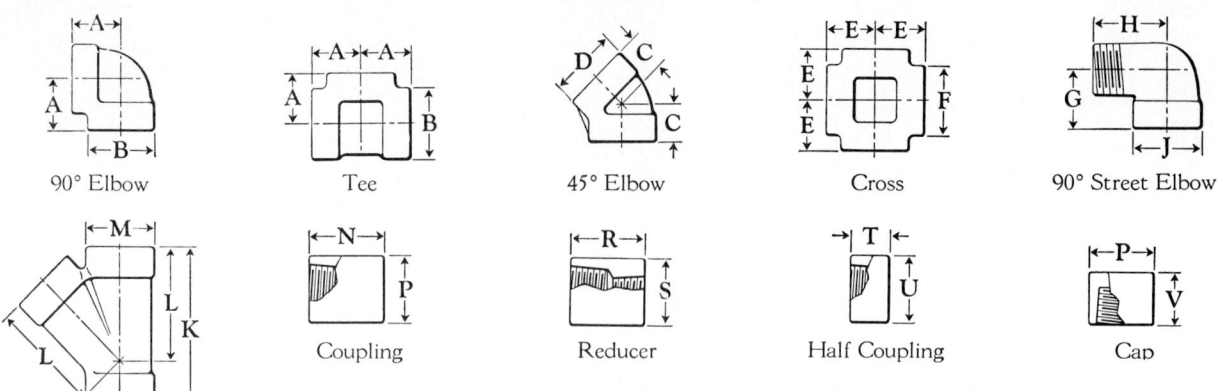

90° Elbow · Tee · 45° Elbow · Cross · 90° Street Elbow
Coupling · Reducer · Half Coupling · Cap
45° Y-Bend

Dimensions of reducing sizes are the same as those of the straight size corresponding to the largest opening.

Size	A	B	C	D	E	F	G	H	J	K	L	M	N	P	R	S	T	U	V
2000-Pound W.O.G. Fittings																		
1/8	13/16	29/32	11/16	15/16	31/32	1													
1/4	13/16	29/32	3/4	11/32	31/32	1				2 5/16	1 5/8	13/16							
3/8	31/32	1 1/32	3/4	1 1/32	31/32	1				2 11/16	1 7/8	1							
1/2	1 1/8	1 5/16	7/8	1 5/16	1 1/8	1 5/16				3	2 1/8	1 5/16							
3/4	1 5/16	1 1/2	1	1 1/2	1 5/16	1 1/2				3 9/16	2 9/16	1 1/2							
1	1 1/2	1 13/16	1 1/8	1 13/16	1 1/2	1 13/16				4 1/8	3	1 13/16							
1 1/4	1 3/4	2 7/32	1 5/16	2 7/32	1 3/4	2 3/16				4 13/16	3 1/2	2 3/16							
1 1/2	2	2 15/32	1 7/16	2 15/32	2	2 7/16				5 3/8	3 15/16	2 7/16							
2	2 3/8	3	1 11/16	3	2 3/8	2 31/32				6 7/16	4 3/4	2 31/32							
2 1/2	3	3 5/8	2 1/16	4	3 1/4	4													
3	3 3/8	4 5/16	2 1/2	4 5/8	3 3/8	4 5/8													
4	4 3/16	5 3/4	3 1/8	5 3/4	4 3/16	5 3/4													
3000-Pound W.O.G. Fittings																		
1/8	13/16	29/32	5/8	7/8	31/32	1							1 1/8	19/32		3/4	5/8	3/4	1/2
1/4	31/32	1 1/32	3/4	1 1/32	31/32	1	7/8	1 1/4	1				1 3/8	3/4	1 3/8	3/4	11/16	3/4	11/16
3/8	1 1/8	1 5/16	7/8	1 5/16	1 1/8	1 5/16	1	1 1/2	1 1/4				1 5/8	15/16	1 1/2	7/8	3/4	7/8	3/4
1/2	1 5/16	1 1/2	1	1 1/2	1 5/16	1 1/2	1 1/4	1 5/8	1 1/2	3 9/16	2 9/16	1 1/2	1 7/8	1 1/8	1 7/8	1 1/8	15/16	1 1/8	1 1/4
3/4	1 1/2	1 13/16	1 1/8	1 13/16	1 1/2	1 13/16	1 3/8	1 7/8	1 3/4	4 1/8	3	1 13/16	2 1/8	1 3/8	2	1 3/8	1	1 3/8	1 1/4
1	1 3/4	2 7/32	1 5/16	2 7/32	1 3/4	2 3/16	1 3/4	2 1/4	2	4 13/16	3 1/2	2 3/16	2 3/8	1 3/4	2 3/8	1 3/4	1 3/16	1 3/4	1 1/2
1 1/4	2	2 15/32	1 7/16	2 15/32	2	2 7/16	2	2 5/8	2 7/16	5 3/8	3 15/16	2 7/16	2 7/8	2 1/4	2 5/8	2 1/4	1 5/16	2 1/4	1 5/8
1 1/2	2 3/8	3	1 11/16	3	2 3/8	2 31/32	2 1/8	2 13/16	2 3/4	6 7/16	4 3/4	2 31/32	2 7/8	2 1/2	3 1/8	2 1/2	1 9/16	2 1/2	1 5/8
2	2 1/2	3 9/16	1 23/32	3 5/16	2 1/2	3 5/16	2 1/2	3 5/16	3 5/16				3 3/8	3	3 3/8	3	1 11/16	3	
2 1/2	3 3/8	4 3/8	2 1/16	4	3 1/4	4							3 5/8	3 5/8	3 5/8	3 5/8	1 13/16	3 5/8	2 3/8
3	3 3/4	4 3/4	2 1/2	4 5/8	3 3/8	4 5/8							4 1/4	4 1/4	4 1/4	4 1/4	2 1/8	4 1/4	2 9/16
3 1/2	4 1/2	6	3 1/8	5 3/4	4 3/16	5 3/4							4 1/2	4 3/4	4 1/2	4 3/4	2 1/4	4 3/4	2 5/8
4	4 1/2	6	3 1/8	5 3/4	4 3/16	5 3/4							4 3/4	5 1/2	4 3/4	5 1/2	2 3/8	5 1/2	2 11/16
5														6 1/2					3 1/16
6														8					3 3/16
6000-Pound W.O.G. Fittings																		
1/8	31/32	1 1/32	3/4	1 1/16	31/32	1							1 1/4	7/8			5/8	3/4	
1/4	1 1/8	1 5/16	7/8	1 5/16	1 1/8	1 5/16	1	1 1/2	1 1/4				1 3/8	1	1 3/8	1	11/16	1	
3/8	1 5/16	1 1/2	1	1 1/2	1 5/16	1 1/2	1 1/8	1 5/8	1 1/2	3 9/16	2 9/16	1 1/2	1 1/2	1 1/4	1 1/2	1 1/4	3/4	1 1/4	
1/2	1 1/2	1 13/16	1 1/8	1 13/16	1 1/2	1 13/16	1 3/8	1 7/8	1 3/4	4 1/8	3	1 13/16	1 7/8	1 1/2	1 7/8	1 1/2	15/16	1 1/2	
3/4	1 3/4	2 7/32	1 5/16	2 3/16	1 3/4	2 3/16	1 3/4	2 1/4	2	4 13/16	3 1/2	2 3/16	2	1 3/4	2	1 3/4	1	1 3/4	
1	2	2 15/32	1 11/32	2 7/16	2	2 7/16	2	2 5/8	2 7/16	5 3/8	3 15/16	2 7/16	2 3/8	2 1/4	2 3/8	2 1/4	1 3/16	2 1/4	
1 1/4	2 3/8	3	1 11/16	2 31/32	2 3/8	2 31/32	2 1/8	2 13/16	2 3/4	6 7/16	4 3/4	2 31/32	2 5/8	2 1/2	2 5/8	2 1/2	1 5/16	2 1/2	
1 1/2	2 1/2	3 9/16	1 23/32	3 5/16	2 1/2	3 5/16	2 1/2	3 5/16	3 5/16				3 1/8	3	3 1/8	3	1 9/16	3	
2	3 3/8	4 3/8	2 1/16	4	3 1/4	4							3 3/8	3 5/8	3 3/8	3 5/8	1 11/16	3 5/8	
2 1/2	3 3/4	4 3/4	2 1/2	4 5/8	3 3/8	4 5/8							3 5/8	4 1/4	3 5/8	4 1/4	1 13/16	4 1/4	
3	4 3/16	5 3/4	3 1/8	5 3/4	4 3/16	5 3/4							4 1/4	5	4 1/4	5	2 1/8	5	
3 1/2	4 1/2	6	3 1/8	5 3/4	4 3/16	5 3/4							4 1/2	5 3/4	4 1/2	5 3/4	2 1/4	5 3/4	
4													4 3/4	6 1/4	4 3/4	6 1/4	2 3/8	6 1/4	

3
FLANGES

Fig. 3-1

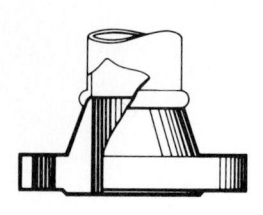

Welding Neck Flanges are distinguished from other types by their long tapered hub and gentle transition of thickness in the region of the butt weld joining them to the pipe. The long tapered hub provides an important reinforcement of the flange proper from the standpoint of strength and resistance to dishing. The smooth transition from flange thickness to pipe wall thickness effected by the taper is extremely beneficial under conditions of repeated bending, caused by line expansion or other variable forces, and produces an endurance strength of welding neck flanged assemblies[1] equivalent to that of a butt welded joint between pipes, which, in practice, is the same as that of unwelded pipe. Thus this type of flange is preferred for every severe service condition, whether this results from high pressure or from sub-zero or elevated temperature, and whether loading conditions are substantially constant or fluctuate between wide limits; welding neck flanges are particularly recommended for handling explosive, flammable or costly liquids, where loss of tightness or local failure may be accompanied by disastrous consequences.

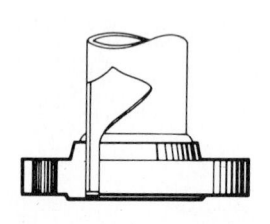

Slip-On Flanges continue to be preferred to welding neck flanges by many users on account of their initially lower cost, the reduced accuracy required in cutting the pipe to length, and the somewhat greater ease by alignment of the assembly; however, their final installed cost is probably not much, if any, less than that of welding neck flanges. Their calculated strength under internal pressure is of the order of two-thirds that of welding neck flanges, and their life under fatigue is about one-third that of the latter. For these reasons, slip-on flanges are limited to sizes NPS ½ to 2½ in the 1500 standard.

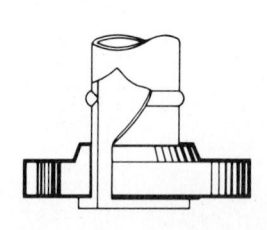

Lap Joint Flanges are primarily employed with lap joint stubs, the combined initial cost of the two items being approximately one-third higher than that of comparable welding neck flanges. Their pressure-holding ability is little, if any, better than that of slip-on flanges and the fatigue life of the assembly is only one-tenth that of welding neck flanges. The chief use of lap joint flanges in carbon or low alloy steel piping systems is in services necessitating frequent dismantling for inspection and cleaning and where the ability to swivel flanges and to align bolt holes materially simplifies the erection of large diameter or unusually stiff piping Their use at points where severe bending stress occurs should be avoided.

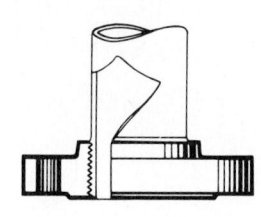

Screwed Flanges made of steel are confined to special applications. Their chief merit lies in the fact that they can be assembled without welding; this explains their use in extremely high pressure services, particularly at or near atmospheric temperature, where alloy steel is essential for strength and where the necessary post-weld heat treatment is impractical. Screwed flanges are unsuited for conditions involving temperature or bending stresses of any magnitude, particularly under cyclic conditions, where leakage through the threads may occur in relatively few cycles of heating or stress; seal welding is sometimes employed to overcome this, but cannot be considered as entirely satisfactory.

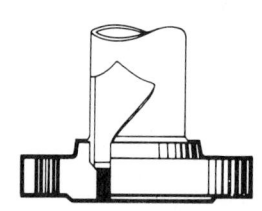

Socket Welding Flanges were initially developed for use on small-size high pressure piping. Their initial cost is about 10% greater than that of slip-on flanges; when provided with an internal weld as illustrated, their static strength is equal to, but their fatigue strength 50% greater than double-welded slip-on flanges. Smooth, pocketless bore conditions can readily be attained (by grinding the internal weld) without having to bevel the flange face and, after welding, to reface the flange as would be required with slip-on flanges. The internally welded socket type flange is becoming increasingly popular in chemical process piping for this reason.

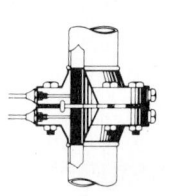

Orifice Flanges are widely used in conjunction with orifice meters for measuring the rate of flow of liquids and gases. They are basically the same as standard welding neck, slip-on and screwed flanges except for the provision of radial, tapped holes in the flange ring for meter connections and additional bolts to act as jack screws to facilitate separating the flanges for inspection or replacement of the orifice plate. In choosing the type of orifice flange, the considerations affecting the choice of welding neck, slip-on and screwed standard flanges apply with equal force.

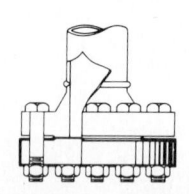

Blind Flanges are used to blank off the ends of piping, valves and pressure vessel openings. From the standpoint of internal pressure and bolt loading, blind flanges, particularly in the larger sizes, are the most highly stressed of all American Standard flange types. However, since the maximum stresses in a blind flange are bending stresses at the center, they can safely be permitted to be higher than in other types of flanges. Where temperature is a service factor, or repeated severe water hammer, consideration should be given to closures made of welding neck flanges and caps.

---

[1] In tests of all types of flanged assemblies, fatigue failure invariably occurred in the pipe or in an unusually weak weld, never in the flange proper. The type of flange, however, and particularly the method of attachment, greatly influence the number of cycles required to cause fracture.

Fig. 3-2 Selection and application of forged steel companion flanges (*Courtesy ITT Grinnell Corporation.*)

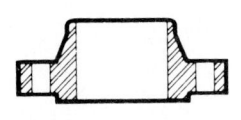

The Raised Face is the most common facing employed with steel flanges; it is $\frac{1}{16}''$ high for Class 150 and Class 300 flanges and $\frac{1}{4}''$ high for all other pressure classes. The facing is machine-tool finished with spiral or concentric grooves (approximately $\frac{1}{64}''$ deep on approximately $\frac{1}{32}''$ centers) to bite into and hold the gasket. Because both flanges of a pair are identical, no stocking or assembly problems are involved in its use. Raised face flanges generally are installed with soft flat ring composition gaskets. The width of the gasket is usually less than the width of the raised face. Faces for use with metal gaskets preferably are smooth finished.

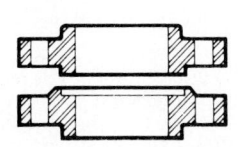

Male-and-Female Facings are standardized in both large and small types. The female face is $\frac{3}{16}''$ deep and the male face $\frac{1}{4}''$ high and both are usually smooth finished since the outer diameter of the female face acts to locate and retain the gasket. The width of the large male and female gasket contact surface, like the raised face, is excessive for use with metal gaskets. The small male and female overcomes this but provides too narrow a gasket surface for screwed flanges assembled with standard weight pipe.

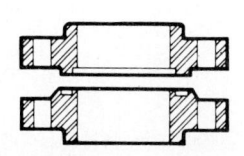

Tongue-and-Groove Facings are also standardized in both large and small types. They differ from male-and-female in that the inside diameters of tongue and groove do not extend to the flange bore, thus retaining the gasket on both its inner and outer diameter; this removes the gasket from corrosive or erosive contact with the line fluid. The small tongue-and-groove construction provides the minimum area of flat gasket it is advisable to use, thus resulting in the minimum bolting load for compressing the gasket and the highest joint efficiency possible with flat gaskets.

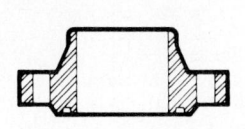

Ring Joint Facing is the most expensive standard facing but also the most efficient, partly because the internal pressure acts on the ring to increase the sealing force. Both flanges of a pair are alike, thus reducing the stocking and assembling problem found with both made-and-female and tongue-and-groove joints. Because the surfaces the gasket contacts are below the flange face, the ring joint facing is least likely of all facings to be damaged in handling or erecting. The flat bottom groove is standard.

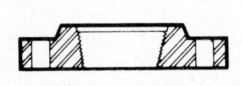

Flat Faces are a variant of raised faces, sometimes formed by machining off the $\frac{1}{64}''$ raised face of Class 150 and Class 300 flanges. Their chief use is for mating with Class 125 and Class 250 cast iron valves and fittings. A flat-faced steel flange permits employing a gasket whose outer diameter equals that of the flange or is tangent to the bolt holes. In this manner the danger of cracking the cast iron flange when the bolts are tightened is avoided.

Flat Ring Gaskets are made of numerous materials such as paper, cloth, rubber, compressed asbestos, ingot iron, nickel, copper, aluminum and other metals, as well as combinations of metals and nonmetals. Their thicknesses normally range from $\frac{1}{64}''$ to $\frac{1}{8}''$. Widths range from $\frac{1}{4}''$ upwards; narrow gaskets are preferable since they require lower bolt loads for joint tightness but they must not be too narrow lest the bolt load crush them, if non-metallic, or indent them into the flange face, if metallic. Paper, cloth and rubber gaskets should not be used for temperatures over 250F. Asbestos may be employed up to 650F or somewhat higher, while ferrous or nickel-base metal gaskets are generally satisfactory for the maximum temperature the flanges themselves will withstand.

Serrated Gaskets are flat, metal gaskets having concentric grooves machined into their face. With the contact area reduced to a few concentric lines, the required bolt load is greatly reduced as compared with an unserrated gasket and hence an efficient joint is obtained for applications where soft gaskets are unsuited; serrated gaskets are used with smooth-finished flange faces.

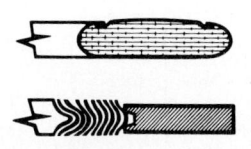

Laminated Gaskets made of metal and soft filler (the filler usually being soft asbestos sheet) are also used; in some gaskets (jacketed type) the laminations are in the plane of the flange face; in others (spiral wound type) the laminations are formed in an axial or edgewise direction. Laminated gaskets with asbestos filler are considered suitable for about 100F higher temperatures than plain asbestos gaskets and require less bolt load to compress them than solid metal gaskets, hence tend to make high-pressure, high-temperature joints more efficient than those using flat, solid metal gaskets.

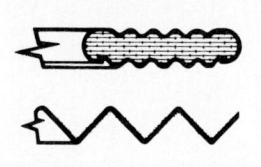

Corrugated Gaskets are a less commonly used type intermediate in stiffness between flat non-metallic and metallic gaskets. The ridges of the corrugations again tend to concentrate the gasket loading along concentric rings. This type is available plain, but to prevent crushing of the corrugations it is preferable to use the asbestos filled or asbestos inserted varieties.

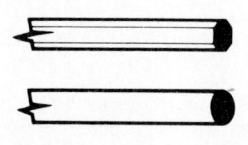

Ring Joint Gaskets are available in two types, octagonal and oval cross-section, both of which are standardized, but the former is considered the superior type. Either may be employed with flat bottom grooves which are now standard. Such rings are almost always made of metal, usually of the softest carbon steel or iron obtainable. In very high temperature or severe corrosion service, they may be made of alloy steel in which case they should be heat-treated to make them as much softer than the flanges proper as possible. For relatively low temperatures, rings made of plastic may be employed to resist corrosion or insulate the joint from electric currents.

Fig. 3-3 Selection and application of facings and gaskets for steel flanges (*Courtesy ITT Grinnell Corporation.*)

FLANGES

TABLE 3-1 FACING DIMENSIONS FOR STEEL FLANGES (ANSI B16.5-1973)
(COURTESY ITT GRINNELL CORPORATION.)

FOR 150, 300, 400, 600, 900, 1500 AND 2500 LB. S.P. STEEL FLANGES

(Other than Ring Joints)

American National Standards Institute

ANSI B16.5—1973

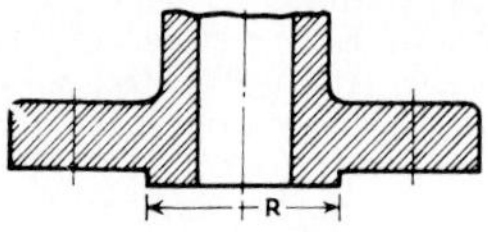

RAISED FACE

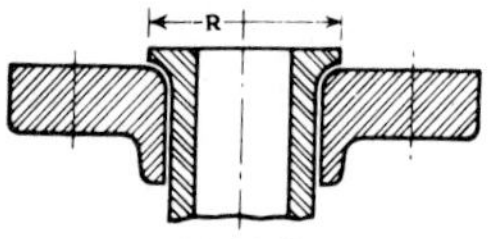

LAPPED

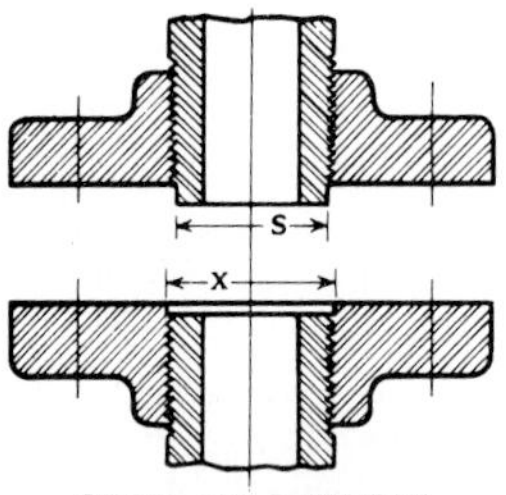

SMALL MALE-FEMALE
ON END OF PIPE

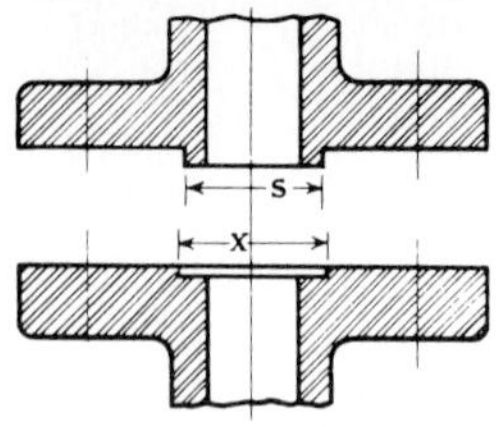

SMALL MALE - FEMALE

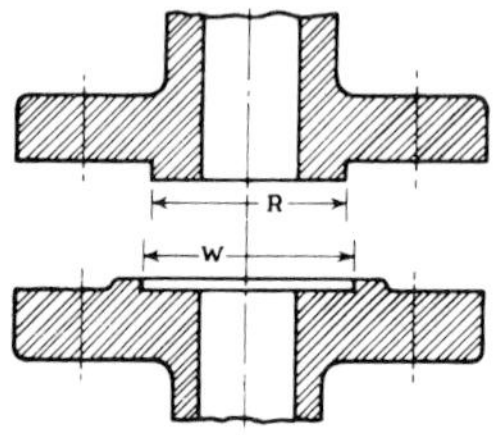

LARGE MALE - FEMALE

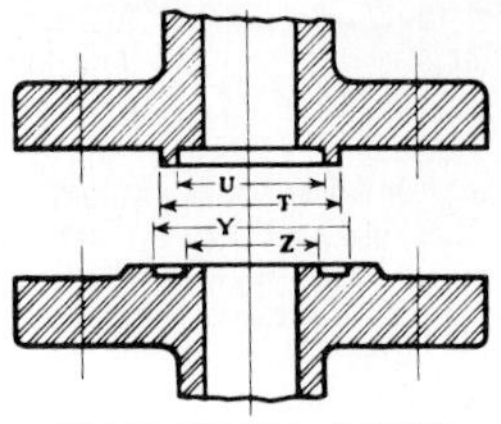

SMALL TONGUE - GROOVE

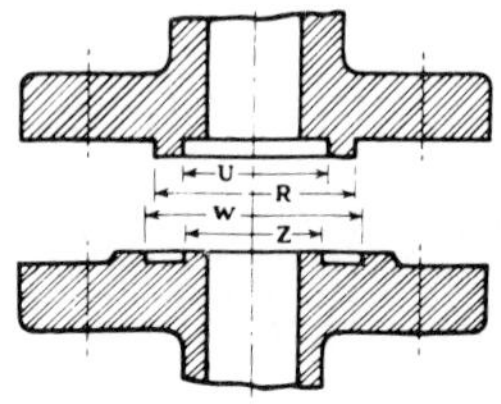

LARGE TONGUE - GROOVE

Nominal Pipe Size	Outside Diameter[3]: Raised Face, Lapped, Large Male and Large Tongue[5] R	Outside Diameter[3]: Small Male[4,5] S	Outside Diameter[3]: Small Tongue[5] T	ID of Large and Small Tongue[3,5] U	Outside Diameter[3]: Large Female and Large Groove[5] W	Outside Diameter[3]: Small Female[4,5] X	Outside Diameter[3]: Small Groove[5] Y	ID of Large and Small Groove[3,5] Z	Height: Raised Face 150 and 300 Lb. St'ds[1]	Height: Raised Face, Large and Small Male and Tongue 400, 600, 900, 1500, and 2500 Lb. St'ds[2]	Depth of Groove or Female
1/2	1 3/8	23/32	1 3/8	1	1 7/16	25/32	1 7/16	15/16	1/16	1/4	3/16
3/4	1 11/16	15/16	1 11/16	1 5/16	1 3/4	1	1 3/4	1 1/4	1/16	1/4	3/16
1	2	1 3/16	1 7/8	1 1/2	2 1/16	1 1/4	1 15/16	1 7/16	1/16	1/4	3/16
1 1/4	2 1/2	1 1/2	2 1/4	1 7/8	2 9/16	1 9/16	2 5/16	1 13/16	1/16	1/4	3/16
1 1/2	2 7/8	1 3/4	2 1/2	2 1/8	2 15/16	1 13/16	2 9/16	2 1/16	1/16	1/4	3/16
2	3 5/8	2 1/4	3 1/4	2 7/8	3 11/16	2 5/16	3 5/16	2 13/16	1/16	1/4	3/16
2 1/2	4 1/8	2 11/16	3 3/4	3 3/8	4 3/16	2 3/4	3 13/16	3 5/16	1/16	1/4	3/16
3	5	3 5/16	4 5/8	4 1/4	5 1/16	3 3/8	4 11/16	4 3/16	1/16	1/4	3/16
3 1/2	5 1/2	3 13/16	5 1/8	4 3/4	5 9/16	3 7/8	5 3/16	4 11/16	1/16	1/4	3/16
4	6 3/16	4 5/16	5 11/16	5 3/16	6 1/4	4 3/8	5 3/4	5 1/8	1/16	1/4	3/16
5	7 5/16	5 3/8	6 13/16	6 5/16	7 3/8	5 7/16	6 7/8	6 1/4	1/16	1/4	3/16
6	8 1/2	6 3/8	8	7 1/2	8 9/16	6 7/16	8 1/16	7 7/16	1/16	1/4	3/16
8	10 5/8	8 3/8	10	9 3/8	10 11/16	8 7/16	10 1/16	9 5/16	1/16	1/4	3/16
10	12 3/4	10 1/2	12	11 1/4	12 13/16	10 9/16	12 1/16	11 3/16	1/16	1/4	3/16
12	15	12 1/2	14 1/4	13 1/2	15 1/16	12 9/16	14 5/16	13 7/16	1/16	1/4	3/16
14 OD	16 1/4	13 3/4	15 1/2	14 3/4	16 5/16	13 13/16	15 9/16	14 11/16	1/16	1/4	3/16
16 OD	18 1/2	15 3/4	17 5/8	16 3/4	18 9/16	15 13/16	17 11/16	16 11/16	1/16	1/4	3/16
18 OD	21	17 3/4	20 1/8	19 1/4	21 1/16	17 13/16	20 3/16	19 3/16	1/16	1/4	3/16
20 OD	23	19 3/4	22	21	23 1/16	19 13/16	22 1/16	20 15/16	1/16	1/4	3/16
24 OD	27 1/4	23 3/4	26 1/4	25 1/4	27 5/16	23 13/16	26 5/16	25 3/16	1/16	1/4	3/16

Notes: All dimensions given in inches.

[1]Regular facing for 150 and 300 lb. steel flanged fittings and companion flange standards is a 1/16 in. raised face included in the minimum flange thickness dimensions.

[2]Regular facing for 400, 600, 900, 1500 and 2500 lb. flange standards is a 1/4 in. raised face not included in minimum flange thickness dimensions.

[3]Tolerance of plus or minus 0.016 in. (1/64 in.) is allowed on the inside and outside diameters of all facings.

[4]For small male and female joints care should be taken in the use of these dimensions to insure that the inside diameter of fitting on pipe is small enough to permit sufficient bearing surface to prevent the crushing of the gasket. This applies particularly on lines where the joint is made on the end of the pipe. Inside diameter of fitting should match inside diameter of pipe as specified by purchaser. Several companion flanges for small male and female joints are furnished with plain face and are threaded with American Standard Locknut Thread (NPSL).

[5]Gaskets for male-female and tongue-groove joints shall cover the bottom of the recess with minimum clearances taking into account the tolerances prescribed in Note 3.

TABLE 3-2 AMERICAN STANDARD RING JOINT FACINGS AND RINGS (*COURTESY CRANE COMPANY.*)

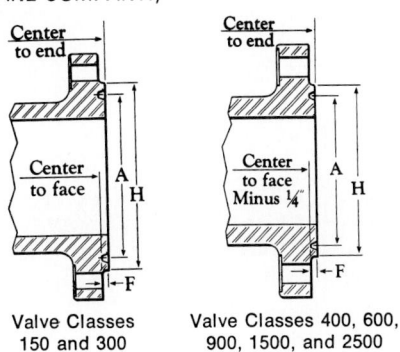

Valve Classes 150 and 300

Valve Classes 400, 600, 900, 1500, and 2500

Oval rings fit grooves having either a flat or round bottom; octagonal rings only fit grooves having a flat bottom.

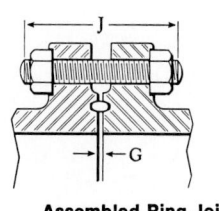

"Z" represents pipe thickness (Cranelap Joints)

"G" is approximate clearance with stud bolts tight.

Assembled Ring Joint

*Dimension "J" does not apply to Cranelap Joints; see "Stud bolts" on next page.

†**Caution:** 3-inch Class 300 and 600 Cranelap Ring Joints use Ring No. R 30, having a pitch diameter of 4.625 inches. When 3-inch Class 300 or 600 ring joint valves are to be bolted to Cranelap joints, orders must specify; they will be machined special.

Class	Size	Ring No.	A	B	C	D	E	F	G	H	J*
Class 150	1	R15	1.875	0.312	0.562	0.500	0.344	0.250	0.16	2.50	3.25
	1¼	R17	2.250	0.312	0.562	0.500	0.344	0.250	0.16	2.88	3.25
	1½	R19	2.562	0.312	0.562	0.500	0.344	0.250	0.16	3.25	3.50
	2	R22	3.250	0.312	0.562	0.500	0.344	0.250	0.16	4.00	3.75
	2½	R25	4.000	0.312	0.562	0.500	0.344	0.250	0.16	4.75	4.00
	3	R29	4.500	0.312	0.562	0.500	0.344	0.250	0.16	5.25	4.25
	3½	R33	5.188	0.312	0.562	0.500	0.344	0.250	0.16	6.06	4.25
	4	R36	5.875	0.312	0.562	0.500	0.344	0.250	0.16	6.75	4.25
	5	R40	6.750	0.312	0.562	0.500	0.344	0.250	0.16	7.62	4.50
	6	R43	7.625	0.312	0.562	0.500	0.344	0.250	0.16	8.62	4.50
	8	R48	9.750	0.312	0.562	0.500	0.344	0.250	0.16	10.75	4.75
	10	R52	12.000	0.312	0.562	0.500	0.344	0.250	0.16	13.00	5.25
	12	R56	15.000	0.312	0.562	0.500	0.344	0.250	0.16	16.00	5.25
	14	R59	15.625	0.312	0.562	0.500	0.344	0.250	0.12	16.75	5.75
	16	R64	17.875	0.312	0.562	0.500	0.344	0.250	0.12	19.00	6.00
	18	R68	20.375	0.312	0.562	0.500	0.344	0.250	0.12	21.50	6.50
	20	R72	22.000	0.312	0.562	0.500	0.344	0.250	0.12	23.50	6.75
	24	R76	26.500	0.312	0.562	0.500	0.344	0.250	0.12	28.00	7.50

Class	Size	Ring No.	A	B	C	D	E	F	G Class 300	G Class 400	G Class 600	H	J* Class 300	J* Class 400	J* Class 600
Class 300† 400 600†	½	R11	1.344	0.250	0.438	0.375	0.281	0.219	0.12	....	0.12	2.00	3.00	....	3.00
	¾	R13	1.688	0.312	0.562	0.500	0.344	0.250	0.16	....	0.16	2.50	3.50	....	3.50
	1	R16	2.000	0.312	0.562	0.500	0.344	0.250	0.16	....	0.16	2.75	3.75	....	3.75
	1¼	R18	2.375	0.312	0.562	0.500	0.344	0.250	0.16	....	0.16	3.12	3.75	....	4.00
	1½	R20	2.688	0.312	0.562	0.500	0.344	0.250	0.16	....	0.16	3.56	4.25	....	4.25
	2	R23	3.250	0.438	0.688	0.625	0.469	0.312	0.22	....	0.19	4.25	4.25	....	4.50
	2½	R26	4.000	0.438	0.688	0.625	0.469	0.312	0.22	....	0.19	5.00	4.75	....	5.00
	†3	†R31	†4.875	0.438	0.688	0.625	0.469	0.312	0.22	....	0.19	5.75	5.00	....	5.25
	3½	R34	5.188	0.438	0.688	0.625	0.469	0.312	0.22	....	0.19	6.25	5.25	....	5.75
	4	R37	5.875	0.438	0.688	0.625	0.469	0.312	0.22	0.22	0.19	6.88	5.25	5.75	6.00
	5	R41	7.125	0.438	0.688	0.625	0.469	0.312	0.22	0.22	0.19	8.25	5.50	6.00	6.75
	6	R45	8.312	0.438	0.688	0.625	0.469	0.312	0.22	0.22	0.19	9.50	5.75	6.25	7.00
	8	R49	10.625	0.438	0.688	0.625	0.469	0.312	0.22	0.22	0.19	11.88	6.25	7.00	7.75
	10	R53	12.750	0.438	0.688	0.625	0.469	0.312	0.22	0.22	0.19	14.00	7.00	7.75	8.75
	12	R57	15.000	0.438	0.688	0.625	0.469	0.312	0.22	0.22	0.19	16.25	7.50	8.25	9.00
	14	R61	16.500	0.438	0.688	0.625	0.469	0.312	0.22	0.22	0.19	18.00	7.75	8.50	9.50
	16	R65	18.500	0.438	0.688	0.625	0.469	0.312	0.22	0.22	0.19	20.00	8.25	9.00	10.25
	18	R69	21.000	0.438	0.688	0.625	0.469	0.312	0.22	0.22	0.19	22.62	8.50	9.25	11.00
	20	R73	23.000	0.500	0.750	0.688	0.531	0.375	0.22	0.22	0.19	25.00	9.00	10.00	11.75
	24	R77	27.250	0.625	0.875	0.812	0.656	0.438	0.25	0.25	0.22	29.50	10.25	11.25	13.25

FLANGES

TABLE 3-2 (CONTD.)

Class	Size	Ring No.	A	B	C	D	E	F	G	H	J*
Class 900	3	R31	4.875	0.438	0.688	0.625	0.469	0.312	0.16	6.12	6.00
	4	R37	5.875	0.438	0.688	0.625	0.469	0.312	0.16	7.12	7.00
	5	R41	7.125	0.438	0.688	0.625	0.469	0.312	0.16	8.50	7.75
	6	R45	8.312	0.438	0.688	0.625	0.469	0.312	0.16	9.50	7.75
	8	R49	10.625	0.438	0.688	0.625	0.469	0.312	0.16	12.12	9.00
	10	R53	12.750	0.438	0.688	0.625	0.469	0.312	0.16	14.25	9.50
	12	R57	15.000	0.438	0.688	0.625	0.469	0.312	0.16	16.50	10.25
	14	R62	16.500	0.625	0.875	0.812	0.656	0.438	0.16	18.38	11.25
	16	R66	18.500	0.625	0.875	0.812	0.656	0.438	0.16	20.62	11.75
	18	R70	21.000	0.750	1.000	0.938	0.781	0.500	0.19	23.38	13.50
	20	R74	23.000	0.750	1.000	0.938	0.781	0.500	0.19	25.50	14.25
	24	R78	27.250	1.000	1.312	1.250	1.062	0.625	0.22	30.38	17.75
Class 1500	½	R12	1.562	0.312	0.562	0.500	0.344	0.250	0.16	2.38	4.25
	¾	R14	1.750	0.312	0.562	0.500	0.344	0.250	0.16	2.62	4.50
	1	R16	2.000	0.312	0.562	0.500	0.344	0.250	0.16	2.81	5.00
	1¼	R18	2.375	0.312	0.562	0.500	0.344	0.250	0.16	3.19	5.00
	1½	R20	2.688	0.312	0.562	0.500	0.344	0.250	0.16	3.62	5.50
	2	R24	3.750	0.438	0.688	0.625	0.469	0.312	0.12	4.88	5.75
	2½	R27	4.250	0.438	0.688	0.625	0.469	0.312	0.12	5.38	6.25
	3	R35	5.375	0.438	0.688	0.625	0.469	0.312	0.12	6.62	7.00
	4	R39	6.375	0.438	0.688	0.625	0.469	0.312	0.12	7.62	7.75
	5	R44	7.625	0.438	0.688	0.625	0.469	0.312	0.12	9.00	9.75
	6	R46	8.312	0.500	0.750	0.688	0.531	0.375	0.12	9.75	10.50
	8	R50	10.625	0.625	0.875	0.812	0.656	0.438	0.16	12.50	12.00
	10	R54	12.750	0.625	0.875	0.812	0.656	0.438	0.16	14.62	13.75
	12	R58	15.000	0.875	1.125	1.062	0.906	0.562	0.19	17.25	15.50
	14	R63	16.500	1.000	1.312	1.250	1.062	0.625	0.22	19.25	17.00
	16	R67	18.500	1.125	1.438	1.375	1.188	0.688	0.31	21.50	18.50
	18	R71	21.000	1.125	1.438	1.375	1.188	0.688	0.31	24.12	20.50
	20	R75	23.000	1.250	1.562	1.500	1.312	0.688	0.38	26.50	22.50
	24	R79	27.250	1.375	1.750	1.625	1.438	0.812	0.44	31.25	25.75
Class 2500	½	R13	1.688	0.312	0.562	0.500	0.344	0.250	0.16	2.56	5.25
	¾	R16	2.000	0.312	0.562	0.500	0.344	0.250	0.16	2.88	5.25
	1	R18	2.375	0.312	0.562	0.500	0.344	0.250	0.16	3.25	5.75
	1¼	R21	2.844	0.438	0.688	0.625	0.469	0.312	0.12	4.00	6.50
	1½	R23	3.250	0.438	0.688	0.625	0.469	0.312	0.12	4.50	7.25
	2	R26	4.000	0.438	0.688	0.625	0.469	0.312	0.12	5.25	7.50
	2½	R28	4.375	0.500	0.750	0.688	0.531	0.375	0.12	5.88	8.25
	3	R32	5.000	0.500	0.750	0.688	0.531	0.375	0.12	6.62	9.25
	4	R38	6.188	0.625	0.875	0.812	0.656	0.438	0.16	8.00	10.75
	5	R42	7.500	0.750	1.000	0.938	0.781	0.500	0.16	9.50	12.75
	6	R47	9.000	0.750	1.000	0.938	0.781	0.500	0.16	11.00	14.50
	8	R51	11.000	0.875	1.125	1.062	0.906	0.562	0.19	13.38	16.00
	10	R55	13.500	1.125	1.438	1.375	1.188	0.688	0.25	16.75	20.50
	12	R60	16.000	1.250	1.562	1.500	1.312	0.688	0.31	19.50	22.50

TABLE 3-3 NUMBER FOR RING-JOINT GASKETS AND GROOVES
(COURTESY LUNKENHEIMER COMPANY, CINCINNATI, OH 45214.)

American National Standards Institute ANSI B16.20 — 1973

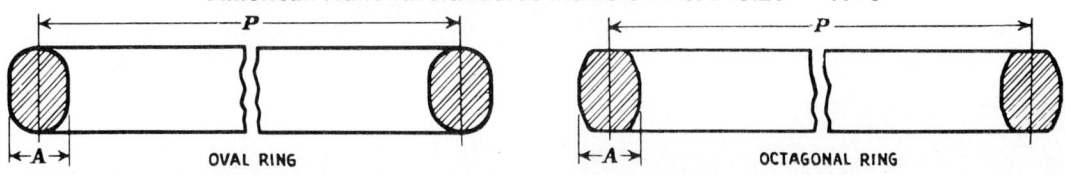

OVAL RING / OCTAGONAL RING

Number	Pitch Diameter P	Width of Ring A	Number	Pitch Diameter P	Width of Ring A	Number	Pitch Diameter P	Width of Ring A	Number	Pitch Diameter P	Width of Ring A
R 11	1 11/32	× 1/4	R 29	4 1/2	× 5/16	R 46	8 5/16	× 1/2	R 63	16 1/2	× 1
R 12	1 9/16	× 5/16	R 30	4 5/8	× 7/16	R 47	9	× 3/4	R 64	17 7/8	× 5/16
R 13	1 11/16	× 5/16	R 31	4 7/8	× 7/16	R 48	9 3/4	× 5/16	R 65	18 1/2	× 7/16
R 14	1 3/4	× 5/16	R 32	5	× 1/2	R 49	10 5/8	× 7/16	R 66	18 1/2	× 5/8
R 15	1 7/8	× 5/16	R 33	5 3/16	× 5/16	R 50	10 5/8	× 5/8	R 67	18 1/2	× 1 1/8
R 16	2	× 5/16	R 34	5 3/16	× 7/16	R 51	11	× 7/8	R 68	20 3/8	× 5/16
R 17	2 1/4	× 5/16	R 35	5 3/8	× 7/16	R 52	12	× 5/16	R 69	21	× 7/16
R 18	2 3/8	× 5/16	R 36	5 7/8	× 5/16	R 53	12 3/4	× 7/16	R 70	21	× 3/4
R 19	2 9/16	× 5/16	R 37	5 7/8	× 7/16	R 54	12 3/4	× 5/8	R 71	21	× 1 1/8
R 20	2 11/16	× 5/16	R 38	6 3/16	× 5/8	R 55	13 1/2	× 1 1/8	R 72	22	× 5/16
R 21	2 27/32	× 7/16	R 39	6 3/8	× 7/16	R 56	15	× 5/16	R 73	23	× 1/2
R 22	3 1/4	× 5/16	R 40	6 3/4	× 5/16	R 57	15	× 7/16	R 74	23	× 3/4
R 23	3 1/4	× 7/16	R 41	7 1/8	× 7/16	R 58	15	× 7/8	R 75	23	× 1 1/4
R 24	3 3/4	× 7/16	R 42	7 1/2	× 3/4	R 59	15 5/8	× 5/16	R 76	26 1/2	× 5/16
R 25	4	× 5/16	R 43	7 5/8	× 5/16	R 60	16	× 1 1/4	R 77	27 1/4	× 5/8
R 26	4	× 7/16	R 44	7 5/8	× 7/16	R 61	16 1/2	× 7/16	R 78	27 1/4	× 1
R 27	4 1/4	× 7/16	R 45	8 5/16	× 7/16	R 62	16 1/2	× 5/8	R 79	27 1/4	× 1 3/8
R 28	4 3/8	× 1/2									

Notes: The edge of each flange and the outside circumference of each ring shall carry corresponding identification marks: i.e., R 11, R 45, etc.
This standard shows only flat bottom grooves, because oval and octagon rings may be used. The former round bottom groove requires the use of an oval gasket.
All dimensions given in inches.

TABLE 3-4 EXTRA HEAVY CAST IRON COMPANION FLANGES AND BOLTS (FOR WORKING PRESSURES UP TO 250 PSI STEAM, 400 PSI WOG) *(COURTESY ITT GRINNELL CORPORATION.)*

PIPE SIZES	DIAM OF FLANGES	DIAM OF BOLT CIRCLE	NO. OF BOLTS	DIAM OF BOLTS	LENGTH OF BOLTS
1	4 7/8	3 1/2	4	5/8	2 1/4
1 1/4	5 1/4	3 7/8	4	5/8	2 1/2
1 1/2	6 1/8	4 1/2	4	3/4	2 1/2
2	6 1/2	5	8	5/8	2 1/2
2 1/2	7 1/2	5 7/8	8	3/4	3
3	8 1/4	6 5/8	8	3/4	3 1/4
3 1/2	9	7 1/4	8	3/4	3 1/4
4	10	7 7/8	8	3/4	3 1/2
5	11	9 1/4	8	3/4	3 3/4
6	12 1/2	10 5/8	12	3/4	3 3/4
8	15	13	12	7/8	4 1/4
10	17 1/2	15 1/4	16	1	5
12	20 1/2	17 3/4	16	1 1/8	5 1/2
14 O.D.	23	20 1/4	20	1 1/8	5 3/4
16 O.D.	25 1/2	22 1/2	20	1 1/4	6
18 O.D.	28	24 3/4	24	1 1/4	6 1/4
20 O.D.	30 1/2	27	24	1 1/4	6 3/4
24 O.D.	36	32	24	1 1/2	7 1/2
30 O.D.	43	39 1/4	28	1 3/4	8 1/2
36 O.D.	50	46	32	2	9 1/2
42 O.D.	57	52 3/4	36	2	10
48 O.D.	65	60 3/4	40	2	11

FLANGES

TABLE 3-5 WELD NECK FLANGES (*COURTESY ITT GRINNELL CORPORATION.*)

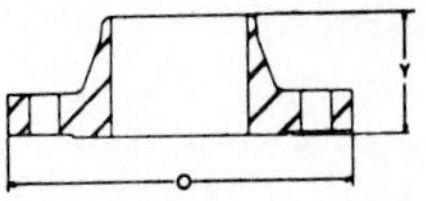

NOM PIPE SIZE	150 LB. OUTSIDE DIAM OF FLANGE O	150 LB. LENGTH THRU HUB Y (1)	300 LB. OUTSIDE DIAM OF FLANGE O	300 LB. LENGTH THRU HUB Y (1)	400 LB. OUTSIDE DIAM OF FLANGE O	400 LB. LENGTH THRU HUB Y (2)	600 LB. OUTSIDE DIAM OF FLANGE O	600 LB. LENGTH THRU HUB Y (2)
1/2	3 1/2	1 7/8	3 3/4	2 1/16	For sizes 3 1/2 and smaller use 600 Lb. Standard		3 3/4	2 1/16
3/4	3 7/8	2 1/16	4 5/8	2 1/4			4 5/8	2 1/4
1	4 1/4	2 3/16	4 7/8	2 7/16			4 7/8	2 7/16
1 1/4	4 5/8	2 1/4	5 1/4	2 9/16			5 1/4	2 5/8
1 1/2	5	2 7/16	6 1/8	2 11/16			6 1/8	2 3/4
2	6	2 1/2	6 1/2	2 3/4			6 1/2	2 7/8
2 1/2	7	2 3/4	7 1/2	3			7 1/2	3 1/8
3	7 1/2	2 3/4	8 1/4	3 1/8			8 1/4	3 1/4
3 1/2	8 1/2	2 13/16	9	3 3/16			9	3 3/8
4	9	3	10	3 3/8	10	3 1/2	10 3/4	4
5	10	3 1/2	11	3 7/8	11	4	13	4 1/2
6	11	3 1/2	12 1/2	3 7/8	12 1/2	4 1/16	14	4 5/8
8	13 1/2	4	15	4 3/8	15	4 5/8	16 1/2	5 1/4
10	16	4	17 1/2	4 5/8	17 1/2	4 7/8	20	6
12	19	4 1/2	20 1/2	5 1/8	20 1/2	5 3/8	22	6 1/8
14	21	5	23	5 5/8	23	5 7/8	23 3/4	6 1/2
16	23 1/2	5	25 1/2	5 3/4	25 1/2	6	27	7
18	25	5 1/2	28	6 1/4	28	6 1/2	29 1/4	7 1/4
20	27 1/2	5 11/16	30 1/2	6 3/8	30 1/2	6 5/8	32	7 1/2
22	29 1/2	5 7/8	33	6 1/2	33	6 3/4	34 1/4	7 3/4
24	32	6	36	6 5/8	36	6 7/8	37	8
26	34 1/4	5	38 1/4	7 1/4	38 1/4	7 5/8	40	8 3/4
30	38 3/4	5 1/8	43	8 1/4	43	8 5/8	44 1/2	9 3/4
34	43 3/4	5 5/16	47 1/2	9 1/8	47 1/2	9 1/2	49	10 5/8
36	46	5 3/8	50	9 1/2	50	9 7/8	51 3/4	11 1/8
42	53	5 5/8	57	10 7/8	57	11 3/8	58 3/4	12 3/4

(1) The 1/16″ raised face **is** included in "Length thru Hub 'Y'."
(2) The 1/4″ raised face **is not** included in "Length thru Hub 'Y'."

TABLE 3-6 SLIP-ON, THREADED, AND SOCKET-TYPE FLANGE DIMENSIONS (*COURTESY ITT GRINNELL CORPORATION.*)

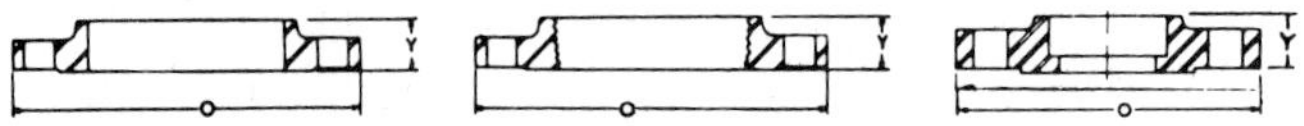

NOM PIPE SIZE	150 LB. OUTSIDE DIAM OF FLANGE O	150 LB. LENGTH THRU HUB Y (1)	300 LB. OUTSIDE DIAM OF FLANGE O	300 LB. LENGTH THRU HUB Y (1)	400 LB. OUTSIDE DIAM OF FLANGE O	400 LB. LENGTH THRU HUB Y (2)	600 LB. OUTSIDE DIAM OF FLANGE O (2)	600 LB. LENGTH THRU HUB Y (2)
1/2	3 1/2	5/8	3 3/4	7/8	For sizes 3 1/2 and smaller use 600 Lb. Standard		3 3/4	7/8
3/4	3 7/8	5/8	4 5/8	1			4 5/8	1
1	4 1/4	11/16	4 7/8	1 1/16			4 7/8	1 1/16
1 1/4	4 5/8	13/16	5 1/4	1 1/16			5 1/4	1 1/8
1 1/2	5	7/8	6 1/8	1 3/16			6 1/8	1 1/4
2	6	1	6 1/2	1 5/16			6 1/2	1 7/16
2 1/2	7	1 1/8	7 1/2	1 1/2			7 1/2	1 5/8
3	7 1/2	1 3/16	8 1/4	1 11/16			8 1/4	1 13/16
3 1/2	8 1/2	1 1/4	9	1 3/4			9	1 15/16
4	9	1 5/16	10	1 7/8	10	2‡	10 3/4	2 1/8‡
5	10	1 7/16	11	2‡	11	2 1/8‡	13	2 3/8‡
6	11	1 9/16	12 1/2	2 1/16‡	12 1/2	2 1/4‡	14	2 5/8‡
8	13 1/2	1 3/4	15	2 7/16‡	15	2 11/16‡	16 1/2	3‡
10	16	1 15/16	17 1/2	2 5/8‡	17 1/2	2 7/8‡	20	3 3/8‡
12	19	2 3/16	20 1/2	2 7/8‡	20 1/2	3 1/8‡	22	3 5/8‡
14	21	2 1/4	23	3‡	23	3 5/16‡	23 3/4	3 11/16‡
16	23 1/2	2 1/2	25 1/2	3 1/4‡	25 1/2	3 11/16‡	27	4 3/16‡
18	25	2 11/16	28	3 1/2‡	28	3 7/8‡	29 1/4	4 5/8‡
20	27 1/2	2 7/8	30 1/2	3 3/4‡	30 1/2	4‡	32	5‡
22	29 1/2	3 1/8†‡	33	4†‡	33	4 1/4†‡	34 1/4	5 1/4†‡
24	32	3 1/4	36	4 3/16‡	36	4 1/2‡	37	5 1/2‡
26	34 1/4	3 3/8†‡	38 1/4	7 1/4†‡	38 1/4	7 5/8†‡	40	8 3/4†‡
30	38 3/4	3 1/2†‡	43	8 1/4†‡	43	8 5/8†‡	44 1/2	9 3/4†‡
34	43 3/4	3 11/16†‡	47 1/2	9 1/8†‡	47 1/2	9 1/2†‡	49	10 5/8†‡
36	46	3 3/4†‡	50	9 1/2†‡	50	9 7/8†‡	51 3/4	11 1/8†‡
42	53	4†‡	57	10 7/8†‡	57	11 3/8†‡	58 3/4	12 3/4†‡

*Not available in Slip-On type.
†Not available in Threaded type.
‡Not available in Socket type.

TABLE 3-7 LAP JOINT FLANGES (*COURTESY ITT GRINNELL CORPORATION.*)

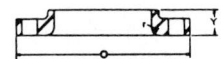

NOM PIPE SIZE	150 LB. OUTSIDE DIAM OF FLANGE O	150 LB. LENGTH THRU HUB Y(1)	300 LB. OUTSIDE DIAM OF FLANGE O	300 LB. LENGTH THRU HUB Y(1)	400 LB. OUTSIDE DIAM OF FLANGE O	400 LB. LENGTH THRU HUB Y(2)	600 LB. OUTSIDE DIAM OF FLANGE O	600 LB. LENGTH THRU HUB Y(2)
1/2	3 1/2	5/8	3 3/4	7/8			3 3/4	7/8
3/4	3 7/8	5/8	4 5/8	1	For sizes 3 1/2 and smaller use 600 Lb. Standard		4 5/8	1
1	4 1/4	11/16	4 7/8	1 1/16			4 7/8	1 1/16
1 1/4	4 5/8	13/16	5 1/4	1 1/16			5 1/4	1 1/8
1 1/2	5	7/8	6 1/8	1 3/16			6 1/8	1 1/4
2	6	1	6 1/2	1 5/16			6 1/2	1 7/16
2 1/2	7	1 1/8	7 1/2	1 1/2			7 1/2	1 5/8
3	7 1/2	1 3/16	8 1/4	1 11/16			8 1/4	1 13/16
3 1/2	8 1/2	1 1/4	9	1 3/4			9	1 15/16
4	9	1 5/16	10	1 7/8	10	2	10 3/4	2 1/8
5	10	1 7/16	11	2	11	2 1/8	13	2 3/8
6	11	1 9/16	12 1/2	2 1/16	12 1/2	2 1/4	14	2 5/8
8	13 1/2	1 3/4	15	2 7/16	15	2 11/16	16 1/2	3
10	16	1 15/16	17 1/2	3 3/8	17 1/2	4	20	4 3/8
12	19	2 3/16	20 1/2	4	20 1/2	4 1/4	22	4 5/8
14	21	3 1/8	23	4 3/8	23	4 5/8	23 3/4	5
16	23 1/2	3 7/16	25 1/2	4 3/4	25 1/2	5	27	5 1/2
18	25	3 13/16	28	5 1/8	28	5 3/8	29 1/4	6
20	27 1/2	4 1/16	30 1/2	5 1/2	30 1/2	5 3/4	32	6 1/2
24	32	4 3/8	36	6	36	6 1/4	37	7 1/4

(1) The 1/16" raised face **is** included in "Length thru Hub 'Y'."
(2) The 1/4" raised face **is not** included in "Length thru Hub 'Y'."

TABLE 3-8 BLIND FLANGE DIMENSIONS (*COURTESY ITT GRINNELL CORPORATION.*)

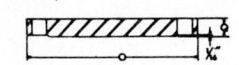

NOM PIPE SIZE	150 LB. OUTSIDE DIAM OF FLANGE O	150 LB. THICKNESS Q(1)	300 LB. OUTSIDE DIAM OF FLANGE O	300 LB. THICKNESS Q(1)	400 LB. OUTSIDE DIAM OF FLANGE O	400 LB. THICKNESS Q(2)	600 LB. OUTSIDE DIAM OF FLANGE O	600 LB. THICKNESS Q(2)
1/2	3 1/2	7/16	3 3/4	9/16			3 3/4	9/16
3/4	3 7/8	1/2	4 5/8	5/8			4 5/8	5/8
1	4 1/4	9/16	4 7/8	11/16	For sizes 3 1/2 and smaller use 600 Lb. Standard		4 7/8	11/16
1 1/4	4 5/8	5/8	5 1/4	3/4			5 1/4	13/16
1 1/2	5	11/16	6 1/8	13/16			6 1/8	7/8
2	6	3/4	6 1/2	7/8			6 1/2	1
2 1/2	7	7/8	7 1/2	1			7 1/2	1 1/8
3	7 1/2	15/16	8 1/4	1 1/8			8 1/4	1 1/4
3 1/2	8 1/2	15/16	9	1 3/16			9	1 3/8
4	9	15/16	10	1 1/4	10	1 3/8	10 3/4	1 1/2
5	10	15/16	11	1 3/8	11	1 1/2	13	1 3/4
6	11	1	12 1/2	1 7/16	12 1/2	1 5/8	14	1 7/8
8	13 1/2	1 1/8	15	1 5/8	15	1 7/8	16 1/2	2 3/16
10	16	1 3/16	17 1/2	1 7/8	17 1/2	2 1/8	20	2 1/2
12	19	1 1/4	20 1/2	2	20 1/2	2 1/4	22	2 5/8
14	21	1 3/8	23	2 1/8	23	2 3/8	23 3/4	2 3/4
16	23 1/2	1 7/16	25 1/2	2 1/4	25 1/2	2 1/2	27	3
18	25	1 9/16	28	2 3/8	28	2 5/8	29 1/4	3 1/4
20	27 1/2	1 11/16	30 1/2	2 1/2	30 1/2	2 3/4	32	3 1/2
22	29 1/2	1 13/16	33	2 5/8	33	2 7/8	34 1/4	3 3/4
24	32	1 7/8	36	2 3/4	36	3	37	4
26	34 1/4	2	38 1/4	3 1/8	38 1/4	3 1/2	40	4 1/4
30	38 3/4	2 1/8	43	3 3/8	43	4	44 1/2	4 1/2
34	43 3/4	2 5/16	47 1/2	4	47 1/2	4 3/8	49	4 3/4
36	46	2 3/8	50	4 1/8	50	4 1/2	51 3/4	4 7/8
42	53	2 5/8	57	4 5/8	57	5 1/8	58 3/4	5 1/2

(1) The 1/16" raised face **is** included in "thickness 'Q'."
(2) The 1/4" raised face **is not** included in "thickness 'Q'."

FLANGES

TABLE 3-9 TEMPLATES FOR DRILLING CLASSES 400, 600, AND 900 STEEL (*COURTESY CRANE COMPANY.*)

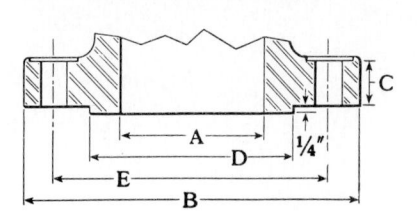

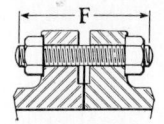

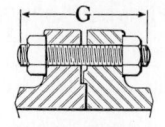

Classes 400, 600, and 900 Steel

Dimensions, in Inches

Class	Nominal pipe size	A†	B	C	D	E	Stud Bolts No.	Stud Bolts Dia.	F	G
Class 400 Steel	4	4.00	10.00	1.38	6.19	7.88	8	⅞	5.50	5.25
	5	5.00	11.00	1.50	7.31	9.25	8	⅞	5.75	5.50
	6	6.00	12.50	1.62	8.50	10.62	12	⅞	6.00	5.75
	8	8.00	15.00	1.88	10.62	13.00	12	1	6.75	6.50
	10	10.00	17.50	2.12	12.75	15.25	16	1⅛	7.50	7.25
	12	12.00	20.50	2.25	15.00	17.75	16	1¼	8.00	7.75
	14	13.12	23.00	2.38	16.25	20.25	20	1¼	8.25	8.00
	16	15.00	25.50	2.50	18.50	22.50	20	1⅜	8.75	8.50
	18	17.00	28.00	2.62	21.00	24.75	24	1⅜	9.00	8.75
	20	18.88	30.50	2.75	23.00	27.00	24	1½	9.75	9.50
	24	22.62	36.00	3.00	27.25	32.00	24	1¾	10.75	10.50
Class 600 Steel	½	0.50	3.75	0.56	1.38	2.62	4	½	3.25	3.00
	¾	0.75	4.62	0.62	1.69	3.25	4	⅝	3.50	3.25
	1	1.00	4.88	0.69	2.00	3.50	4	⅝	3.75	3.50
	1¼	1.25	5.25	0.81	2.50	3.88	4	⅝	4.00	3.75
	1½	1.50	6.12	0.88	2.88	4.50	4	¾	4.25	4.00
	2	2.00	6.50	1.00	3.62	5.00	8	⅝	4.25	4.00
	2½	2.50	7.50	1.12	4.12	5.88	8	¾	4.75	4.50
	3	3.00	8.25	1.25	5.00	6.62	8	¾	5.00	4.75
	3½	3.50	9.00	1.38	5.50	7.25	8	⅞	5.50	5.25
	4	4.00	10.75	1.50	6.19	8.50	8	⅞	5.75	5.50
	5	5.00	13.00	1.75	7.31	10.50	8	1	6.50	6.25
	6	6.00	14.00	1.88	8.50	11.50	12	1	6.75	6.50
	8	7.88	16.50	2.19	10.62	13.75	12	1⅛	7.75	7.50
	10	9.75	20.00	2.50	12.75	17.00	16	1¼	8.50	8.25
	12	11.75	22.00	2.62	15.00	19.25	20	1¼	8.75	8.50
	14	12.88	23.75	2.75	16.25	20.75	20	1⅜	9.25	9.00
	16	14.75	27.00	3.00	18.50	23.75	20	1½	10.00	9.75
	18	16.50	29.25	3.25	21.00	25.75	20	1⅝	10.75	10.50
	20	18.25	32.00	3.50	23.00	28.50	24	1⅝	11.50	11.25
	24	22.00	37.00	4.00	27.25	33.00	24	1⅞	13.00	12.75
	30	...	44.50	4.50	33.75	40.25	28	2	...	...
Class 900 Steel	3	2.88	9.50	1.50	5.00	7.50	8	⅞	5.75	5.50
	4	3.88	11.50	1.75	6.19	9.25	8	1⅛	6.75	6.50
	5	4.75	13.75	2.00	7.31	11.00	8	1¼	7.50	7.25
	6	5.75	15.00	2.19	8.50	12.50	12	1⅛	7.75	7.50
	8	7.50	18.50	2.50	10.62	15.50	12	1⅜	8.75	8.50
	10	9.38	21.50	2.75	12.75	18.50	16	1⅜	9.25	9.00
	12	11.12	24.00	3.12	15.00	21.00	20	1⅜	10.00	9.75
	14	12.25	25.25	3.38	16.25	22.00	20	1½	10.75	10.50
	16	14.00	27.75	3.50	18.50	24.25	20	1⅝	11.25	11.00
	18	15.75	31.00	4.00	21.00	27.00	20	1⅞	12.75	12.50
	20	17.50	33.75	4.25	23.00	29.50	20	2	13.50	13.50
	24	21.00	41.00	5.50	27.25	35.50	20	2½	17.25	17.00

†Dimension "A" applies to valves or fittings.

TABLE 3-10 TEMPLATES FOR DRILLING. CLASSES 1500 AND 2500 STEEL (*COURTESY CRANE COMPANY.*)

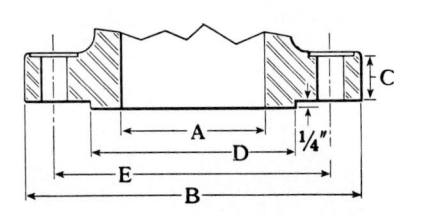

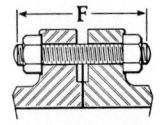

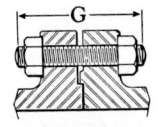

Stud Bolt Length "G" also applies for Tongue to Groove Flanged Joint

Classes 1500 and 2500 Steel

Dimensions, in Inches

Class	Size	A†	B	C	D	E	Stud Bolts No.	Stud Bolts Dia.	F	G
	½	0.50	4.75	0.88	1.38	3.25	4	¾	4.25	4.00
	¾	0.56	5.12	1.00	1.69	3.50	4	¾	4.50	4.25
	1	0.88	5.88	1.12	2.00	4.00	4	⅞	5.00	4.75
	1¼	1.12	6.25	1.12	2.50	4.38	4	⅞	5.00	4.75
	1½	1.38	7.00	1.25	2.88	4.88	4	1	5.50	5.25
	2	1.88	8.50	1.50	3.62	6.50	8	⅞	5.75	5.50
	2½	2.25	9.62	1.62	4.12	7.50	8	1	6.25	6.00
	3	2.75	10.50	1.88	5.00	8.00	8	1⅛	7.00	6.75
Class 1500 Steel	4	3.62	12.25	2.12	6.19	9.50	8	1¼	7.75	7.50
	5	4.38	14.75	2.88	7.31	11.50	8	1½	9.75	9.50
	6	5.38	15.50	3.25	8.50	12.50	12	1⅜	10.25	10.00
	8	7.00	19.00	3.62	10.62	15.50	12	1⅝	11.50	11.25
	10	8.75	23.00	4.25	12.75	19.00	12	1⅞	13.25	13.00
	12	10.38	26.50	4.88	15.00	22.50	16	2	14.75	14.50
	14	11.38	29.50	5.25	16.25	25.00	16	2¼	16.00	15.75
	16	13.00	32.50	5.75	18.50	27.75	16	2½	17.50	17.25
	18	14.62	36.00	6.38	21.00	30.50	16	2¾	19.50	19.00
	20	16.38	38.75	7.00	23.00	32.75	16	3	21.50	21.00
	24	19.62	46.00	8.00	27.25	39.00	16	3½	24.50	24.00
	½	0.44	5.25	1.19	1.38	3.50	4	¾	5.25	5.00
	¾	0.56	5.50	1.25	1.69	3.75	4	¾	5.25	5.00
	1	0.75	6.25	1.38	2.00	4.25	4	⅞	5.75	5.50
	1¼	1.00	7.25	1.50	2.50	5.12	4	1	6.25	6.00
	1½	1.12	8.00	1.75	2.88	5.75	4	1⅛	7.00	6.75
Class 2500 Steel	2	1.50	9.25	2.00	3.62	6.75	8	1	7.25	7.00
	2½	1.88	10.50	2.25	4.12	7.75	8	1⅛	8.00	7.75
	3	2.25	12.00	2.62	5.00	9.00	8	1¼	9.00	8.75
	4	2.88	14.00	3.00	6.19	10.75	8	1½	10.25	10.00
	5	3.62	16.50	3.62	7.31	12.75	8	1¾	12.00	11.75
	6	4.38	19.00	4.25	8.50	14.50	8	2	13.75	13.50
	8	5.75	21.75	5.00	10.62	17.25	12	2	15.25	15.00
	10	7.25	26.50	6.50	12.75	21.75	12	2½	19.50	19.25
	12	8.62	30.00	7.25	15.00	24.38	12	2¾	21.50	21.25

†Dimension "A" applies to valves or fittings.

FLANGES

TABLE 3-11 TEMPLATES FOR DRILLING. CLASSES 25 and 125 CAST IRON (*COURTESY CRANE COMPANY.*)

"Dimensions, in Inches"

Nominal pipe size	Class 25 Cast Iron						Class 125 Cast Iron						
	Diameter of flange	Thickness of flange	Diameter of bolt circle	Number of bolts	Diameter of bolts	Length of bolts	Diameter of flange	Thickness of flange	Diameter of bolt circle	Number of bolts	Diameter of bolts	Length of bolts	Length of stud bolts with 2 nuts
1	...	...	...	...	...	...	4.25	0.44	3.12	4	½	1.75	...
1¼	...	...	...	...	...	...	4.62	0.50	3.50	4	½	2.00	...
1½	...	...	...	...	...	...	5.00	0.56	3.88	4	½	2.00	...
2	...	...	...	...	...	...	6.00	0.62	4.75	4	⅝	2.25	...
2½	...	...	...	...	...	...	7.00	0.69	5.50	4	⅝	2.50	...
3	...	...	...	...	...	...	7.50	0.75	6.00	4	⅝	2.50	...
3½	...	...	...	...	...	...	8.50	0.81	7.00	8	⅝	2.75	...
4	9.00	0.75	7.50	8	⅝	2.50	9.00	0.94	7.50	8	⅝	3.00	...
5	10.00	0.75	8.50	8	⅝	2.50	10.00	0.94	8.50	8	¾	3.00	...
6	11.00	0.75	9.50	8	⅝	2.50	11.00	1.00	9.50	8	¾	3.25	...
8	13.50	0.75	11.75	8	⅝	2.50	13.50	1.12	11.75	8	¾	3.50	...
10	16.00	0.88	14.25	12	⅝	2.75	16.00	1.19	14.25	12	⅞	3.75	...
12	19.00	1.00	17.00	12	⅝	3.00	19.00	1.25	17.00	12	⅞	3.75	...
14	21.00	1.12	18.75	12	¾	3.50	21.00	1.38	18.75	12	1	4.25	...
16	23.50	1.12	21.25	16	¾	3.50	23.50	1.44	21.25	16	1	4.50	...
18	25.00	1.25	22.75	16	¾	3.75	25.00	1.56	22.75	16	1⅛	4.75	...
20	27.50	1.25	25.00	20	¾	3.75	27.50	1.69	25.00	20	1⅛	5.00	...
24	32.00	1.38	29.50	20	¾	4.00	32.00	1.88	29.50	20	1¼	5.50	...
30	38.75	1.50	36.00	28	⅞	4.50	38.75	2.12	36.00	28	1¼	6.25	...
36	46.00	1.62	42.75	32	⅞	4.75	46.00	2.38	42.75	32	1½	7.00	8.75
42	53.00	1.75	49.50	36	1	5.25	53.00	2.62	49.50	36	1½	7.50	9.25
48	59.50	2.00	56.00	44	1	5.75	59.50	2.75	56.00	44	1½	7.75	9.50
54	66.25	2.25	62.75	44	1	6.25	66.25	3.00	62.75	44	1¾	8.50	10.50
60	73.00	2.25	69.25	52	1⅛	6.25	73.00	3.12	69.25	52	1¾	8.75	10.75
72	86.50	2.50	82.50	60	1⅛	6.75	86.50	3.50	82.50	60	1¾	9.50	11.50
84	99.75	2.75	95.50	64	1¼	7.50	99.75	3.88	95.50	64	2	10.50	12.75
96	113.25	3.00	108.50	68	1¼	8.00	113.25	4.25	108.50	68	2¼	11.50	14.00

TABLE 3-12 TEMPLATES FOR DRILLING. CLASSES 250 AND 800 HYDRAULIC CAST IRON (*COURTESY CRANE COMPANY.*)

Class 250 Cast Iron

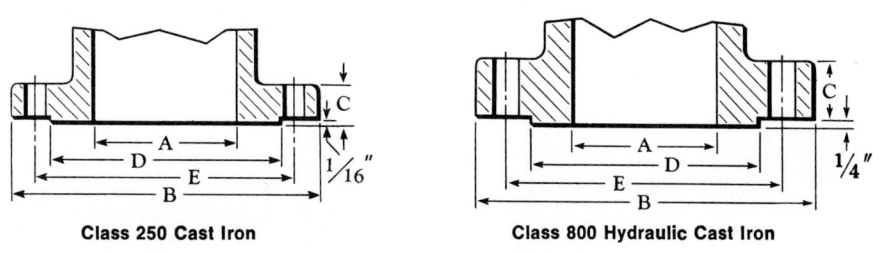

Class 800 Hydraulic Cast Iron

Dimensions, in Inches

Class	Nominal pipe size	A (Valve or fitting)	B	C	D	E	No. of bolts	Dia. of bolts	Length of bolts (See Note X)	Length of stud bolts with 2 nuts (See Note X)
	1	1.00	4.88	0.69	2.69	3.50	4	⅝	2.50	...
	1¼	1.25	5.25	0.75	3.06	3.88	4	⅝	2.50	...
	1½	1.50	6.12	0.81	3.56	4.50	4	¾	2.75	...
	2	2.00	6.50	0.88	4.19	5.00	8	⅝	2.75	...
	2½	2.50	7.50	1.00	4.94	5.88	8	¾	3.25	...
	3	3.00	8.25	1.12	5.69	6.62	8	¾	3.50	...
	3½	3.50	9.00	1.19	6.31	7.25	8	¾	3.50	...
	4	4.00	10.00	1.25	6.94	7.88	8	¾	3.75	...
	5	5.00	11.00	1.38	8.31	9.25	8	¾	4.00	...
	6	6.00	12.50	1.44	9.69	10.62	12	¾	4.00	...
Class 250 Cast Iron	8	8.00	15.00	1.62	11.94	13.00	12	⅞	4.50	...
	10	10.00	17.50	1.88	14.06	15.25	16	1	5.25	...
	12	12.00	20.50	2.00	16.44	17.75	16	1⅛	5.50	...
	14	13.25	23.00	2.12	18.94	20.25	20	1⅛	6.00	...
	16	15.25	25.50	2.25	21.06	22.50	20	1¼	6.25	...
	18	17.00	28.00	2.38	23.31	24.75	24	1¼	6.50	...
	20	19.00	30.50	2.50	25.56	27.00	24	1¼	6.75	...
	24	23.00	36.00	2.75	30.31	32.00	24	1½	7.50	9.50
	30	29.00	43.00	3.00	37.19	39.25	28	1¾	8.50	10.50
	36	34.50	50.00	3.38	43.69	46.00	32	2	9.50	11.75
	42	40.25	57.00	3.67	50.44	52.75	36	2	10.00	12.50
	48	46.00	65.00	4.00	58.44	60.75	40	2	10.75	13.00

	Nominal pipe size	A (Valve or fitting)	B	C	D	E	No. of bolts	Dia. of bolts	Length of bolts (See Note Y)	(See Note Z)
Class 800 Hydraulic Cast Iron	2	2.00	6.50	1.25	3.62	5.00	8	⅝	3.75	3.50
	2½	2.50	7.50	1.38	4.12	5.88	8	¾	4.25	4.00
	3	3.00	8.25	1.50	5.00	6.62	8	¾	4.50	4.25
For sizes not listed, use Class 600 steel data	3½	3.50	9.00	1 62	5.50	7.25	8	⅞	5.00	4.75
	4	4.00	10.75	1.88	6.19	8.50	8	⅞	5.50	5.25
	5	5.00	13.00	2.12	7.31	10.50	8	1	6.00	5.75
	6	6.00	14.00	2.25	8.50	11.50	12	1	6.25	6.00
	8	7.88	16.50	2.50	10.62	13.75	12	1⅛	7.00	6.75
	10	9.75	20.00	2.88	12.75	17.00	16	1¼	7.75	7.50
	12	11.75	22.00	3.00	15.00	19.25	20	1¼	8.00	7.75

FLANGES 47

TABLE 3-14 BOLTING DIMENSIONS FOR 300-LB FLANGES (*COURTESY ITT GRINNELL CORPORATION.*)

NOM PIPE SIZE	300 LB. STEEL FLANGES				
	DIAM OF BOLT CIRCLE	DIAM OF BOLTS	NO. OF BOLTS	LENGTH OF STUDS 1/16" RAISED FACE	BOLT LENGTH
½	2⅝	½	4	2½	2
¾	3¼	⅝	4	2¾	2½
1	3½	⅝	4	3	2½
1¼	3⅞	⅝	4	3	2¾
1½	4½	¾	4	3½	3
2	5	⅝	8	3¼	3
2½	5⅞	¾	8	3¾	3¼
3	6⅝	¾	8	4	3½
3½	7¼	¾	8	4¼	3¾
4	7⅞	¾	8	4¼	3¾
5	9¼	¾	8	4½	4
6	10⅝	¾	12	4¾	4¼
8	13	⅞	12	5¼	4¾
10	15¼	1	16	6	5¼
12	17¾	1⅛	16	6½	5¾
14	20¼	1⅛	20	6¾	6
16	22½	1⅛	20	7¼	6½
18	24¾	1¼	24	7½	6¾
20	27	1¼	24	8	7
22	29¼	1½	24	8¾	7½
24	32	1½	24	9	7¾
26	34½	1⅝	28	10	8¾
30	39¼	1¾	28	11¼	10
34	43½	1⅞	28	12¼	10¾
36	46	2	32	12¾	11¼
42	52¾	2	36	13¾	13½

Bolting arrangement for 250 lb. cast iron flanges are the same as shown for 300 lb. steel flanges.

TABLE 3-13 BOLTING DIMENSIONS FOR 150-LB FLANGES (*COURTESY ITT GRINNELL CORPORATION.*)

NOM PIPE SIZE	150 LB. STEEL FLANGES				
	DIAM OF BOLT CIRCLE	DIAM OF BOLTS	NO. OF BOLTS	LENGTH OF STUDS 1/16" RAISED FACE	BOLT LENGTH
½	2⅜	½	4	2¼	1¾
¾	2¾	½	4	2¼	2
1	3⅛	½	4	2½	2
1¼	3½	½	4	2½	2¼
1½	3⅞	½	4	2¾	2¼
2	4¾	⅝	4	3	2¾
2½	5½	⅝	4	3¼	3
3	6	⅝	4	3½	3
3½	7	⅝	8	3½	3
4	7½	⅝	8	3½	3
5	8½	¾	8	3¾	3¼
6	9½	¾	8	3¾	3¼
8	11¾	¾	8	4	3½
10	14¼	⅞	12	4½	3¾
12	17	⅞	12	4½	4
14	18¾	1	12	5	4¼
16	21¼	1	16	5¼	4½
18	22¾	1⅛	16	5¾	4¾
20	25	1⅛	20	6	5¼
22	27¼	1¼	20	6½	5½
24	29½	1¼	20	6¾	5¾
26	31¾	1¼	24	7	6
30	36	1¼	28	7¼	6¼
34	40½	1½	32	8	7
36	42¾	1½	32	8¼	7
42	49½	1½	36	8¾	7½

Stud lengths for lap joint flanges are equal to lengths shown plus the thickness of two laps of the stub ends.
Bolting arrangement for 125 lb. cast iron flanges are the same as shown for 150 lb. steel flanges.

TABLE 3-15 BOLTING DIMENSIONS FOR 400- AND 600-LB FLANGES (*COURTESY ITT GRINNELL CORPORATION.*)

NOM PIPE SIZE	400 LB. STEEL FLANGES				600 LB. STEEL FLANGES			
	DIAM OF BOLT CIRCLE	DIAM OF BOLTS	NO. OF BOLTS	LENGTH OF STUDS ¼" RAISED FACE	DIAM OF BOLT CIRCLE	DIAM OF BOLTS	NO. OF BOLTS	LENGTH OF STUDS ¼" RAISED FACE
½	2⅝	½	4	3	2⅝	½	4	3
¾	3¼	⅝	4	3¼	3¼	⅝	4	3¼
1	3½	⅝	4	3½	3½	⅝	4	3½
1¼	3⅞	⅝	4	3¾	3⅞	⅝	4	3¾
1½	4½	¾	4	4	4½	¾	4	4
2	5	⅝	8	4	6	⅝	8	4
2½	5⅞	¾	8	4½	5⅞	¾	8	4½
3	6⅝	¾	8	4¾	6⅝	¾	8	4¾
3½	7¼	⅞	8	5¼	7¼	⅞	8	5¼
4	7⅞	⅞	8	5¼	8½	⅞	8	5½
5	9¼	⅞	8	6½	10½	1	8	6¼
6	10⅝	⅞	12	5¾	11½	1	12	6½
8	13	1	12	6½	13¾	1⅛	12	7½
10	15¼	1⅛	16	7¼	17	1¼	16	8¼
12	17¾	1¼	16	7¾	19¼	1¼	20	8½
14	20¼	1¼	20	8	20¾	1⅜	20	9
16	22½	1⅜	20	8½	23¾	1½	20	9¾
18	24¾	1⅜	24	8¾	25¾	1⅝	20	10½
20	27	1½	24	9½	28½	1⅝	24	11¼
22	29¼	1⅝	24	10	30⅝	1¾	24	12
24	32	1¾	24	10½	33	1⅞	24	12¾
26	34½	1¾	28	11½	36	1⅞	28	13¼
30	39¼	2	28	13	40¼	2	28	14
34	43½	2	28	13¾	44½	2¼	28	15
36	46	2	32	14	47	2½	28	15¾
42	52¾	2½	32	16¼	53¾	2¾	28	17½

Stud lengths for lap joint flanges are equal to lengths shown minus ½" plus the thickness of two laps of the stub ends.

TABLE 3-16 STANDARD CAST IRON COMPANION FLANGES AND BOLTS (FOR WORKING PRESSURES UP TO 125 PSI STEAM, 175 PSI WOG) (*COURTESY ITT GRINNELL CORPORATION.*)

SIZE	DIAM OF FLANGE	BOLT CIRCLE	NO. OF BOLTS	SIZE OF BOLTS	LENGTH OF BOLTS
¾	3½	2½	4	⅜	1⅜
1	4¼	3⅛	4	½	1½
1¼	4⅝	3½	4	½	1½
1½	5	3⅞	4	½	1¾
2	6	4¾	4	⅝	2
2½	7	5½	4	⅝	2¼
3	7½	6	4	⅝	2½
3½	8½	7	8	⅝	2½
4	9	7½	8	⅝	2¾
5	10	8½	8	¾	3
6	11	9½	8	¾	3
8	13½	11¾	8	¾	3¼
10	16	14¼	12	⅞	3½
12	19	17	12	⅞	3¾
14	21	18¾	12	1	4¼
16	23½	21¼	16	1	4¼

FLANGES

TABLE 3-17 TEMPLATES FOR DRILLING. BRONZE AND CORROSION-RESISTANT STANDARDS (*COURTESY CRANE COMPANY.*)

Dimensions, in Inches

Nominal pipe size	Diameter of flange	Thickness of Flange — Bronze	Thickness of Flange — Corrosion Resistant	Diameter of bolt circle	Number of bolts	Diameter of bolts	Length of bolts	Length of stud bolts with 2 nuts
Class 150 Bronze ANSI or Corrosion-Resistant MSS Standards								
*¼	2.50	...	0.34	1.69	4	⅜	1.12	1.50
*⅜	2.50	...	0.34	1.69	4	⅜	1.12	1.50
½	3.50	0.31	0.38	2.38	4	½	1.25	1.88
¾	3.88	0.34	0.41	2.75	4	½	1.50	1.88
1	4.25	0.38	0.44	3.12	4	½	1.50	2.00
1¼	4.62	0.41	0.50	3.50	4	½	1.50	2.00
1½	5.00	0.44	0.56	3.88	4	½	1.50	2.12
2	6.00	0.50	0.62	4.75	4	⅝	1.75	2.50
2½	7.00	0.56	0.62	5.50	4	⅝	2.00	2.62
3	7.50	0.62	0.62	6.00	4	⅝	2.00	2.75
*3½	8.50	0.69	...	7.00	8	⅝	2.25	2.88
4	9.00	0.69	0.69	7.50	8	⅝	2.25	2.88
*5	10.00	0.75	...	8.50	8	¾	2.50	3.25
6	11.00	0.81	0.81	9.50	8	¾	2.50	3.38
8	13.50	0.94	0.94	11.75	8	¾	2.75	3.62
10	16.00	1.00	1.00	14.25	12	⅞	3.25	4.12
12	19.00	1.06	1.06	17.00	12	⅞	3.25	4.25
Class 300 Bronze ANSI Standard								
½	3.75	0.50	...	2.62	4	½	1.75	2.25
¾	4.62	0.53	...	3.25	4	⅝	2.00	2.50
1	4.88	0.59	...	3.50	4	⅝	2.00	2.62
1¼	5.25	0.62	...	3.88	4	⅝	2.00	2.75
1½	6.12	0.69	...	4.50	4	¾	2.25	3.12
2	6.50	0.75	...	5.00	8	⅝	2.25	3.00
2½	7.50	0.81	...	5.88	8	¾	2.50	3.38
3	8.25	0.91	...	6.62	8	¾	2.75	3.50
3½	9.00	0.97	...	7.25	8	¾	3.00	3.62
4	10.00	1.06	...	7.88	8	¾	3.00	3.88
5	11.00	1.12	...	9.25	8	¾	3.25	4.00
6	12.50	1.19	...	10.62	12	¾	3.25	4.12
8	15.00	1.38	...	13.00	12	⅞	3.75	4.75

*Sizes ¼ and ⅜-inch are not included in ANSI B16.24.
Sizes 3½ and 5-inch are not included in MSS SP-42.

4

THREADS

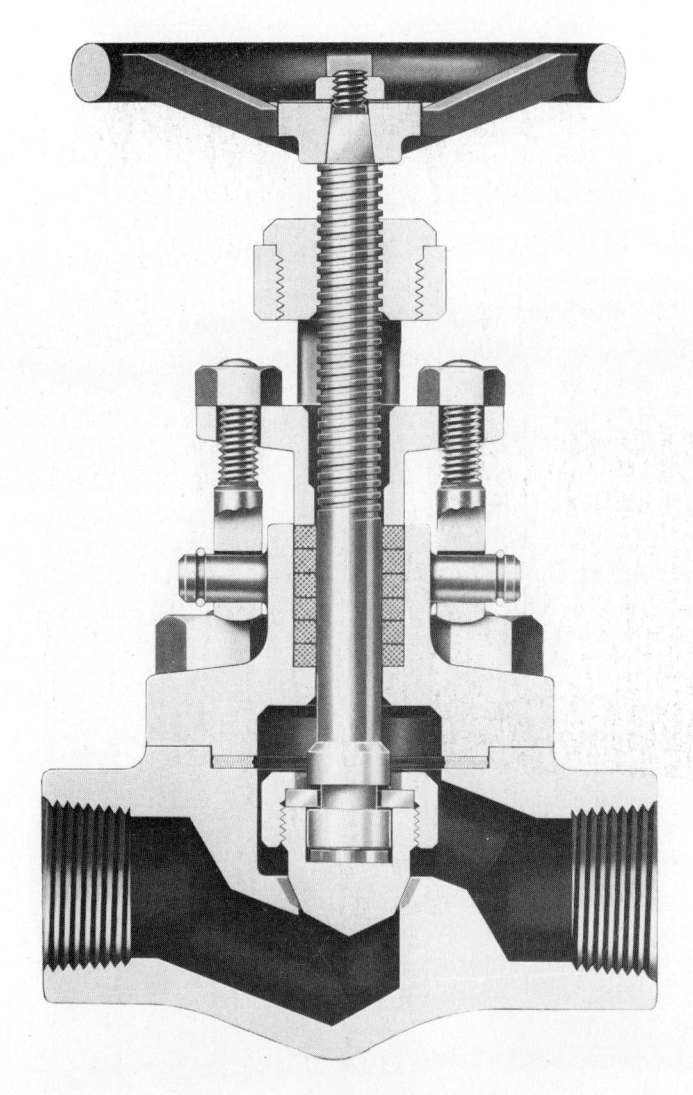

Fig. 4-1 *(Courtesy Jenkins Bros.)*

THREADED PIPE JOINTS

Standardized Threads

Several standards have been established covering pipe threads for various purposes. The oldest and probably most commonly used is the American National Standard for pipe threads, known as ANSI B2.1. There are also the American Petroleum Institute Standards No. 5A, 6A, and 5L covering oil field tubular goods, such as line pipe and casing threads.

The British Standard taper pipe thread system, in accordance with British Standard No. 21, is used in Great Britain. The form of thread is that of the Whitworth system; the sides of the thread form an angle of 55 degrees with each other, and the crests and roots of the threads are rounded to a radius equal to 0.1373 × the pitch of the thread. The total taper is 0.75 in. per foot, the same as for American National Standard taper pipe threads.

The number of threads per inch are as follows:

Pipe Size	Threads
1/8"	28 per inch
1/4, 3/8"	19 per inch
1/2, 3/4"	14 per inch
1 to 6"	11 per inch
8, 10"	10 per inch
12" and up	8 per inch

Thread Assembling

In making up threaded pipe joints, it is very important that the threads in both parts be thoroughly cleaned. Any threads which may have become burred or bent should be straightened or removed and afterward a good grade of lubricant should be applied to the threads. The lubricant reduces the friction, which allows the two parts to be pulled up further, resulting in a more effective pipe joint. Pipe joints should not be screwed together too rapidly in order to avoid raising the temperature of the two parts to a high degree.

Leaky Joints

Leaky joints can usually be traced either to faulty threading or an improper lubricant. Frequently the trouble lies in the thread on the pipe, which may have been cut with dull or improperly adjusted threading tools, resulting in wavy, shaved, rough, or chewed threads.

Wavy threads are noticeable both to the eye and touch, due to circumferential waves or longitudinal flats of slightly helical form rather than the desired true circular form. Shaved threads appear to have been threaded with two dies, one not matching the other, giving double thread appearance at the start of the thread. Rough or chewed threads are noticeably rough and torn.

Should the threads have any of these defects, it is possible that leaky joints might result.

Normal Engagement

The normal amount of engagement required between male and female threads to make a tight joint is given in Table 4-1. The dimensions are based on parts being threaded to the American National Standard for pipe threads or the API Standard for line pipe threads and have been established from tests made under practical working conditions. No allowance was made for variations in tapping or threading.

In order to obtain the thread engagements listed in the table, it is necessary to vary the torque or power applied according to the size, weight, and kind of material, as well as the type of lubrication used. For example, it requires considerably less power to make up a threaded joint using a light bronze valve than a high-pressure steel valve.

Threaded fittings must be clean-cut and uniform to provide strength, tightness, and durability in service. The American Standard pipe tapered thread is tapered 1/16 in. (0.16 cm) per inch (3/4 in. per ft) to ensure a tight joint at the fitting. The crest of the thread is flattened, and the root is filled so that the depth of the thread is equal to 0.80P (80 percent of the pitch). The number of threads per inch for a given nominal diameter can be obtained from Table 4-2. The American Standard tapered

TABLE 4-1 NORMAL THREAD ENGAGEMENT (*COURTESY CRANE COMPANY.*)

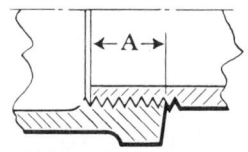

American National Standard and API Line Pipe Threads
Normal Thread Engagement

Pipe Size	A	Pipe Size	A	Pipe Size	A
1/8"	0.25"	1 1/4"	0.69"	4"	1.12"
1/4	0.38	1 1/2	0.69	5	1.25
3/8	0.38	2	0.75	6	1.31
1/2	0.50	2 1/2	0.94	8	1.44
3/4	0.56	3	1.00	10	1.62
1	0.69	3 1/2	1.06	12	1.75

TABLE 4-2 AMERICAN NATIONAL STANDARD TAPER PIPE THREADS (NPT)
(COURTESY CRANE COMPANY.)

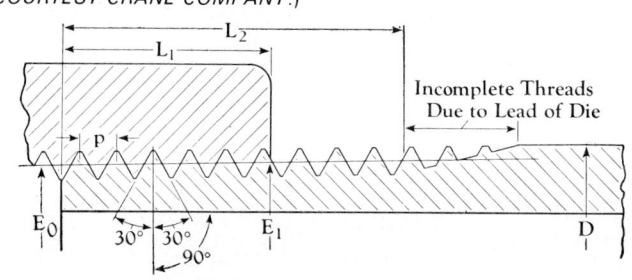

$E_0 = D - (0.050D + 1.1)p$
*$E_1 = E_0 + 0.0625 L_1$
$L_2 = (0.80D + 6.8)p$

p = Pitch
Depth of thread = $0.80p$
Total Taper ¾-inch per Foot

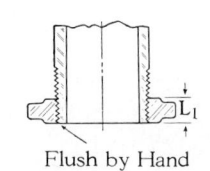

Tolerance on Product
One turn large or small from notch on plug gauge or face of ring gauge.

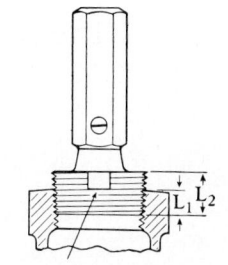

Notch flush with face of fitting. If chamfered, notch flush with bottom of chamfer

Dimensions in Inches

Nominal pipe size	D Outside diameter of pipe	Number of threads per inch	p Pitch of thread	E_0 Pitch diameter at end of external thread	E_1† Pitch diameter at end of internal thread	L_1¶ Normal engagement by hand between external and internal threads	L_2§ Length of effective external thread	Height of thread
1/16	0.3125	27	0.03704	0.27118	0.28118	0.160	0.2611	0.02963
⅛	0.405	27	0.03704	0.36351	0.37360	0.1615	0.2639	0.02963
¼	0.540	18	0.05556	0.47739	0.49163	0.2278	0.4018	0.04444
⅜	0.675	18	0.05556	0.61201	0.62701	0.240	0.4078	0.04444
½	0.840	14	0.07143	0.75843	0.77843	0.320	0.5337	0.05714
¾	1.050	14	0.07143	0.96768	0.98887	0.339	0.5457	0.05714
1	1.315	11.5	0.08696	1.21363	1.23863	0.400	0.6828	0.06957
1¼	1.660	11.5	0.08696	1.55713	1.58338	0.420	0.7068	0.06957
1½	1.900	11.5	0.08696	1.79609	1.82234	0.420	0.7235	0.06957
2	2.375	11.5	0.08696	2.26902	2.29627	0.436	0.7565	0.06957
2½	2.875	8	0.12500	2.71953	2.76216	0.682	1.1375	0.10000
3	3.500	8	0.12500	3.34062	3.38850	0.766	1.2000	0.10000
3½	4.000	8	0.12500	3.83750	3.88881	0.821	1.2500	0.10000
4	4.500	8	0.12500	4.33438	4.38712	0.844	1.3000	0.10000
5	5.563	8	0.12500	5.39073	5.44929	0.937	1.4063	0.10000
6	6.625	8	0.12500	6.44609	6.50597	0.958	1.5125	0.10000
8	8.625	8	0.12500	8.43359	8.50003	1.063	1.7125	0.10000
10	10.750	8	0.12500	10.54531	10.62094	1.210	1.9250	0.10000
12	12.750	8	0.12500	12.53281	12.61781	1.360	2.1250	0.10000
14 O.D.	14.000	8	0.12500	13.77500	13.87262	1.562	2.2500	0.10000
16 O.D.	16.000	8	0.12500	15.76250	15.87575	1.812	2.4500	0.10000
18 O.D.	18.000	8	0.12500	17.75000	17.87500	2.000	2.6500	0.10000
20 O.D.	20.000	8	0.12500	19.73750	19.87031	2.125	2.8500	0.10000
24 O.D.	24.000	8	0.12500	23.71250	23.86094	2.375	3.2500	0.10000

† Also pitch diameter at gauging notch.
§ Also length of plug gauge.
¶ Also length of ring gauge, and length from gauging notch to small end of plug gauge.
* For the ⅛-27 and ¼-18 sizes ... E_1 approx. = $D - (0.05 D + 0.827) p$

Above information extracted from American National Standard for Pipe Threads, ANSI B2.1.

THREADS

TABLE 4-3 NORMAL ENGAGEMENT FOR TIGHT JOINTS (*COURTESY ITT GRINNELL CORPORATION.*)

(does not apply to companion flanges)

length of pipe entering fitting

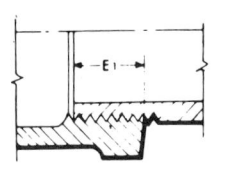

American Standard and API line pipe threads

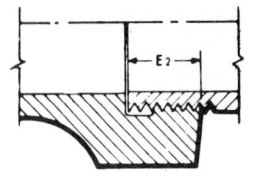

shoulder type drainage fitting threads

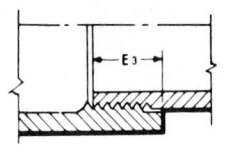

railing fitting thread assembly

dimensions, in.

size	1/8	1/4	3/8	1/2	3/4	1	1 1/4	1 1/2	2	2 1/2	3	3 1/2	4	5	6	8	10	12
E1, E2	1/4	3/8	3/8	1/2	9/16	11/16	11/16	11/16	3/4	15/16	1	1 1/16	1 1/8	1 1/4	1 5/16	1 7/16	1 5/8	1 3/4
E3	......	......	......	1/2	1/2	5/8	11/16	3/4	3/4									

thread or pipe thread is very similar to the American Standard thread or straight thread.

In general, threaded couplings and pipe are used for piping 2½ in. (6.3 cm) or less in diameter except for high-temperature and -pressure situations. It is also common to use external straight threads for male ends and taper threads for female ends, since piping materials are usually sufficiently ductile that threads can adjust themselves and create a tight joint. Taper threads are recommended by the ASA for most threaded joints, except for pressure-tight joints for couplings, pressure-tight joints for raised-cup fuel or oil filter fittings, loose-fitting mechanical joints for hose couplings, and free-fitting mechanical joints for fixtures. Other than these four situations, the normal American Standard pipe thread is to be used. Table 4-3 shows the normal engagement for tight joints for the length of pipe entering the fitting. The normal engagement specified for American Standard pipe thread is based on parts being threaded to the American Standard pipe thread or the API standard for pipeline threads. ANSI B2.1, which covers pipe threads for various purposes, was one of the first standards established and most commonly used. The API standards 5A, 6A, and 5L cover oil field tubular material such as line pipe and casing threads. Line pipe threads have the same form and taper as American Standard pipe threads.

All internal threads are known as female threads and all external threads are male threads. The length of engagement between male and female threads to make a tight joint is based on the thread being machined to the American Standard for pipe threads or the API standard for line pipe threads which have been established under practical working conditions. For designating various pipe thread types, it is important to call out one of the various designations:

- NPT   American Standard pipe threads
- NPTR American Standard taper pipe threads for rail fittings
- NPSH American Standard straight pipe threads for hose couplings and nipples
- NPSL American Standard pipe threads for locknuts and locknut pipe threads
- NPSI American Standard internal straight pipe or sealer
- NPSM American Standard straight pipe threads for mechanical joints
- NPSC American Standard straight pipe threads and pipe couplings

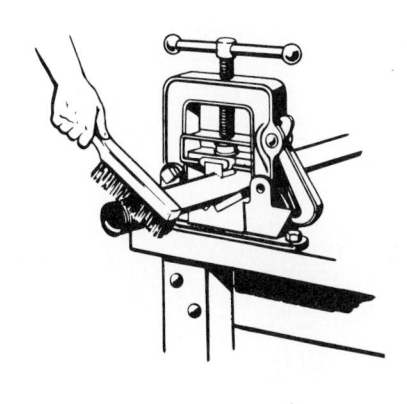

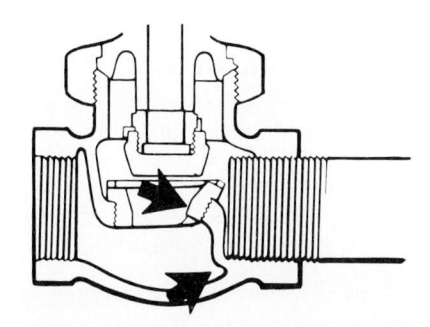

Fig. 4-2 Making up screwed joints (*Courtesy Crane Company.*)

- NPSF American Standard straight pipe threads for pressure-tight joints for use without lubricant

HOW TO MAKE UP A SCREWED JOINT

The screwed joint (threaded male and female ends screwed together) is the most common method of joining pipe. A high degree of standardization is maintained in the threading of piping materials. American Pipe Thread Standards guide the piping industry.

Figure 4-2 shows the normal engagement male and female threads for making up screwed joints.

Use a Wire Brush Since dirt can make a good fit impossible, the first thing to do is to wipe both male and female threads clean. Wire brushing is recommended, especially if threaded pipe has been exposed to weather. Running a tap or die over the threads will usually straighten any that may be damaged and will help to assure a good metal-to-metal joint.

Lubricant is Good—But Not Inside a Line Next, use a bit of good thread lubricant or "pipe dope," but put it on the *male threads only*. This prevents any excess from squeezing into the pipe and causing harm to valve seats or other mechanisms. Lubricant reduces friction when pulling up a joint. It is not expected to seal the joint or to compensate for poor workmanship.

With lubricant applied, start the joint by hand. If the thread engagement feels "right," turn it up as far as it will go. If the threading job is up to standard, and if the threads are free of dirt, the joint is now started right. Further turns with a wrench will complete the job.

Pipe Run in Too Far But be careful with the wrench. Never use an oversized one that will tempt you to "lean on the joint." Too much pull-up can cause damage, especially in the case of valves. Don't try to run all the male threads into the joint. The lead of the die always leaves a few imperfect and unusable male threads which should be left exposed.

THREADS

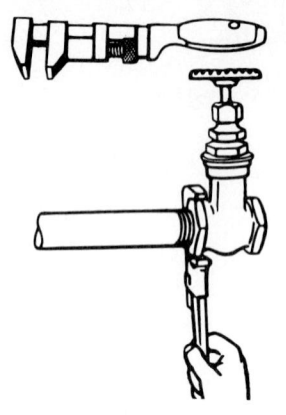

hex and monkey wrench

Your hexagonal or monkey wrench —with its smooth, square jaws— is always your best bet for hex end valves and fittings. It gives a better fit on the part to be turned, and doesn't have the crushing effect of a pipe wrench. Because of the shape of their jaws, hex or monkey wrenches obviously cannot be used on pipe or other round objects.

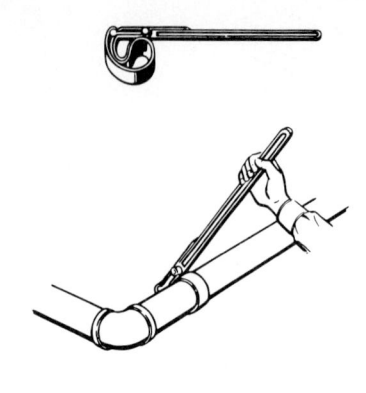

strap wrench

Strap wrenches are used mainly when working with plated or polished materials to avoid marring the surface. They also come in handy in tight places where you can't get in with a pipe wrench.

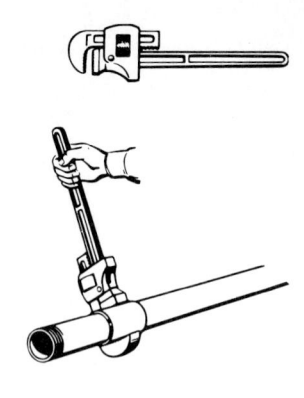

pipe wrench

Designed for use on pipe and screwed end fittings only. Not as efficient as a monkey wrench on hexagonal ends. The harder you pull a pipe wrench, the tighter it squeezes. This squeezing action can do great damage to a valve. It may cripple or even crush its body.

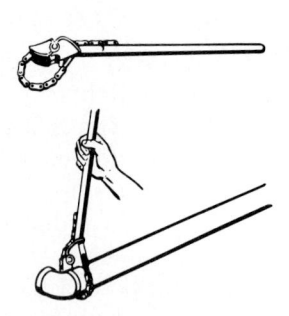

pipe tongs or chain pipe wrench

Pipe tongs are wrenches of the chain strap and lever type. They are generally used for larger sizes of pipe—about 3-inches and above —although they are available for sizes down to $\frac{1}{8}$-inch.

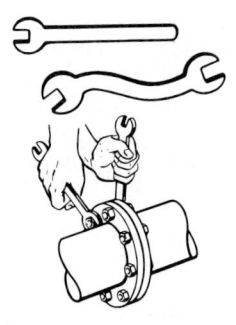

open end wrenches

Open end or socket wrenches are most suitable and fastest to work with when pulling up flange bolts and nuts. Getting the right size is the most important consideration here. Slippage of the wrench can cause bruised knuckles and can wear nut and bolt heads round.

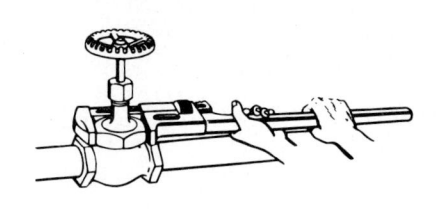

too much leverage is dangerous

Avoid bearing down on the joint with an oversize wrench or a wrench with an extension handle. It often gives you more leverage than you need, and may result in pulling up a joint so tight that the fitting will be cracked, the valve twisted out of shape, or the pipe run clear into the seats.

Fig. 4-3 Wrench types and proper use (*Courtesy Crane Company.*)

TABLE 4-4   AMERICAN STANDARD STRAIGHT THREADS (*COURTESY ITT GRINNELL CORPORATION.*)

Information abstracted from American Standard Pipe Threads **ANSI B2.1**.

■ Straight thread gages are used to gage mechanical joint straight pipe threads.

The actual pitch diameters of the tapped hole will be slightly smaller than the values given.

♦ American Standard taper pipe thread plug gages are used to gage straight pipe threads in couplings with the gaging notch coming flush with the edge of the thread or with the bottom of chamfer, if chamfered, allowing a tolerance of one and one half turns large or small.

basic dimensions (inches)

nominal pipe size	threads per inch	pitch of thread	straight pipe threads♦ in pipe couplings (pressure tight joints) internal max	min	straight pipe threads for mechanical joints■ (free fitting) external max	min	internal max	min	straight pipe threads for locknut connections (loose fitting mechanical joints) external max	min	internal max	min
⅛	27	0.0370	0.3782	0.3713	0.3748	0.3713	0.3782	0.3748	0.3840	0.3805	0.3898	0.3863
¼	18	0.0556	0.4951	0.4847	0.4899	0.4847	0.4951	0.4899	0.5038	0.4986	0.5125	0.5073
⅜	18	0.0556	0.6322	0.6218	0.6270	0.6218	0.6322	0.6270	0.6409	0.6357	0.6496	0.6444
½	14	0.0714	0.7851	0.7717	0.7784	0.7717	0.7851	0.7784	0.7963	0.7896	0.8075	0.8008
¾	14	0.0714	0.9956	0.9822	0.9889	0.9822	0.9956	0.9889	1.0067	1.0000	1.0179	1.0112
1	11½	0.0870	1.2468	1.2305	1.2386	1.2305	1.2468	1.2386	1.2064	1.2523	1.2739	1.2658
1¼	11½	0.0870	1.5915	1.5752	1.5834	1.5752	1.5915	1.5834	1.6051	1.5970	1.6187	1.6106
1½	11½	0.0870	1.8305	1.8142	1.8223	1.8142	1.8305	1.8223	1.8441	1.8360	1.8576	1.8495
2	11½	0.0870	2.3044	2.2881	2.2963	2.2881	2.3044	2.2963	2.3180	2.3099	2.3315	2.3234
2½	8	0.1250	2.7739	2.7505	2.7622	2.7505	2.7739	2.7622	2.7934	2.7817	2.8129	2.8012
3	8	0.1250	3.4002	3.3768	3.3885	3.3768	3.4002	3.3885	3.4198	3.4081	3.4393	3.4276
3½	8	0.1250	3.9005	3.8771	3.8888	3.8771	3.9005	3.8888	3.9201	3.9084	3.9396	3.9279
4	8	0.1250	4.3988	4.3754	4.3871	4.3754	4.3988	4.3871	4.4184	4.4067	4.4379	4.4262
5	8	0.1250			5.4493	5.4376	5.4610	5.4493	5.4805	5.4688	5.5001	5.4884
6	8	0.1250			6.5060	6.4943	6.5177	6.5060	6.5372	6.5255	6.5567	6.5450
8	8	0.1250							8.5313	8.5196	8.5508	8.5391
10	8	0.1250							10.6522	10.6405	10.6717	10.6600
12	8	0.1250							12.6491	12.6374	12.6686	12.6569

TABLE 4-5 BRITISH STANDARD TAPER PIPE THREADS (*COURTESY ITT GRINNELL CORPORATION.*)

Reproduced from British Standard No. 21—1957

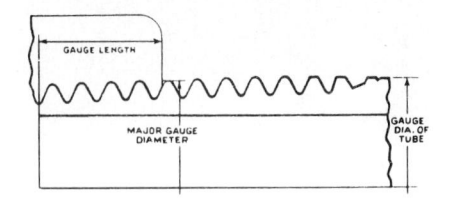

Nominal Size	Max. Outside Diameter	No. of Threads per Inch	Depth of Thread	Major	Effective Pitch Diameter	Minor	Gauge Length
⅛	0.412	28	0.0229	0.383	0.3601	0.3372	0.1563
¼	0.550	19	0.0337	0.518	0.4843	0.4506	0.2367
⅜	0.688	19	0.0337	0.656	0.6223	0.5886	0.2500
½	0.859	14	0.0457	0.825	0.7793	0.7336	0.3214
¾	1.075	14	0.0457	1.041	0.9953	0.9496	0.3750
1	1.351	11	0.0582	1.309	1.2508	1.1926	0.4091
1¼	1.692	11	0.0582	1.650	1.5918	1.5336	0.5000
1½	1.924	11	0.0582	1.882	1.8238	1.7656	0.5000
2	2.403	11	0.0582	2.347	2.2888	2.2306	0.6250
2½	3.021	11	0.0582	2.960	2.9018	2.8436	0.6875
3	3.526	11	0.0582	3.460	3.4018	3.3436	0.8125
3½	4.021	11	0.0582	3.950	3.8918	3.8336	0.8750
4	4.526	11	0.0582	4.450	4.3918	4.3336	1.0000
5	5.536	11	0.0582	5.450	5.3918	5.3336	1.1250
6	6.541	11	0.0582	6.450	6.3918	6.3336	1.1250

Notes: Included thread angle 55°, formed at right angle to the pipe axis, and rounded equally top and bottom.
All dimensions given in inches.

TABLE 4-6 DRILL SIZES FOR PIPE TAPS (*COURTESY ITT GRINNELL CORPORATION.*)

SIZE OF TAP	NUMBER OF THREADS PER INCH	DIAM. OF DRILL	SIZE OF TAP	NUMBER OF THREADS PER INCH	DIAM. OF DRILL
⅛	27	11/32	2	11½	2 3/16
¼	18	7/16	2½	8	2 9/16
⅜	18	37/64	3	8	3 3/16
½	14	23/32	3½	8	3 11/16
¾	14	59/64	4	8	4 3/16
1	11½	1 5/32	4½	8	4 ¾
1¼	11½	1½	5	8	5 5/16
1½	11½	1 49/64	6	8	6 5/16

TABLE 4-7 TAP AND DRILL SIZES (AMERICAN STANDARD COURSE) (*COURTESY ITT GRINNELL CORPORATION.*)

SIZE OF DRILL	SIZE OF TAP	THREADS PER INCH	SIZE OF DRILL	SIZE OF TAP	THREADS PER INCH
7	¼	20	49/64	⅞	9
F	5/16	18	53/64	15/16	9
5/16	⅜	16	⅞	1	8
U	7/16	14	63/64	1⅛	7
27/64	½	13	1 7/64	1¼	7
31/64	9/16	12	1 13/64	1⅜	6
17/32	⅝	11	1 11/32	1½	6
19/32	11/16	11	1 29/64	1⅝	5½
21/32	¾	10	1 9/16	1¾	5
23/32	13/16	10	1 11/16	1⅞	5
			1 25/32	2	4½

5
VALVES

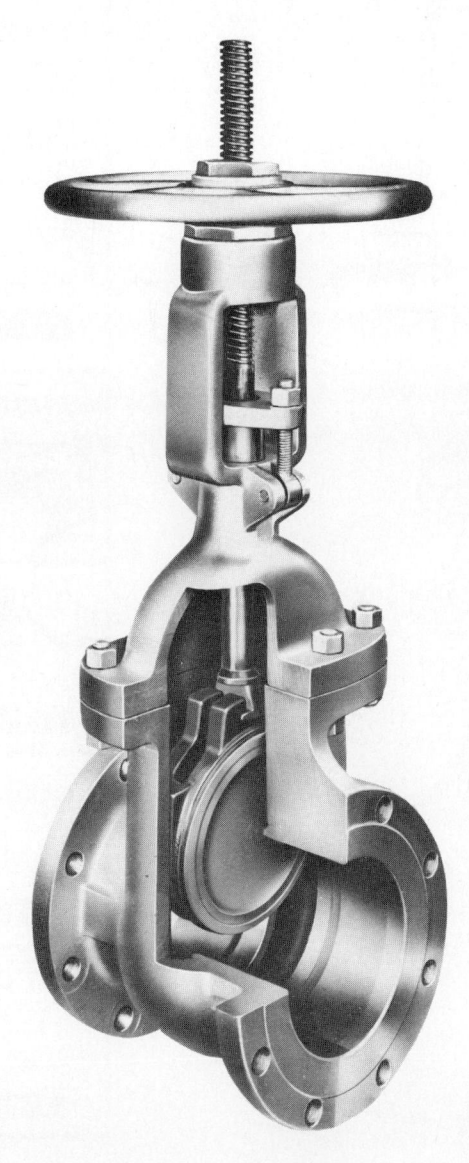

Fig. 5-1 (*Courtesy Lunkenheimer Company, Cincinnati, OH 45214.*)

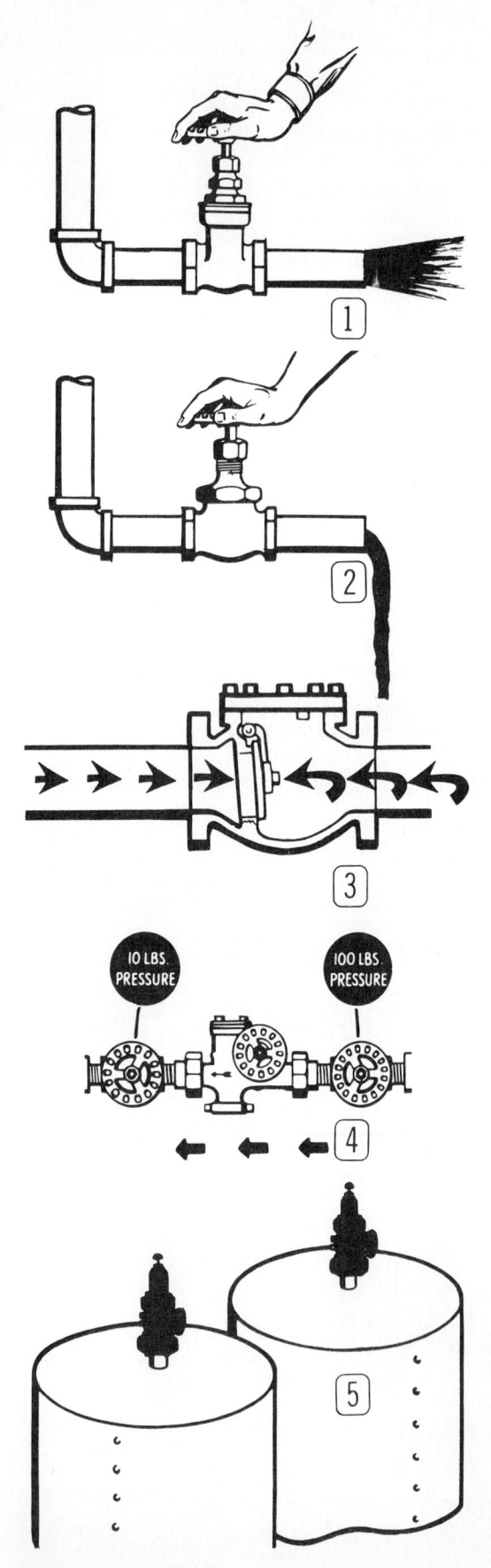

Fig. 5-2 Principal functions of valves (*Courtesy Crane Company.*)

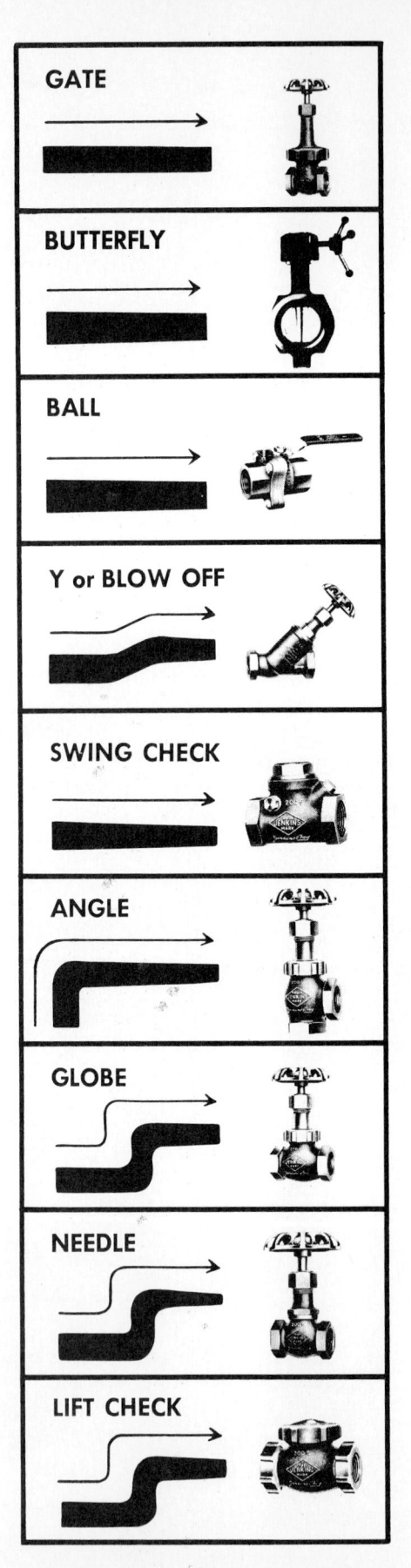

Fig. 5-3 Valve flow characteristics (*Courtesy Jenkins Bros.*)

CHAPTER 5

THE PRINCIPAL FUNCTIONS OF VALVES

Starting and Stopping Flow This is the service for which valves are most generally used—starting and stopping flow. Gate valves are excellently suited for such service. Their seating design, when open, permits fluid to move through the valve in a straight line with minimum restriction of flow and loss of pressure at the valve. The gate principle is not practical for throttling.

Regulating or Throttling Flow Regulating or throttling flow is done most efficiently with globe or angle valves. Their seating design causes a change in direction of flow through valve body, thereby increasing resistance to flow at the valve. Globe and angle valve disk construction permits closer regulation of flow. These valves are seldom used in sizes above 12 in., owing to the difficulty of opening and closing the larger valves against pressure.

Preventing Backflow Check valves perform the single function of checking or preventing reversal of flow in piping. They come in two basic types: swing and lift checks. Flow keeps these valves open, and gravity and reversal of flow close them automatically. As a general rule, swing checks are used with gate valves and lift checks with globe valves.

Regulating Pressure Pressure regulators are used in lines where it is necessary to reduce incoming pressure to the required service pressure. They not only reduce pressure but maintain it at the point desired. Reasonable fluctuations of inlet pressure to a regulator valve do not affect outlet pressure for which it is set.

Relieving Pressure Boilers and other equipment subject to damage from excessive pressures should be equipped with safety valves. They usually are spring-loaded valves which open automatically when pressure exceeds limit for which the valve is set. These valves are known as safety valves and relief valves. Safety valves are generally used for steam, air, or other gases. Relief valves are usually used for liquids.

VALVE FLOW CHARACTERISTICS

The amount of fluid permitted to pass through a valve varies with its basic pattern. Generally, the greater the degree of pressure control, the greater the restriction of flow, pressure drop, and energy loss. Many formulas and equations have been developed to determine precisely pressure drops and energy losses for specific sizes of valves, types of valves, fluids, and flow conditions.

To assist in making average calculations, the diagrams in Fig. 5-3 indicate the relative capacity and direction of flow permitted by several basic valve patterns, and Table 5-1 gives the approxi-

TABLE 5-1   FLOW RESISTANCE (*COURTESY JENKINS BROS.*)

VALVE TYPE

VALVE SIZE, IN.	GATE	BLV	BFV	Y	SWING CHECK	ANGLE	GLOBE
½	0.4	2	—	3.6	4	8	16
¾	0.5	1.5	—	4.5	5	12	22
1	0.6	1.9	—	6.3	7	15	27
1¼	0.8	2.4	—	8.0	9	18	37
1½	0.9	2.8	—	10.0	11	21	44
2	1.2	3	2.2	12.5	14	28	55
2½	1.4	—	2.6	13.5	15	32	65
3	1.6	4.7	2.9	18.0	19	41	80
3½	2.0	—	—	20.0	22	50	100
4	2.2	4.4	3.9	22.5	25	55	120
5	2.9	—	5.4	28.0	32	70	140
6	3.5	9.7	6.3	36	40	80	160
8	4.5	—	6.8	45	50	110	220
10	5.5	—	7.5	58	65	140	280
12	6.5	—	9.0	68	75	160	340
14	8.0	—	—	81	90	190	380
16	9.0	—	—	95	105	220	430
18	10.0	—	—	108	120	250	500
20	12.0	—	—	117	130	270	550
24	14.0	—	—	135	150	380	650

mate flow resistance of full open valves compared to the equivalent feet of schedule 40 pipe.

As indicated, gate valves allow maximum flow, with butterfly, ball, Y, swing check, angle, globe, and lift check patterns following in increasing order of flow resistance.

VALVE ABBREVIATIONS

1. Valve descriptions

AI	All iron
Al	Aluminum
BR	Bronze
CI	Cast iron
Cr	Chromium
Cr 13	Type 410 stainless steel
CS	Cast steel
DI	Ductile iron
FS	Forged steel
HF	Stellite face (hard face)
IBBM	Iron body bronze mounted
M	Monel metal
MI	Malleable iron
Mo	Molybdenum
N	Nickel
NI	Nickel iron
NICU	Nickel copper alloy
PVC	Polyvinyl chloride
SA	Sludge acid
SA	Sulfuric acid
SS	Stainless steel
Tef	Teflon
18-8 Mo	Type 316 stainless steel

2. Ratings

CWP	Cold working pressure
S	Steam pressure
SP	Steam pressure
WOG	Water, oil, gas pressure
WP	Working pressure
WSP	Working steam pressure

3. Operating mechanism

NRS	Nonrising stem
OS&Y	Outside screw and yoke
RS	Rising stem

4. Facings, disks, and joints

DD	Double disk
FF	Flat face
RF	Raised face
RTJ	Ring-type joint

5. End connections

BW	Butt-welded ends
FE	Flanged ends
FFD	Flanged, faced, and drilled
Flg	Flanged ends
Scr	Screwed ends
SE	Screwed ends
SJ	Soldered ends
SW	Socket-welded ends

6. Societies

ANSI	American National Standards Institute
API	American Petroleum Institute
ASME	American Society of Mechanical Engineers
ASTM	American Society of Testing Materials
MSS	Manufacturers Standardization Society of the Valve and Fittings Industry
SAE	Society of Automotive Engineers

7. Measurements

cfm	cubic feet per minute
gpm	gallons per minute
psi	pounds per square inch
rpm	revolutions per minute

SERVICE CONDITIONS

Choosing the right valve and its design options (bonnet, stem, disk) is determined mainly by the conditions under which it will be in operation. The nature of the service, critical or noncritical, is of primary importance. Noncritical services usually include low-pressure/moderate-temperature plumbing and heating systems. These uses are probably more familiar to the reader than the critical services—which entail high pressures, high temperatures, and fire hazards in the power generation and petrochemical industries. Critical services demand close evaluation of valve types and their suitability to the job. Whenever there is danger to people or property and valve operation is critical, extreme caution is taken selecting the valve, its connections, bonnet construction, and safety features.

Every liquid and gas has its own characteristics and subsequent effect on piping materials and equipment. Erosion and corrosion of pipes and valves, crystallization, vaporization of line contents—all are aspects which need to be considered when selecting the valve type. Air, gas, oil, water, and solids suspended in liquid have a specific effect on the system.

A valve's pressure classification is always given in pounds per square inch (psi). Each item is rated for steam pressure (SP) first, then water, oil, and gas (WOG). Sometimes the WOG rating is replaced by a cold working pressure (CWP) rating. The temperature rating will sometimes follow the SP rating (for example, 150 SP 450°F-250 WOG). Many valves have only the SP rating cast on their bodies. For example, a 150 SP-300 WOG bronze globe valve might have 150 SP on its side. Each product is tested and rated by design and allowable stress at a certain temperature. Temperature restrictions are listed in catalogs provided by the valve company. Valves may be used at or below the SP rating, but the temperature must not exceed the rating established for the valve material. This rating takes into account the hazard of controlling high-temperature fluids and has an ample safety factor.

On pipeline valves, it is traditional to mark only the ANSI rating (no SP rating). Thus class 150 is understood to have a CWP (WOG) of 275 psi; while class 300 is understood to have a CWP (WOG) of 720 psi; and class 600 is understood to have a CWP (WOG) of 1440 psi.

The WOG rating designates the maximum nonshock pressure (at atmospheric temperatures) at which the product may be used. This rating is usually cast on the product along with the SP rating if space permits. Exclusive WOG valves have only a WOG rating. One should become familiar with these ratings because a drafter's final job on most drawings is to construct a bill of materials and callout list.

Shock is a critical consideration in the design and placement of valves in a system. Physical shock from *external* sources becomes a concern when the valve is connected to moving or vibrating equipment such as pumps, motors, and compressors. Physical shock due to frequency of operation, maintenance, and the surrounding environment must also be taken into account. A prime consideration in many areas of the world where the service is critical is seismic movement—the impact caused by earthquakes. When referring to *internal* shock, one must consider all movements and stress produced by overpressure in the pipeline. Overpressure is caused by the sudden arresting of flow, which when superimposed on static pressure may overload the valve and also damage fittings and pipes. In these cases, it is advisable to select higher-rated valves for the service or provide a surge relief system.

Thermal shock is caused by rapidly rising and falling temperatures in the line, which cause expansion and contraction of materials. This type of shock is extremely serious in the nuclear power field.

PLACEMENT

Adequate attention must be given to the placement of valves. Ease of access, frequency of operation, accessibility for maintenence, hanger placement, possible temperature losses through the valve—all are considerations. Sometimes the placement will even dictate the type and style of valve selected for the job.

Environmental considerations include the following questions: Is the valve suitable for outside service where it may be exposed to extreme heat, cold, or moisture? Will it be buried in ground which freezes? Can the valve materials withstand the conditions under which it must operate effectively? Will insulating the valve affect its placement and its ability to be operated and maintained? These and other questions are essential to the proper selection and placement of valves.

DESIGN OPTIONS

Bonnet Assemblies

The bonnet is the part of a valve which covers the internal moving parts and through which the stem threads must pass. *Screwed bonnets* are the least expensive type and are used for low-pressure plumbing. They provide easy access to the valve's inner parts but are not advised for frequent dismantling because of wear and tear on the threads and on the valve itself. These bonnets come on smaller valves which are not intended for critical assignments.

Bolted bonnets are easily dismantled, come in larger sizes, and are recommended for critical service. The steel type is preferred in the power and process areas. A variation of the bolted bonnet is the *flanged type with ring joint,* which is used for high temperatures and high pressures. *Flanged with gasket bonnets* are common for 900°F (482.2°C) and below. The *pressure-seal* type is self-sealing and requires periodic adjustment. This bonnet design is good for frequent dismantling.

Welded bonnets for high temperatures and high pressures are used in critical situations that do not involve frequent dismantling. This design is excellent for corrosive materials and provides a tight leakproof joint.

Screwed union bonnets are used for small valves with high-temperature and high-pressure ratings because, unlike screwed bonnets, there is no danger of their being screwed off in normal operation. This type is widely used in petrochemical plants.

Stems

Stem selection is important because of headroom clearance. Stem options are available in several

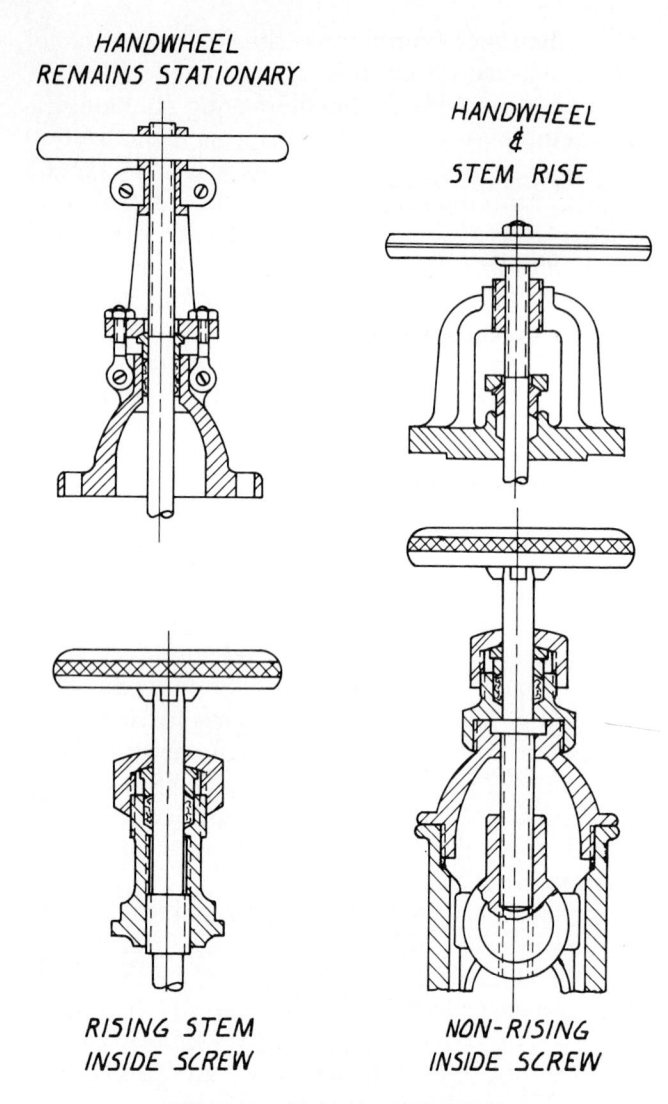

Fig. 5-4 Valve stem variations

designs from which to choose. Frequency of operation and ease in reading disk position are just two considerations involved in choosing an option (Fig. 5-4).

Outside screw and yoke (OS&Y) is the best construction for high temperatures and pressures. Stem threads are not subject to severe thermal shock and are easy to lubricate. Corrosion possibilities are minimal. The need for adequate headroom for operation and maintenance is a major drawback, however. A variation of this style has both a rising stem and stationary handwheel. The stem position indicates disk position. Stem and threads must be protected from external damage. In the second option—*rising stem and stationary handwheel*—the only differences are that disk position is less easily noticed by stem position and the danger of external damage to threads is greater unless they are covered.

Inside screw options feature a *rising stem and inside screw* which is most common in small bronze valves that are 2 in. (5.08 cm) and less. Such valves are used in situations where threads are not damaged by line contents (water, hydrocarbons, steam). Stem position indicates the disk position, and adequate headroom is necessary. The second variation is the *nonrising stem and inside screw*. This option is mainly for valves 4 in. (10.16 cm) and below and is not recommended for critical services. Since line fluids are in constant contact with threads, high-temperature use is undesirable. Two positive aspects of this option are low wear on packing (because only the stem revolves) and less headroom needed. This valve must be equipped with a stem indicator to designate disk position and is usually used as a plumbing valve.

CONSTRUCTION MATERIALS

Valve bodies come in several types of metal and plastic. Bronze, brass, iron, and steel alloys are most common. Selection of the body material depends on the valve's function (shutoff or throttling), the pressure and temperature ratings, line contents and their corrosive effects, line stresses, and fire hazards. One should always check the manufacturer's guide to selection and recommendations before ordering. Valve bodies are often classified by material.

Brass is a copper-zinc alloy used for noncritical service at temperatures below 400°F (204.4°C) and low pressures. Both brass and bronze are copper-based alloys. Check with the manufacturer's data as to their true composition. *Bronze*, a copper-tin alloy, is stronger than brass and used at 500°F (259.9°C) maximum and 300 psi and below. This material is usually found in valves 3 in. and smaller.

Iron comes in a variety of types and compositions. Since manufacturers have different grades, it is important to check the specific valve type one wants and its specifications. *Cast iron* valve bodies come in most sizes. The SP ratings of 250 and a temperature of 450°F (232.2°C) are maximum. This material is an alloy of iron, carbon, silicon, and manganese and comes in a wide variety of grades and compositions. *Malleable iron* valves are excellent for service subject to thermal shock. Their tightness, stiffness, and toughness are desirable characteristics.

Steel valves come either forged or cast, or

they can be fabricated from plate stock. *Cast steel* is used for large valves where service is critical and other materials are inadequate. *Forged steel* valves are drilled and machined from solid stock and hence are not practical for anything but small sizes. They are used for highly corrosive conditions. *Fabricated steel* is used for extremely large valves where cast steel would be impractical because of weight considerations.

TRIM MATERIALS

Valve trim consists of the seat, ring, disk, facing, and stem, all of which come in an assortment of steel and bronze alloys. The trim is as important as the body itself, for without proper trim the valve cannot perform satisfactorily. When selecting the trim, consideration must be given to operating temperature, chemical stability of the line fluid, and tensile properties (hardness and toughness). (See Table 5-2.) Seat-dish facings, their seizing possibilities, and the trim's corrosive resistance must also be taken into account. These data can usually be found in the valve company's catalog.

When specifying valves, one may order in the following way:

1. SP rating
2. Temperature rating (where applicable)
3. WOG rating (CWP)
4. Body material
5. Type (gate, globe) and size
6. Trim (ends, stem variations, bonnet, disk)
7. Catalog figure and catalog page

For example: 150 SP 300°F 250 WOG; 2-in. bronze globe valve, screwed ends, inside screw, nonrising stem, composition disk, screwed bonnet; Fig. 20, p. 150A, Lunkenheimer Valve Company. Normally, only call out SP and WOG, size, type, company, and figure or order number (sometimes only order number).

End connections are also important in valve selection because not all valve types and sizes are adaptable to all means of connection. The drafter must be familiar with the possible end connection variations, which will also influence the symbols

TABLE 5-2 VALVE TRIM SELECTION GUIDE (*COURTESY STOCKHAM VALVES AND FITTINGS.*)

Range	Temperature (°F)	Material
Very high	2000	Refractory metals, ceramics
High	1200 to 1600	High-temperature alloy steels
Intermediate	1000	Carbon steel
	650	Ductile iron
	550	Bronze
	450	Cast iron
	150	PVC plastic
Low	−250	Low-alloy steels, bronze
Very low (cryogenic)	−450	Bronze, austenitic ductile iron, austenitic stainless steels

TABLE 5-3 SERVICE CONDITIONS FOR GATE VALVES (*COURTESY STOCKHAM VALVES AND FITTINGS.*)

Service Conditions	Disk Options: Solid	Disk Options: Split Wedge, Double, or Flexible	Stem Options: Rising Stem	Stem Options: Nonrising Stem
Viscous or gummy fluids	●		●	
Limited operating space	●			●
Varying temperatures (expansion/contraction)		●	●	●
Visual position indication of disk		●	●	
Steam service (and general oil, gas, and water services)	●		●	
Gases, volatile fluids, light liquids		●	●	●
Water mains and distribution		●		●

TABLE 5-4 SELECTION GUIDE FOR DISK VARIATIONS FOR GLOBE VALVES (*COURTESY STOCKHAM VALVES AND FITTINGS.*)

Disk Type	Renewable Disks	Regrinding Disks	Renewable Seats	Hardened Seats to Resist Erosion	Limited Sealing Contact	Close Regulation	Tight *Shutoff* of *Gases*
TFE	●						●
Composition	●						●
Spherical or full-way		●	●†		●		
Plug		●	●†	●†			
Needle					●	●	

†In some cases.

VALVES

TABLE 5-5 SELECTION GUIDE FOR LUBRICATED VERSUS NONLUBRICATED PLUG VALVES (*COURTESY STOCKHAM VALVES AND FITTINGS.*)

Feature	Lubricated	Lift Type	Nonlubricated Nonmetallic Sleeve Type	Eccentric Plug Type
Temperatures above 450°F		●		
Some throttling (with minimum abrasion)	●			●
Slurries		●	●	
Minimum operating torques			●	●
Low operating cost (no lubrication)		●	●	●
Protected body seats		●		
Coking service		●		

for valves—there are different symbols for most of the end connections. In many instances, industries that deal exclusively in one type of joint will establish their own simplified method of drawing end connections for fittings, valves, and other equipment. If only flanged fittings are used, for example, no flanges are drawn or symbolized.

But not all companies do things the same way. A drafter's greatest asset will be an ability to adjust to the job at hand. Most companies have compiled a book of standards they have found useful throughout many years of experience. Employers do not always conform to all the rules and regulations found in drafting books or taught in schools. These are only meant to be guidelines, not laws.

END CONNECTIONS ON VALVES*

Just as a chain is no stronger than its weakest link, a piping system is no better than its connecting links. Specifying the right end connection on valves and fittings is, therefore, of the utmost importance. Be sure you know which type of end connection is best suited to the working conditions of your particular system.

Shown below are the types of end connections most commonly used on valves and fittings, although not all piping materials are available in all types of end connections. Each type has one or more variants, with steel flange joints having the greatest variety.

Screwed Ends Screwed end valves and fittings are by far the most widely used. This type of end connection is found in brass, iron, steel, and alloy piping materials. They are suited for all pressures, but are usually confined to smaller pipe sizes. The larger the pipe size, the more difficult it is to make up the screwed joint.

Welding Ends Welding ends, available in steel valves and fittings only, are used mainly for higher pressure-temperature services. They are recommended for lines not requiring frequent dismantling. There are two types of welding end materials: butt and socket welding. Butt-welding valves and fittings come in all sizes; socket-welding ends are usually limited to smaller sizes.

Brazing Ends Brazing end connections are available on brass materials. The ends of such materials are specially designed for use of brazing alloys to make the joint. When the equipment and brazing material are heated with a welding torch to the temperature required by the alloy, a tight seal is formed between the pipe and valve or fitting. While made in a manner similar to a solder joint, a brazed joint will withstand higher temperatures because of the brazing materials used.

Solder Ends Solder-joint valves and fittings are used with copper tubing for plumbing and heating lines and for many low-pressure industrial services. The joint is soldered by applying heat. Because of close clearance between the tubing and the socket of the fitting or valve, the solder flows into the joint by capillary attraction. The use of soldered joints under temperature is limited because of the low melting point of the solder.

Flared Ends Flared end connections are commonly used on valves and fittings for metal and plastic tubing up to 2-in. diameter. The end of tubing is skirted or flared, and a ring nut is used to make a union-type joint.

Hub Ends Hub end connections are generally limited to valves for water supply and sewage piping. The joint is assembled on the socket principle, with pipe inserted in hub end of valve or fitting, then caulked with oakum and sealed with molten lead.

Flanged Ends Flanged end materials, although made in sizes as small as 1/2 in., are generally used for larger lines because they are easy to assemble and take down. Flanged joints are made with much smaller tools than required for screwed connections of comparable size. Flanges which are separate pieces can be attached to pipe in several ways.

---
*(Courtesy Crane Company.)

Flanged joints are made up by bolting two flanges together with a gasket between their machined faces to assure a tight seal. The proper steps in assembling a flanged joint are shown below.

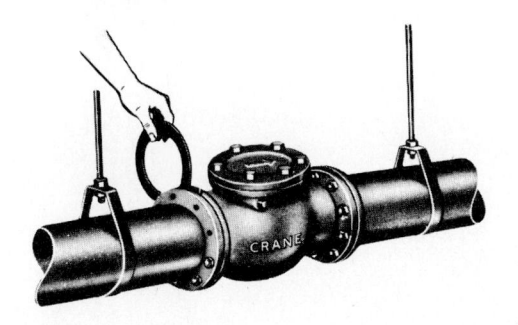

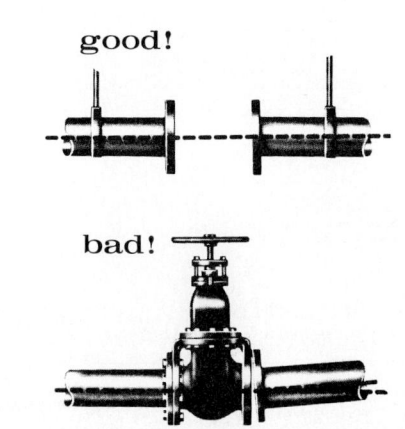

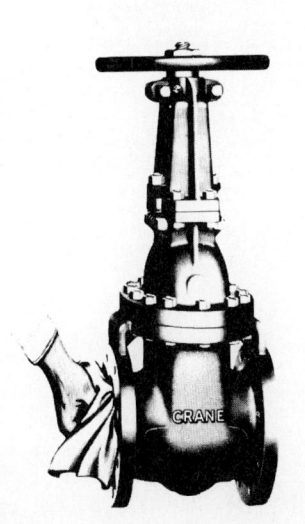

good!

bad!

clean all parts

As with screwed joints, all parts should be clean to assure best results. Use a solvent-soaked rag to remove the rust-preventing grease put on flanges at the factory. Next, clean off all dirt and grit particles. Then wipe off the gasket just to be sure it is clean.

align and support pipe

When the pipe is in place, be sure it is properly supported. For example, a valve cannot hold up an unsupported length of pipe without undergoing great strain. The flanges should be accurately aligned—by checking with a spirit level, both horizontally along the pipe and vertically across the flange faces. It is then ready to be bolted tight.

insert gasket

With the flanges secured in position, slip in half the bolts at the bottom. They will hold the gasket in place. Gaskets should be coated with a little graphite and oil or other recommended lubricant before they are inserted. They are then easier to remove if the joint is opened later. Finally, slip in the rest of the bolts, apply a little thread lubricant to each, and then turn up all nuts by hand, far as they'll go.

it's best to tighten bolts in this order

Now start pulling up with a wrench—not in rotation, but by the crossover method to load the bolts evenly and eliminate any concentrated stresses. Keep repeating by going over and across until the joint is uniformly tight.

Fig. 5-5 How to assemble a flanged joint (*Courtesy Crane Company.*)

VALVES

TABLE 5-6 SELECTION OF VALVES AND VALVE DESIGNS
(*COURTESY ITT GRINNELL CORPORATION.*)

Material	Maximum Temperature (°F)	Recommended Service
Aluminum	800	
Asbestos	Cold	Water
Composition	750	Steam, oil, air, gas
Spiral wound	750	Steam
Metal	600	Steam
Woven	600	Gas
Composition blue	Hot-cold (varies)	Acids
Woven blue	Hot-cold (varies)	Acids
Metallic	1000	Ammonia, gas
Composition		Ammonia
Spiral-wound composition	1000	Steam, air
Graphitic	1000	Gasoline, kerosene
Brass, admiralty	500	
Brass, high	500	
Chrome, moly steel	1200 (varies)	Acids
Chrome, steel	1300 (varies)	Acids
Copper	600	Steam
Cork fiber	212	Cold oil
Everdur	600	
Gold	1200	
Hastelloy	2000	
Inconel	2000	
Iron, ingot	1000	Steam, oil
Lead	212	
Lead sheet	Cold	Ammonia
Magnesium	400	
Monel	1500	Steam
Nickel	1400	
Platinum	2300	
Rubber		
Red	Hot	Water
Red	220	Steam, air, gas
Black	Cold	Water, ammonia
Soft	Cold	Water
Brown	Hot	Water
Silver	1200	
Steel		
Corrugated	1000	Steam
Low carbon	1000	
Sheet alloy		Acids
Stainless	1000	Steam
Stainless (304, 316)	800	
Stainless (347)	1700	
Tantalum	3000	
Tin	212	
Zinc	212	

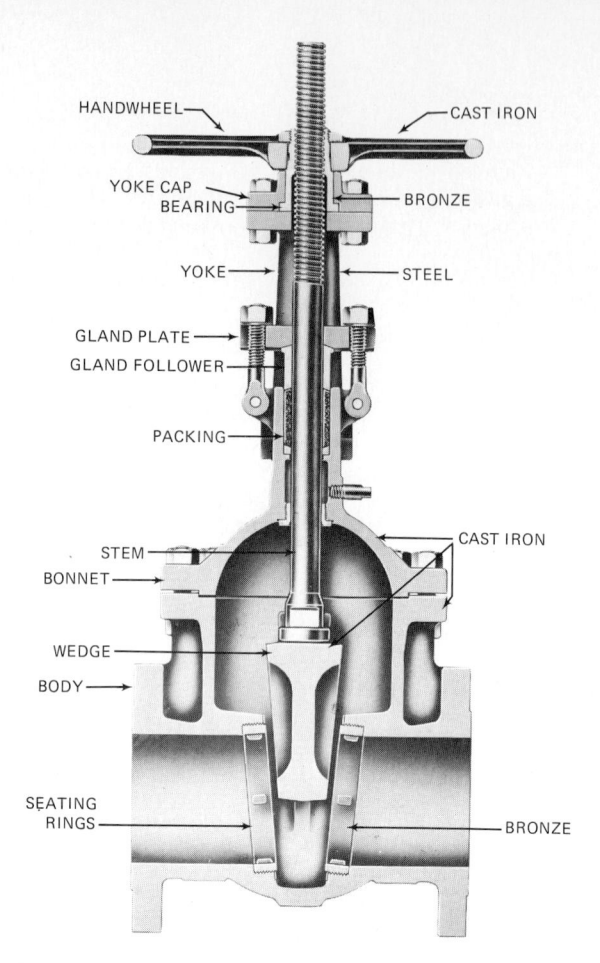

Fig. 5-6 Typical materials on a cast iron gate valve (*Courtesy Jenkins Bros.*)

Fig. 5-7 Split-wedge gate valve (*Courtesy Lunkenheimer Company, Cincinnati, OH 45214.*)

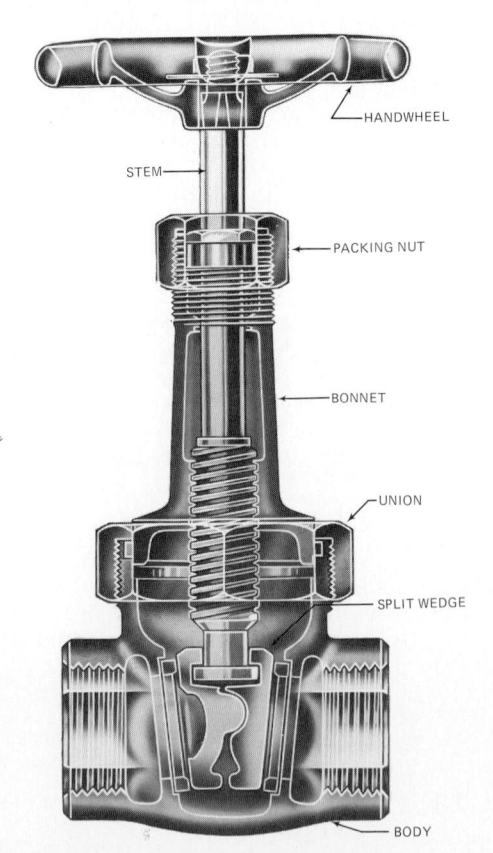

Fig. 5-8 OS & Y globe valve (*Courtesy Jenkins Bros.*)

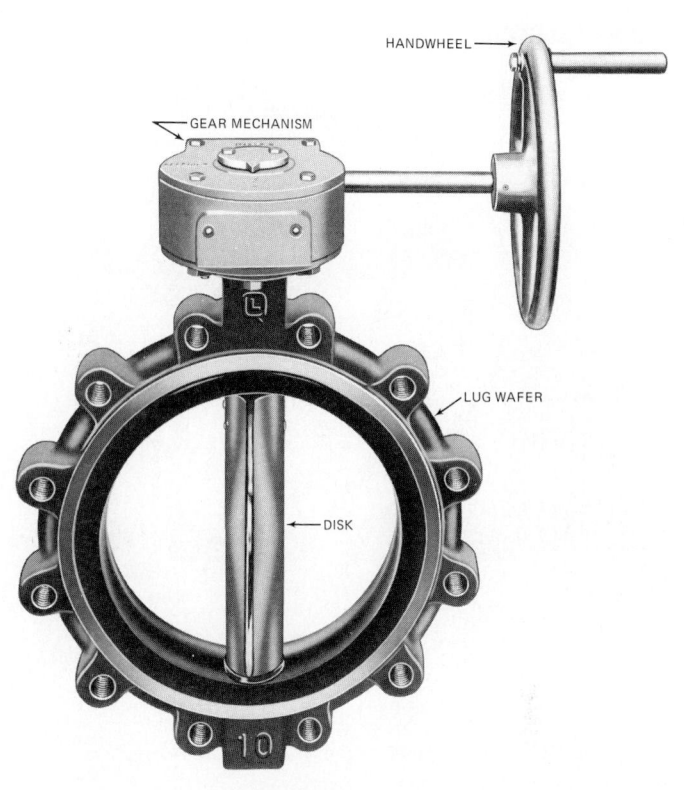

Fig. 5-10 Gear-operated butterfly valve (*Courtesy Lunkenheimer Company, Cincinnati, OH 45214.*)

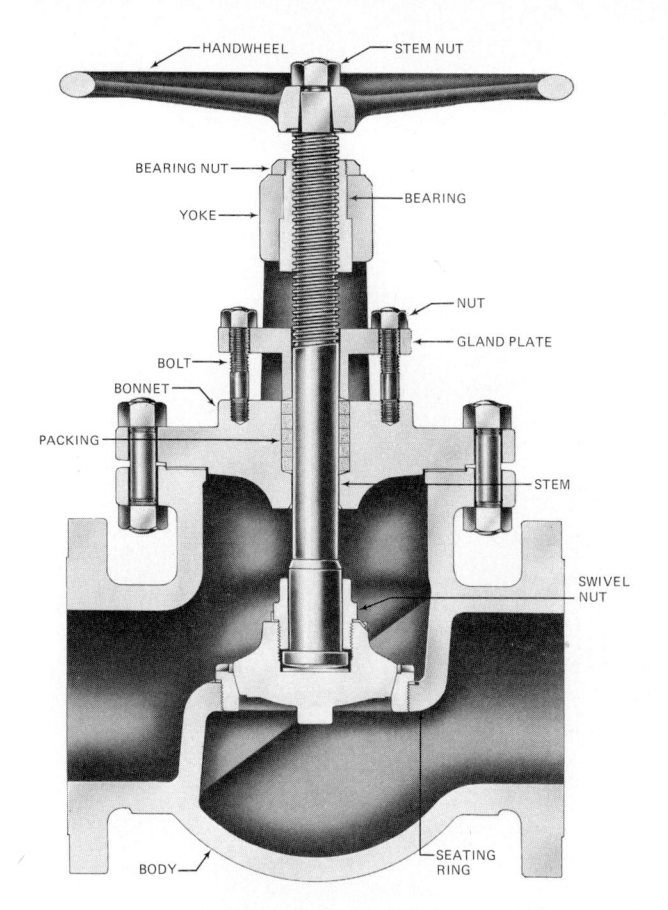

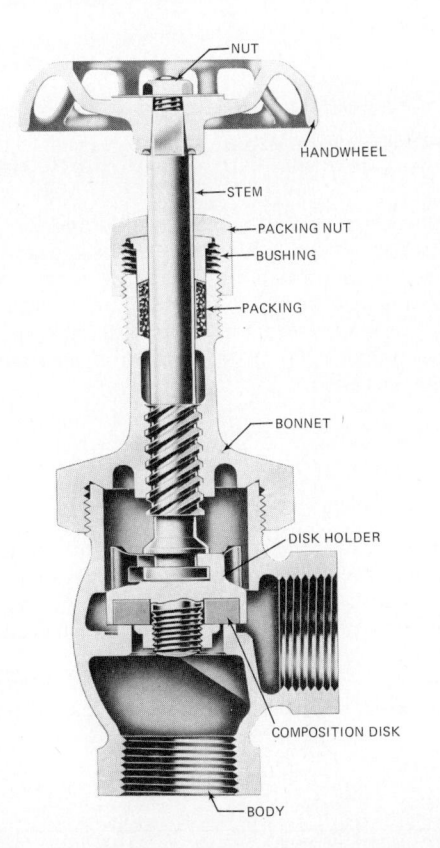

Fig. 5-9 Angle globe valve (*Courtesy Jenkins Bros.*)

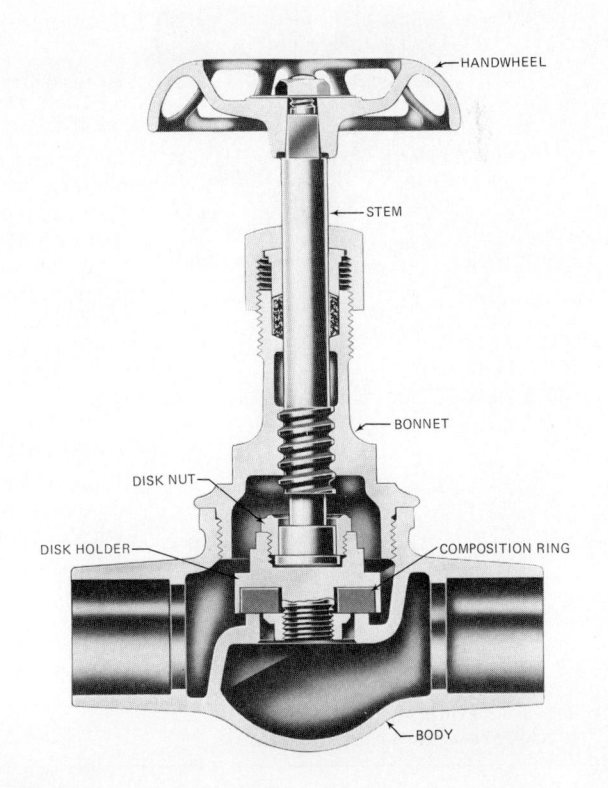

Fig. 5-11 Composition disk globe valve (*Courtesy Jenkins Bros.*)

VALVES 69

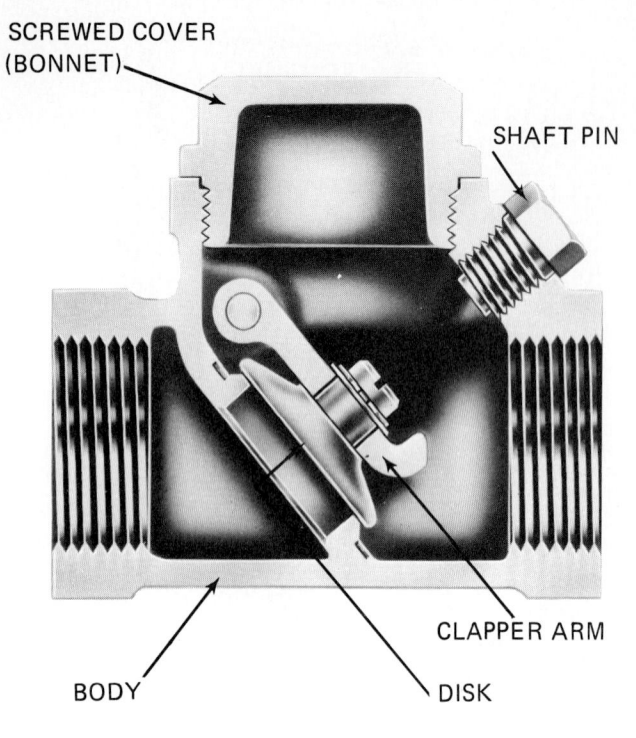

Fig. 5-12 Swing check valve (*Courtesy Jenkins Bros.*)

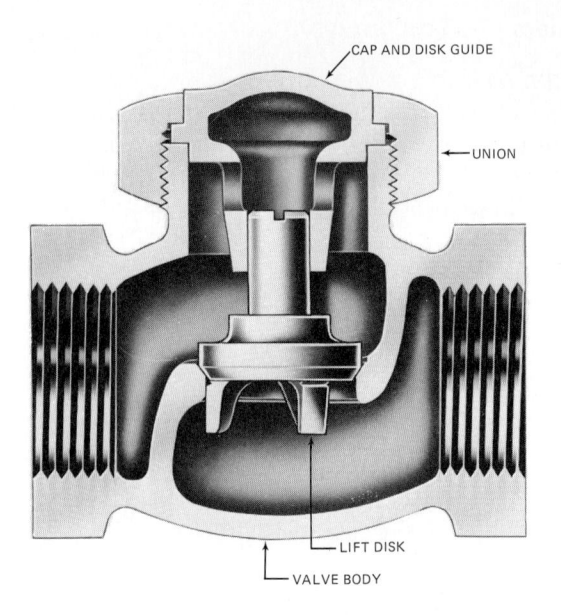

Fig. 5-13 Lift check valve (*Courtesy Jenkins Bros.*)

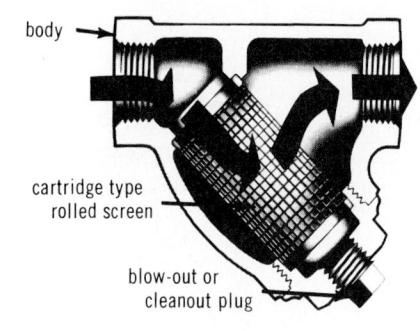

dirt means trouble... stop it with strainers

strainers reduce maintenance cost and lengthen life of valves, regulators, traps, gauges, and all automatic devices in piping

Scale and dirt in piping systems cause endless trouble and frequently serious damage. Such foreign matter lodges on valve seats and starts leakage, gets into the close fitting parts of automatic devices to prevent proper operation, and collects at other points, thereby obstructing flow.

A generous use of strainers or sediment separators is a small but sound investment for any piping system—regardless of how clean the piping materials are when installed.

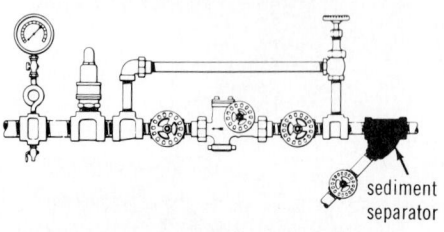

No danger of sediment accumulating in this Pressure Regulator. The strainer on the inlet line will entrap foreign matter in the system. Accumulated sediment should be blown out at regular intervals.

how a strainer works

Line fluid passes through an extra large screening area that assures free fluid flow with minimum pressure loss. Not only do strainers effectively trap all foreign matter, they also furnish a pocket for its accumulation. Periodic blowing out is only maintenance needed. All internal parts are readily accessible.

The blow-off connection can also be used to drain the line or to remove condensate when starting up, particularly in the case of pressure regulators.

Fig. 5-14 Strainer installation (*Courtesy Crane Company.*)

CHAPTER 5

STEAM TRAP HOOKUPS*

General Recommendations

Steam trap piping arrangements must be carefully planned to ensure dependable operation and permit maintenance when required. Hookups for many often encountered applications are shown which illustrate the following recommendations.

- Shutoff valves on both sides of the trap should be gate-type for unrestricted flow.

---
*(Courtesy Sarco.)

- Unions installed on each side of the trap. Uniform face-to-face dimensions permit fast trap replacement and shop rather than field maintenance.
- Pipeline strainers protect all traps, valves, etc., from pipe scale and dirt. They require periodic cleaning to remain effective.
- Steam supply piping to any equipment should be taken from the top of the supply main to obtain dry steam.
- Use a trap on each piece of steam equipment; do not "group trap." Differences in condensing rates cause uneven pressure drops resulting in air-binding and partial flooding when

TABLE 5-7 STEAM TRAP SELECTION (*COURTESY SARCO.*)

Application	Balanced Pressure Thermostatic A	Inverted Bucket B	Application	Float-Thermostatic A	Liquid Expansion Thermostatic B
Drips for Low Pressure Mains & Risers 2" and up			Process Tanks, large		
Drips for Low Pressure Mains & Risers under 2"			Dryers, Coil Type		
Drips for High Pressure Mains (saturated)			Dryers, Fan Type		
Drips for Process Lines			Drying Tumblers		
Steam Lines, Outdoors			Water Heaters		
Tracer Lines			Preheaters, Fuel Oil		
Radiators & Wall Coils			Rotating Cylinders		
Unit Heaters, large			Slashers		
Unit Heaters, small			Dry Cans		
Blast Coils & Stacks			Flat Work Ironers		
Hot Water Heaters			Presses, Laundry		
Storage Tanks			Stocking Forms		
Track Heaters			Presses, Plastics		
Kettles & Pans, slow boiling			Presses, Platen		
Kettles & Pans, quick boiling			Evaporators		
Process Tanks, small			Plating Tanks		

Use strainers ahead of all steam traps.

The alternate recommendations A and B given below are based on the most common operating conditions encountered in industry. They are subject to modification on local conditions such as unusually high or low steam pressures, corrosive condensate, the possibility of water-hammer, etc.

VALVES

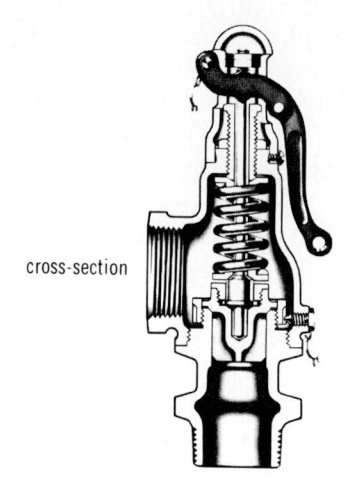

pop safety valves

for steam, air, or gas • do's and dont's when installing

No chips, scale, pipe dope, or other matter should be left in inlet of valve or in adjacent connections.

Always install pop safety valves with the stem in a vertical position. For air or gas service these valves should preferably be installed inverted to allow moisture to collect and seal the seating surfaces. Mount valve directly to tank or vessel without any intervening pipe or fitting.

If discharge piping is used, it should be as short as possible, and preferably larger in size than the outlet of the valve. The height of discharge piping should be adequately supported and *never* imposed on the valve. A loose slip joint in the outlet piping is recommended, since it permits expansion and contraction without unduly affecting the valve.

Wrenches should be used with care so as not to abuse or distort the valve. Periodic testing of the valve by pulling the lever is advised. If not operating properly, valve should be returned to the factory—not serviced in the field.

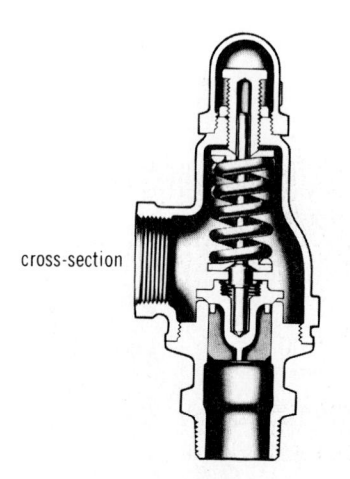

cross-section

relief valves

how they operate • installation and maintenance hints

Spring loaded relief valves should be installed with the stem vertical. For air or gas service, these valves should preferably be installed inverted, to allow moisture to collect and seal the seating surfaces. No chips, scale, pipe dope, or other foreign matter should be left in the inlet of the valve or in the adjacent connections.

Whenever piping is installed in the inlet or outlet of these valves, it must be at least as large as the valve connections. Under no circumstances should it be reduced. This piping should be adequately supported to prevent line strains from causing the valve to leak at the seat.

cross-section

Fig. 5-15 Installation of safety and relief valves (*Courtesy Crane Company.*)

several pieces of equipment are drained into a common trap.

Piping to the Trap

- The steam trap should be located directly below and as close as possible to the equipment being drained. Temperature-operated traps may have several feet of cooling leg ahead of them for condensate accumulation.
- Horizontal piping to the trap should be pitched for gravity drainage and to avoid steam locking of the line. Pipe size must be equal to or larger than the trap connections.
- Put a full pipe size, or larger, collecting leg with a dirt pocket before the trap inlet.
- Equipment which cannot readily be shut down for service requires a bypass line around the trap. A standby trap in the bypass line allows uninterrupted service and prevents steam loss should the bypass valve leak or be left open.

Return Line Piping

- Hot condensate discharged by a trap flashes to a much greater volume of low-pressure steam. To prevent back pressure buildup, return lines must always be larger than the supply piping and sized to handle the total capacity of all traps discharging into them.
- Discharge condensate into the top of the return line.
- Condensate return lines should drain by gravity to a common receiver. The condensate can then be pumped overhead or to the boiler.

6
PIPE SYMBOLS

Fig. 6-1 (*Courtesy Manville Products Corporation.*)

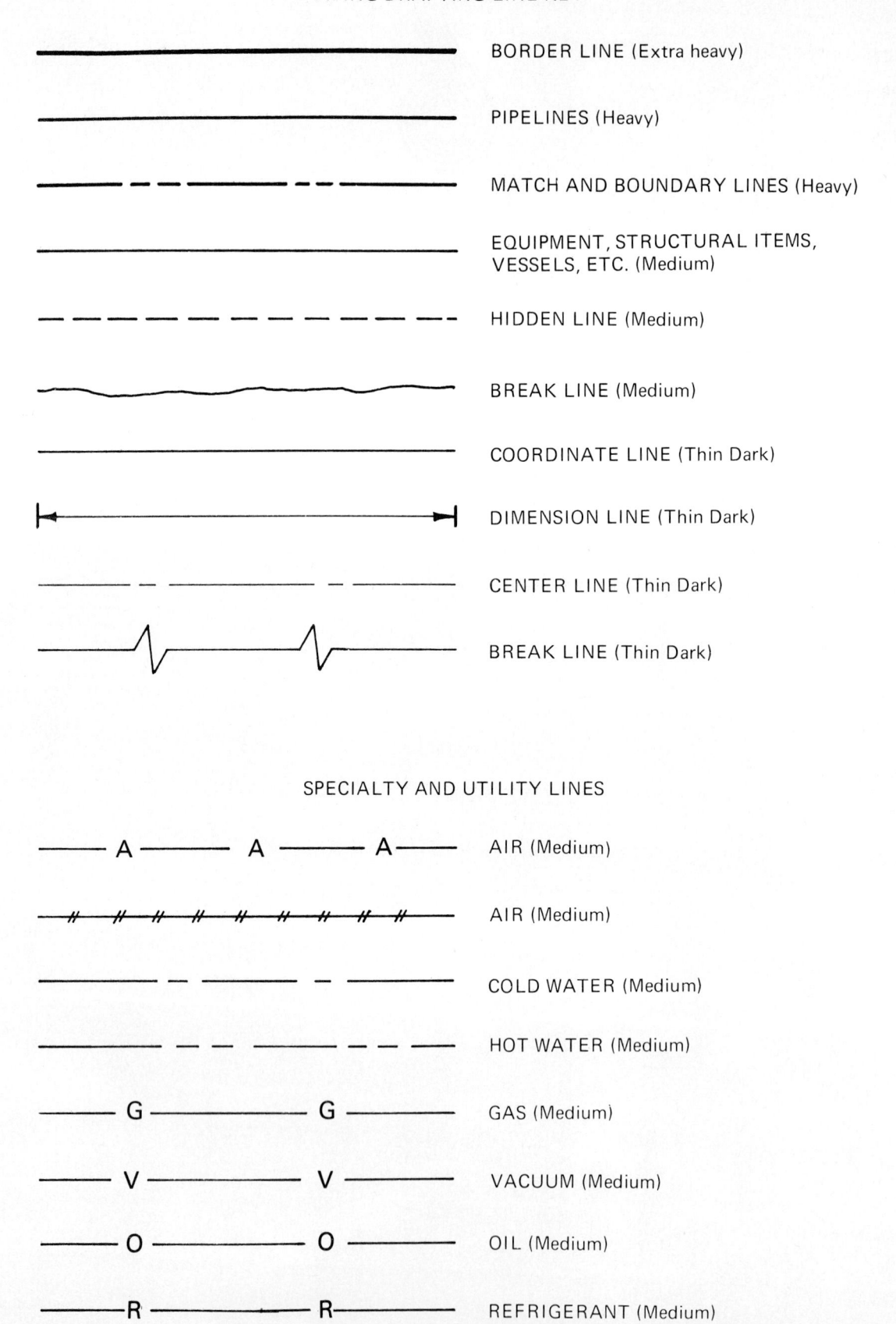

PIPING DRAFTING LINE KEY

————————————— BORDER LINE (Extra heavy)

————————————— PIPELINES (Heavy)

———— — — ———— — — ———— MATCH AND BOUNDARY LINES (Heavy)

————————————— EQUIPMENT, STRUCTURAL ITEMS, VESSELS, ETC. (Medium)

— — — — — — — — — — — HIDDEN LINE (Medium)

~~~~~~~~~~~~~~ BREAK LINE (Medium)

————————————— COORDINATE LINE (Thin Dark)

|⟵—————————————⟶| DIMENSION LINE (Thin Dark)

———— — ———— — ———— CENTER LINE (Thin Dark)

———⌐⌐———⌐⌐——— BREAK LINE (Thin Dark)

SPECIALTY AND UTILITY LINES

——A————A————A—— AIR (Medium)

——//—//—//—//—//—//—//—//—— AIR (Medium)

———— — ———— — ———— COLD WATER (Medium)

———— — — ———— — — ———— HOT WATER (Medium)

——G————G———— GAS (Medium)

——V————V———— VACUUM (Medium)

——O————O———— OIL (Medium)

——R————R———— REFRIGERANT (Medium)

Fig. 6-2 Piping drafting line key

GRAPHICAL SYMBOLS FOR PIPING

AIR CONDITIONING

28 BRINE RETURN	— — — BR — — —
29 BRINE SUPPLY	———— B ————
30 CIRCULATING CHILLED OR HOT-WATER FLOW	———— CH ————
31 CIRCULATING CHILLED OR HOT-WATER RETURN	— — — CHR — — —
32 CONDENSER WATER FLOW	———— C ————
33 CONDENSER WATER RETURN	— — — CR — — —
34 DRAIN	———— D ————
35 HUMIDIFICATION LINE	— - — H — - —
36 MAKE-UP WATER	— - — - — - —
37 REFRIGERANT DISCHARGE	———— RD ————
38 REFRIGERANT LIQUID	———— RL ————
39 REFRIGERANT SUCTION	— - — RS — - —

HEATING

40 AIR-RELIEF LINE	— — — — —
41 BOILER BLOW OFF	———————
42 COMPRESSED AIR	———— A ————
43 CONDENSATE OR VACUUM PUMP DISCHARGE	—o——o——o—
44 FEEDWATER PUMP DISCHARGE	—oo——oo——oo—
45 FUEL-OIL FLOW	———— FOF ————
46 FUEL-OIL RETURN	— — — FOR — — —
47 FUEL-OIL TANK VENT	— - — FOV — - —
48 HIGH-PRESSURE RETURN	—#— —#— —#—
49 HIGH-PRESSURE STEAM	—#— —#— —#—
50 HOT-WATER HEATING RETURN	— — — — —
51 HOT-WATER HEATING SUPPLY	———————

52 LOW-PRESSURE RETURN	— — — — —
53 LOW-PRESSURE STEAM	———————
54 MAKE-UP WATER	— — — — —
55 MEDIUM PRESSURE RETURN	—/— —/— —/—
56 MEDIUM PRESSURE STEAM	—/— —/— —/—

PLUMBING

57 ACID WASTE	———— ACID ————
58 COLD WATER	— - — - — - —
59 COMPRESSED AIR	———— A ————
60 DRINKING-WATER FLOW	— - — - — - —
61 DRINKING-WATER RETURN	— — — — —
62 FIRE LINE	——— F ——— F
63 GAS	——— G ——— G
64 HOT WATER	— - - — - - —
65 HOT-WATER RETURN	— - - — - - —
66 SOIL, WASTE OR LEADER (ABOVE GRADE)	———————
67 SOIL, WASTE OR LEADER (BELOW GRADE)	— — — — —
68 VACUUM CLEANING	——— V ——— V
69 VENT	— — — — —

PNEUMATIC TUBES

70 TUBE RUNS	═══════

SPRINKLERS

71 BRANCH AND HEAD	——o——o—
72 DRAIN	—S— — —S
73 MAIN SUPPLIES	———— S

Fig. 6-3 Special pipeline line symbols

PIPE SYMBOLS

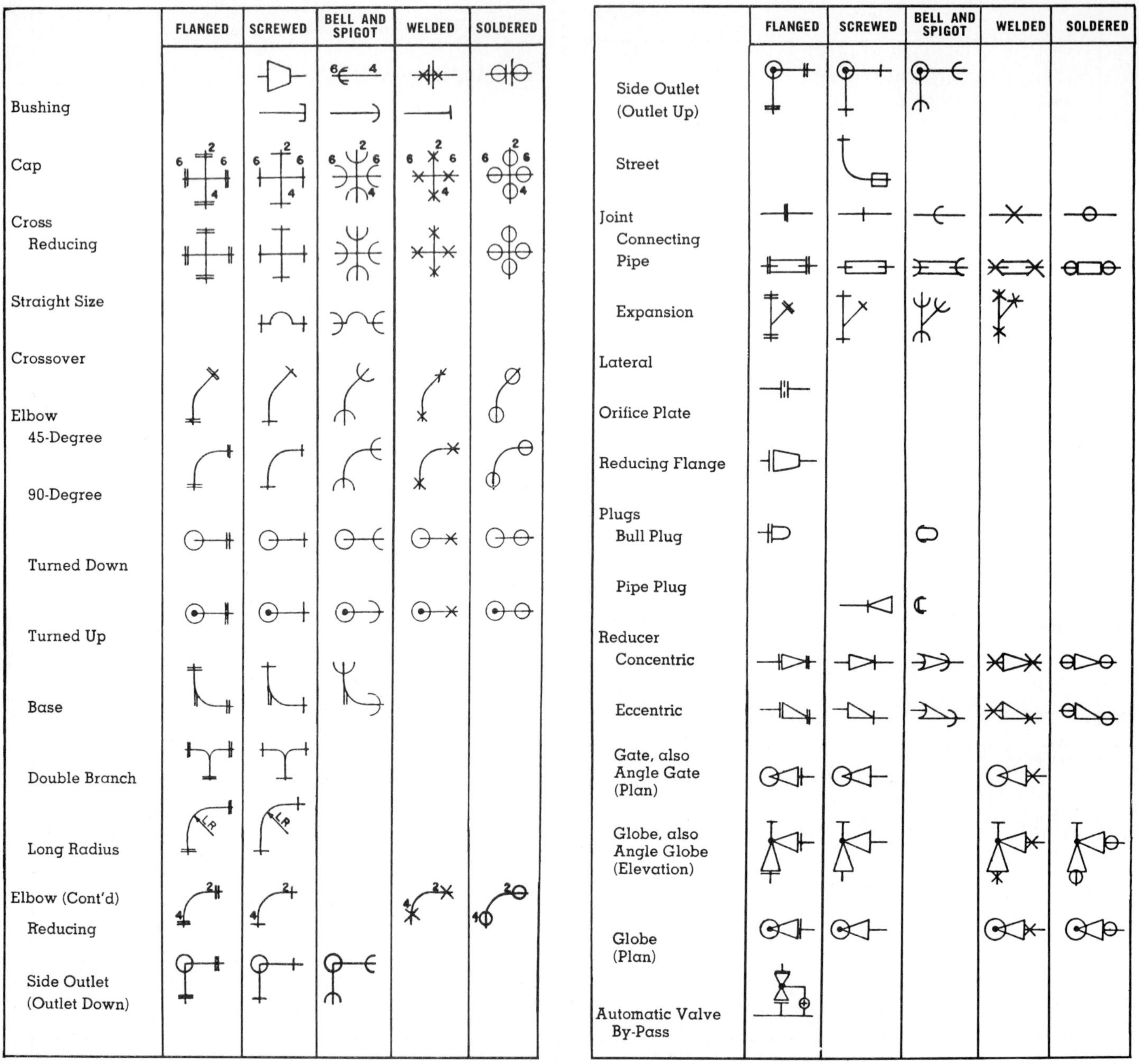

Fig. 6-4 Basic piping fitting symbols (*Courtesy ITT Grinnell Corporation.*)

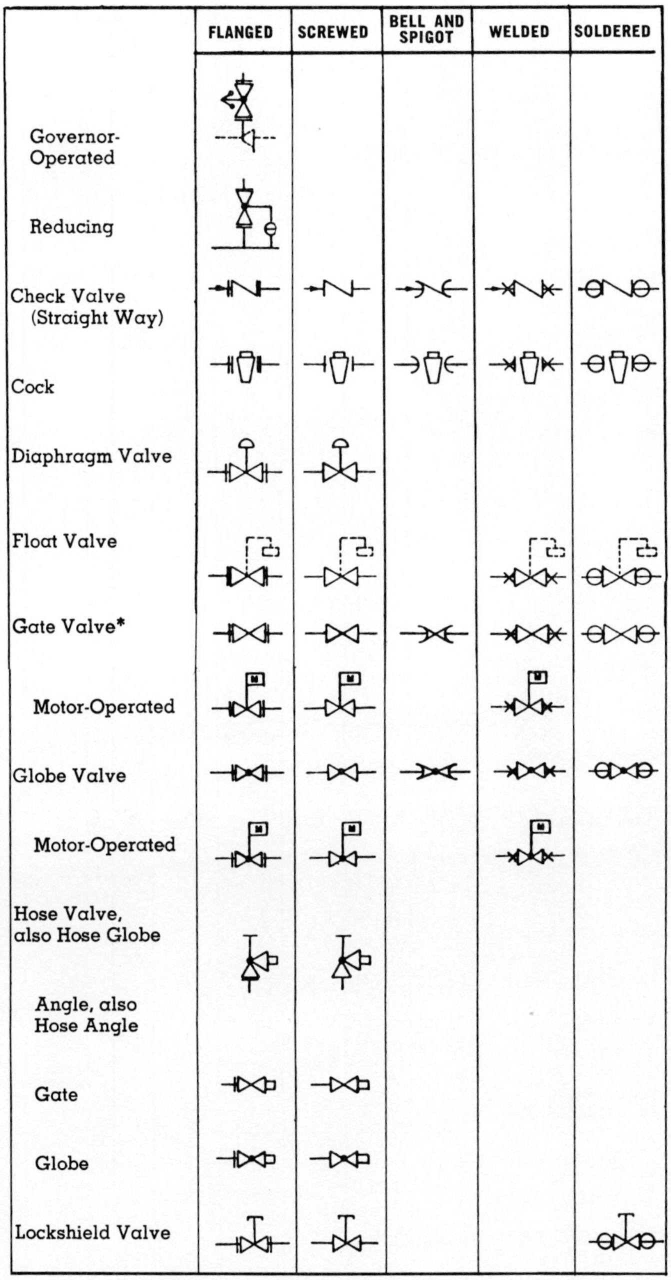

	FLANGED	SCREWED	BELL AND SPIGOT	WELDED	SOLDERED
Governor-Operated					
Reducing					
Check Valve (Straight Way)					
Cock					
Diaphragm Valve					
Float Valve					
Gate Valve*					
Motor-Operated					
Globe Valve					
Motor-Operated					
Hose Valve, also Hose Globe					
Angle, also Hose Angle					
Gate					
Globe					
Lockshield Valve					

	FLANGED	SCREWED	BELL AND SPIGOT	WELDED	SOLDERED
Quick Opening Valve					
Safety Valve					
Sleeve					
Tee Straight Size					
(Outlet Up)					
(Outlet Down)					
Double Sweep					
Reducing					
Single Sweep					
Side Outlet (Outlet Down)					
Side Outlet (Outlet Up)					
Union					
Angle Valve Check, also Angle Check					
Gate, also Angle Gate (Elevation)					

*Also used for General Stop Valve Symbol when amplified by specification.

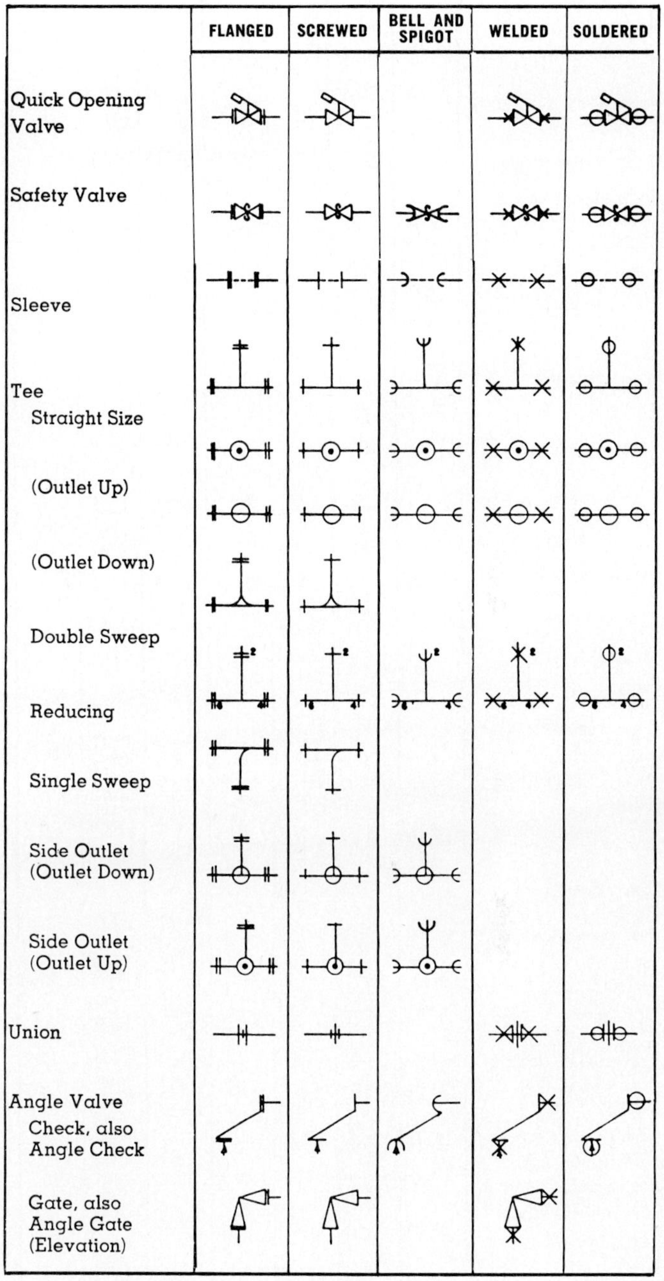

Fig. 6-4 (*contd.*)

PIPE SYMBOLS

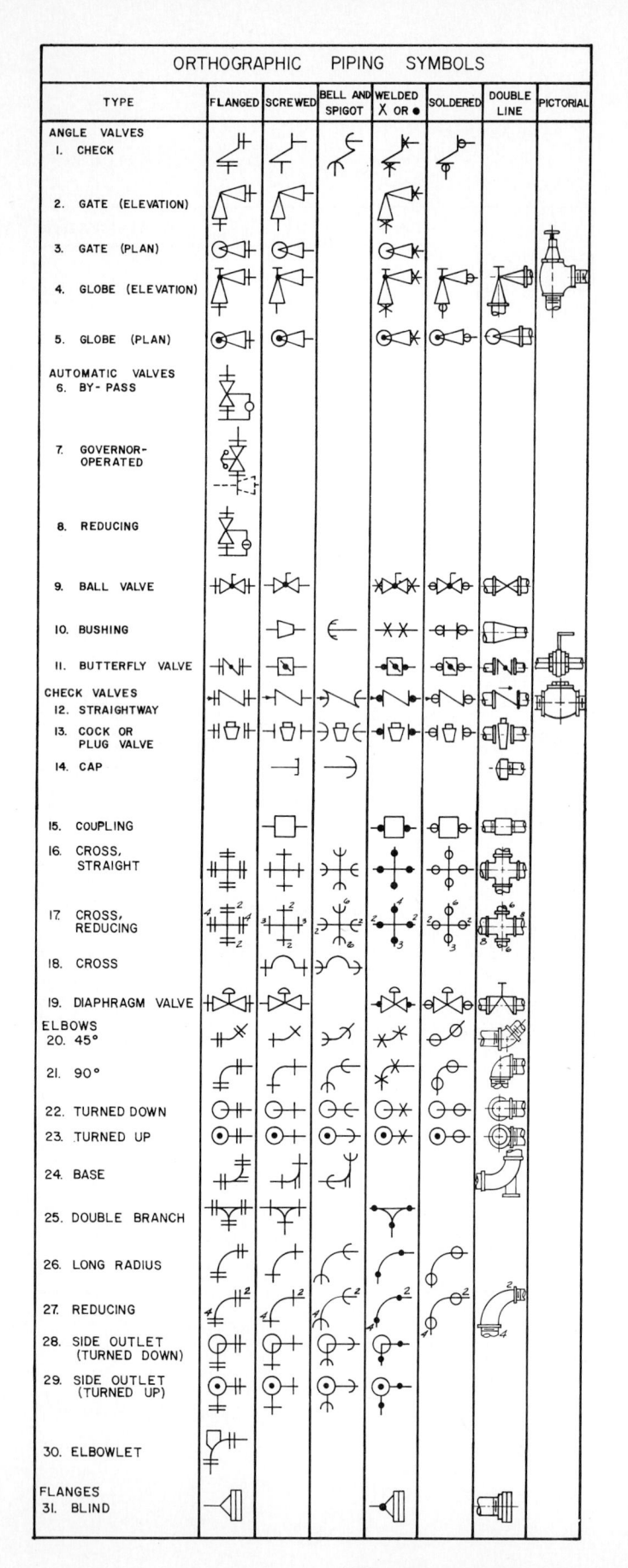

Fig. 6-5 Comprehensive guide to piping drawing symbols

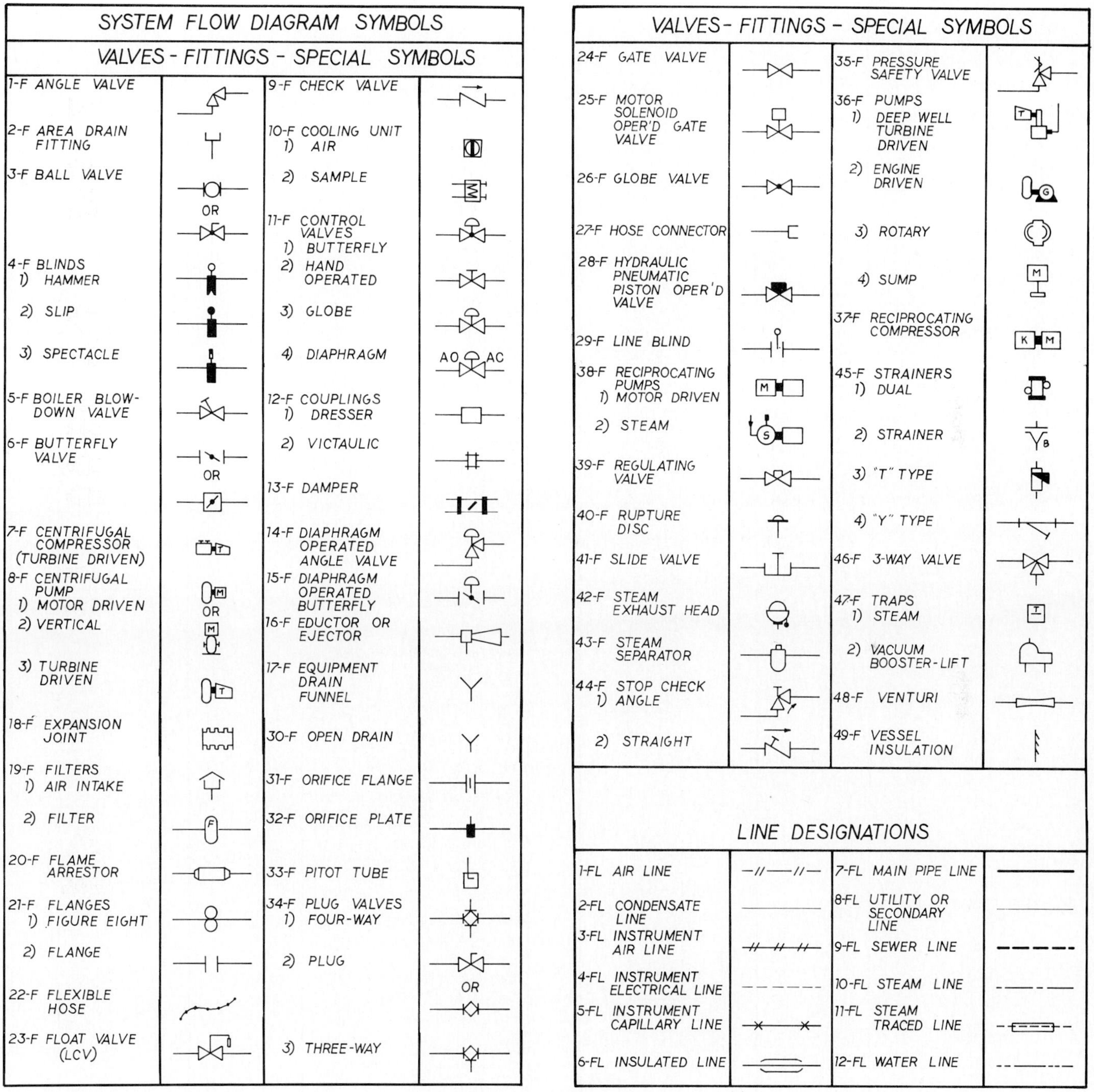

SYSTEM FLOW DIAGRAM SYMBOLS

VALVES - FITTINGS - SPECIAL SYMBOLS

#	Item		#	Item
1-F	ANGLE VALVE		9-F	CHECK VALVE
2-F	AREA DRAIN FITTING		10-F	COOLING UNIT
				1) AIR
3-F	BALL VALVE			2) SAMPLE
	OR		11-F	CONTROL VALVES
				1) BUTTERFLY
4-F	BLINDS			2) HAND OPERATED
	1) HAMMER			3) GLOBE
	2) SLIP			4) DIAPHRAGM
	3) SPECTACLE			
5-F	BOILER BLOW-DOWN VALVE		12-F	COUPLINGS
				1) DRESSER
6-F	BUTTERFLY VALVE			2) VICTAULIC
	OR		13-F	DAMPER
7-F	CENTRIFUGAL COMPRESSOR (TURBINE DRIVEN)		14-F	DIAPHRAGM OPERATED ANGLE VALVE
8-F	CENTRIFUGAL PUMP		15-F	DIAPHRAGM OPERATED BUTTERFLY
	1) MOTOR DRIVEN		16-F	EDUCTOR OR EJECTOR
	2) VERTICAL		17-F	EQUIPMENT DRAIN FUNNEL
	3) TURBINE DRIVEN			
18-F	EXPANSION JOINT		30-F	OPEN DRAIN
19-F	FILTERS		31-F	ORIFICE FLANGE
	1) AIR INTAKE		32-F	ORIFICE PLATE
	2) FILTER			
20-F	FLAME ARRESTOR		33-F	PITOT TUBE
21-F	FLANGES		34-F	PLUG VALVES
	1) FIGURE EIGHT			1) FOUR-WAY
	2) FLANGE			2) PLUG
22-F	FLEXIBLE HOSE			OR
23-F	FLOAT VALVE (LCV)			3) THREE-WAY

VALVES - FITTINGS - SPECIAL SYMBOLS

#	Item		#	Item
24-F	GATE VALVE		35-F	PRESSURE SAFETY VALVE
25-F	MOTOR SOLENOID OPER'D GATE VALVE		36-F	PUMPS
				1) DEEP WELL TURBINE DRIVEN
26-F	GLOBE VALVE			2) ENGINE DRIVEN
27-F	HOSE CONNECTOR			3) ROTARY
28-F	HYDRAULIC PNEUMATIC PISTON OPER'D VALVE			4) SUMP
29-F	LINE BLIND		37-F	RECIPROCATING COMPRESSOR
38-F	RECIPROCATING PUMPS		45-F	STRAINERS
	1) MOTOR DRIVEN			1) DUAL
	2) STEAM			2) STRAINER
39-F	REGULATING VALVE			3) "T" TYPE
40-F	RUPTURE DISC			4) "Y" TYPE
41-F	SLIDE VALVE		46-F	3-WAY VALVE
42-F	STEAM EXHAUST HEAD		47-F	TRAPS
				1) STEAM
43-F	STEAM SEPARATOR			2) VACUUM BOOSTER-LIFT
44-F	STOP CHECK		48-F	VENTURI
	1) ANGLE			
	2) STRAIGHT		49-F	VESSEL INSULATION

LINE DESIGNATIONS

#	Item		#	Item
1-FL	AIR LINE	—//—//—	7-FL	MAIN PIPE LINE
2-FL	CONDENSATE LINE		8-FL	UTILITY OR SECONDARY LINE
3-FL	INSTRUMENT AIR LINE		9-FL	SEWER LINE
4-FL	INSTRUMENT ELECTRICAL LINE		10-FL	STEAM LINE
5-FL	INSTRUMENT CAPILLARY LINE		11-FL	STEAM TRACED LINE
6-FL	INSULATED LINE		12-FL	WATER LINE

Fig. 6-6 Flow diagram symbols

PIPE SYMBOLS

INSTRUMENTATION FLOW DIAGRAM SYMBOLS				NAME	LOCAL	BOARD	IN PRACTICE
NAME	LOCAL	BOARD	IN PRACTICE				
1-IF CONDUCTIVITY RECORDER	CR	CR	CR	19-IF LEVEL RECORDING CONTROLLER	LRC	LRC	LRC
2-IF DENSITY RECORDER	DC	DC	DC	20-IF LEVEL SWITCH	LS	LS	LS
3-IF HAND OPERATED CONTROL VALVE	HCV	HCV		21-IF MOISTURE RECORDER	MR	MR	MR
4-IF HAND ACTUATED PNEUMATIC CONTROLLER	HIC	HIC	HIC	22-IF PRESSURE ALARM	PA	PA	PA
5-IF FLOW ALARM	FA	FA	FA	23-IF PRESSURE CONTROL VALVE	PCV	PCV	PCV
6-IF FLOW ELEMENT	FE	FE	FE	24-IF PRESSURE TEST CONNECTION	PE	PE	PE
7-IF FLOW INDICATOR	FI	FI	FI	25-IF PRESSURE INDICATOR	PI	PI	PI
8-IF FLOW INDICATING CONTROLLER	FIC	FIC	FIC	26-IF PRESSURE INDICATING CONTROLLER	PIC	PIC	PIC
9-IF FLOW RECORDER	FR	FR	FR	27-IF PRESSURE RECORDER	PR	PR	PR
10-IF FLOW RECORDING CONTROLLER	FRC	FRC	FRC	28-IF PRESSURE RECORDING CONTROLLER	PRC	PRC	PRC
11-IF FLOW RATIO RECORDING CONTROLLER	FrRC	FrRC	FrRC	29-IF PRESSURE - DIFFERENTIAL RECORDING CONTROLLER	PdRC	PdRC	PdRC
12-IF DISPLACEMENT FLOW METER	FmI	FmI	FmI	30-IF PRESSURE SAFETY VALVE	PSV	PSV	PSV
13-IF LEVEL ALARM	LA	LA	LA	31-IF SPEED RECORDER	SR	SR	SR
14-IF LEVEL CONTROLLER	LC	LC	LC	32-IF STEAM TRAP	T	T	T
15-IF LEVEL INDICATOR	LI	LI	LI	33-IF TEMPERATURE ALARM	TA	TA	TA
16-IF LEVEL INDICATING CONTROLLER	LIC	LIC	LIC	34-IF TEMPERATURE CONTROLLER	TC	TC	TC
17-IF LEVEL GLASS	LG	LG	LG	35-IF TEMPERATURE INDICATING CONTROLLER	TIC	TIC	TIC
18-IF LEVEL RECORDER	LR	LR	LR				

Fig. 6-7 Instrumentation symbols

NAME	LOCAL	BOARD	IN PRACTICE
36-IF TEMPERATURE ELEMENT	TE	TE	TE
37-IF TEMPERATURE INDICATOR	TI	TI	TI
38-IF TEMPERATURE INDICATING CONTROLLER	TIC	TIC	
39-IF TEMPERATURE RECORDER	TR	TR	TR
40-IF TEMPERATURE RECORDING CONTROLLER	TRC	TRC	TRC
41-IF TEMPERATURE SWITCH	TS	TS	TS
42-IF TEMPERATURE OPERATING VALVE	TV	TV	TV
43-IF TEMPERATURE WELL	TW	TW	TW
44-IF VISCOSITY RECORDER	VR	VR	VR
45-IF WEIGHT RECORDER	WR	WR	WR

SPECIAL INSTRUMENTATION

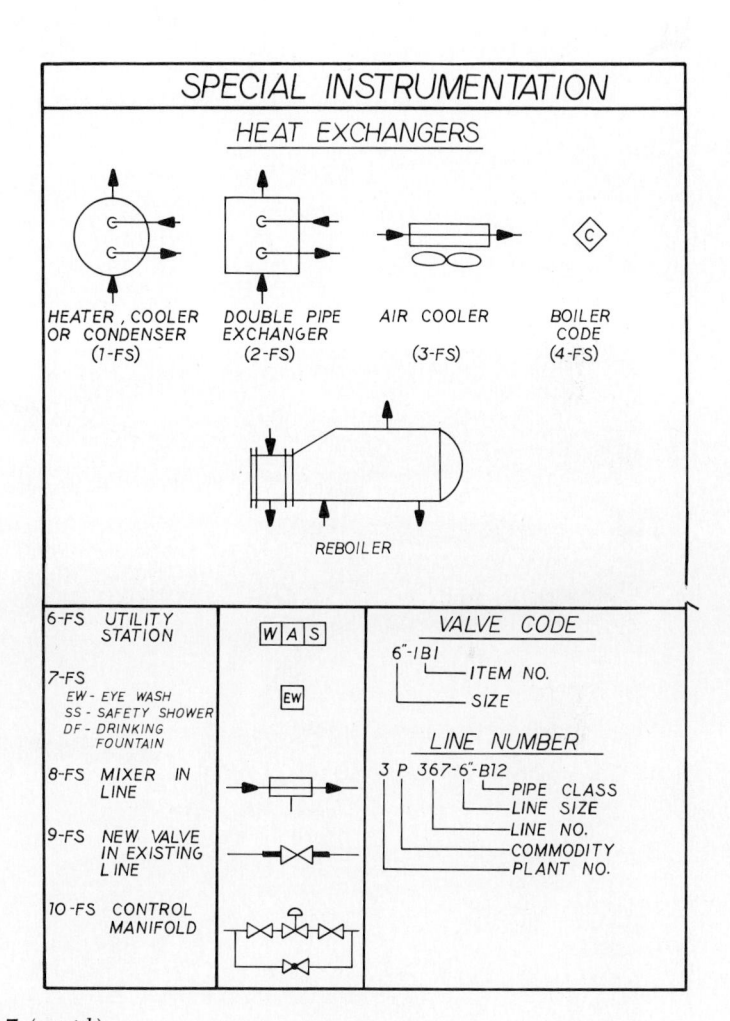

HEAT EXCHANGERS

HEATER, COOLER OR CONDENSER (1-FS)

DOUBLE PIPE EXCHANGER (2-FS)

AIR COOLER (3-FS)

BOILER CODE (4-FS)

REBOILER

6-FS UTILITY STATION — WAS

7-FS
EW - EYE WASH
SS - SAFETY SHOWER
DF - DRINKING FOUNTAIN
— EW

8-FS MIXER IN LINE

9-FS NEW VALVE IN EXISTING LINE

10-FS CONTROL MANIFOLD

VALVE CODE
6"-1B1
— ITEM NO.
— SIZE

LINE NUMBER
3 P 367-6"-B12
— PIPE CLASS
— LINE SIZE
— LINE NO.
— COMMODITY
— PLANT NO.

Fig. 6-7 (*contd.*)

PIPE SYMBOLS

7
PIPE SETUP

Fig. 7-1 (*Courtesy Manville Products Corporation.*)

Proper alignment is important if a piping system is to be correctly fabricated. Poor alignment may result in welding difficulties and a system that does not function properly.

ITT Grinnell welding rings may be employed to assure proper alignment as well as the correct welding gap. In addition to using welding rings, some simple procedures can be followed to assist the pipe fitter. Below and on the following page are alignment procedures commonly used by today's craftsmen.

PIPE-TO-PIPE

1. Level one length of pipe using spirit level
2. Bring lengths together leaving only small welding gap
3. Place spirit level over both pipes as shown and maneuver unpositioned length until both are level
4. Tack weld top and bottom
5. Rotate pipe 90°
6. Repeat procedure

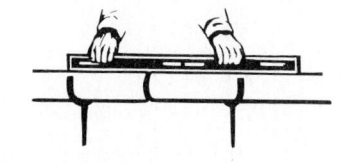

45° ELBOW-TO-PIPE

1. Level pipe using spirit level
2. Place fitting to pipe leaving small welding gap
3. Place 45° spirit level on face of elbow and maneuver elbow until bubble is centered
4. Tack weld in place

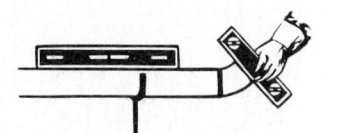

90° ELBOW-TO-PIPE

1. Level pipe using spirit level
2. Place fitting to pipe leaving small welding gap
3. Place spirit level on face of elbow and maneuver elbow until level
4. Tack weld in place

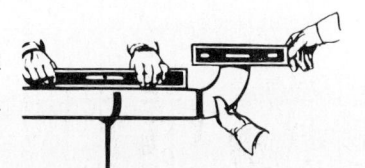

TEE-TO-PIPE

1. Level pipe using spirit level
2. Place tee to pipe leaving small welding gap
3. Place spirit level on face of tee and maneuver tee until level
4. Tack weld in place

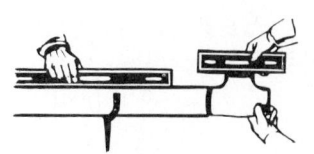

FLANGE-TO-PIPE

1. Bring flange to pipe end leaving small welding gap
2. Align top two holes of flange with spirit level
3. Tack weld in place
4. Center square on face of flange as shown
5. Tack weld in place
6. Check sides in same way

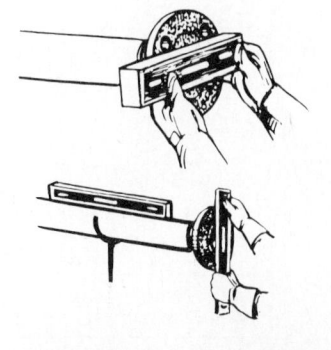

JIG FOR SMALL DIAMETER PIPING

The jig is made from channel iron 3′ 9″ long. Use 1/8″ x 1½″ for pipe sizes 1¼″ thru 3″; 1⅛″ x ¾″ for sizes 1″ or smaller.

1. Cut out 90° notches about 9″ from end.
2. Heat bottom of notch with torch.
3. Bend channel iron to 90° angle and weld sides.
4. Place elbow in jig and saw half thru sides of channel iron as shown. Repeat this step with several elbows so jig may be used for different operations.
5. A used hack saw blade placed in notch as shown will provide proper welding gap.

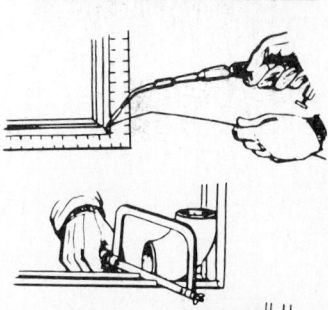

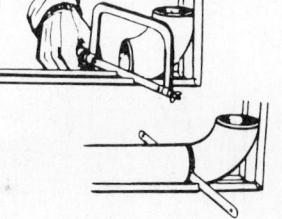

Fig. 7-2 Alignment of pipe (*Courtesy ITT Grinnell Corporation.*)

HEADER AND BRANCH LAYOUT

To lay out a header and its corresponding full-size branch line, scribe a centerline mark at the center point where the branch will be located. Divide and mark the circumference into four equal parts. Point C is located at the intersection of the header and branch centerlines (on the circumference of the pipe, as shown in Fig. 7-3). Points A and B are located along the header's circumference where the branch will be placed. The distance between A and B is equal to the diameter of the branch. A and B can be found by setting off one-half the outside diameter of the branch as measured from the scribed centerline.

A wraparound can then be used to scribe the contour between point A and point C (about the upper half of the pipe's centerline). Repeat the process for the contour between points B and C.

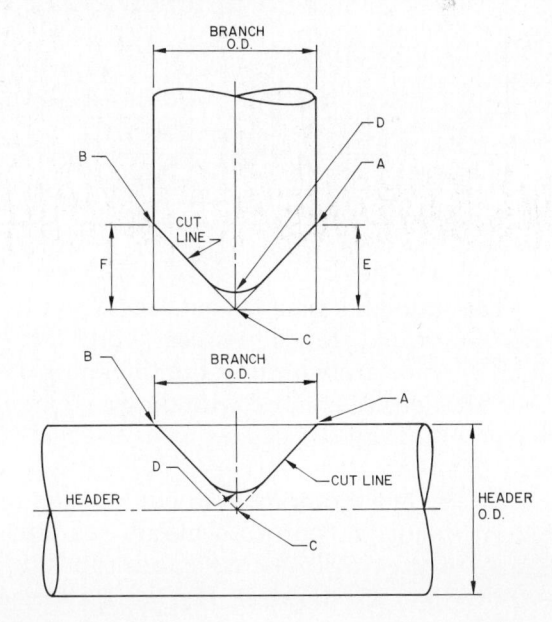

Fig. 7-3 Branch and header layout (*Courtesy Texas Pipe Bending Company.*)

PIPE SETUP

TABLE 7-1 HOW TO CUT ODD-ANGLE ELBOWS (*COURTESY ITT GRINNELL CORPORATION*)

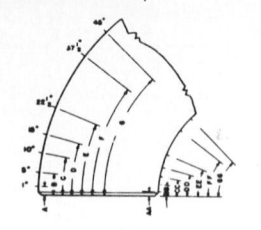

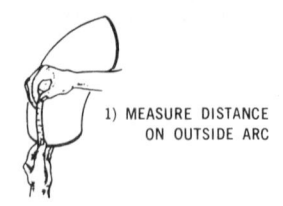

1) MEASURE DISTANCE ON OUTSIDE ARC

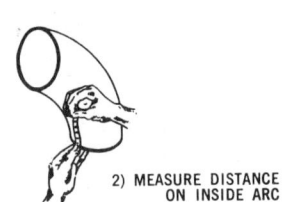

2) MEASURE DISTANCE ON INSIDE ARC

3) WRAP TAPE AROUND ELBOW AND MARK CUTTING LINE

ODD DEGREE LONG RADIUS ELBOWS							
NOM SIZE	\multicolumn{7}{c	}{OUTSIDE ARC}
	A	B	C	D	E	F	G
2	5/64	3/8	23/32	1 3/32	1 21/32	2 3/4	3 9/32
2 1/2	3/32	7/16	29/32	1 11/32	2 1/2	3 3/8	4 1/16
3	7/64	9/16	1 1/8	1 5/8	2 15/32	4 3/32	4 29/32
3 1/2	1/8	5/8	1 9/32	1 29/32	2 27/32	4 3/4	5 11/16
4	9/64	23/32	1 7/16	2 5/32	3 1/4	5 13/32	6 15/32
5	3/16	29/32	1 25/32	2 11/16	4 1/32	6 23/32	8 1/16
6	7/32	1 1/16	2 5/32	3 7/32	4 27/32	8 1/16	9 21/32
8	9/32	1 7/16	2 27/32	4 9/32	6 13/32	10 11/16	12 13/16
10	11/32	1 25/32	3 9/16	5 11/32	8	13 11/32	16
12	7/16	2 1/8	4 1/4	6 3/8	9 9/16	15 31/32	19 5/32
14	1/2	2 7/16	4 7/8	7 5/16	11	18 5/16	22
16	9/16	2 13/16	5 19/32	8 3/8	12 9/16	20 15/16	25 1/4
18	5/8	3 3/8	6 9/32	9 7/16	14 1/8	23 3/16	28 9/32
20	11/16	3 1/2	7	10 15/32	15 23/32	26 3/16	31 13/32
22	3/4	3 27/32	7 11/16	11 17/32	17 9/32	28 13/16	34 9/16
24	27/32	4 3/16	8 3/8	12 9/16	18 27/32	31 13/32	37 11/16
26	29/32	4 17/32	9 3/32	13 5/8	20 13/32	34 1/32	40 27/32
30	1 1/32	5 1/4	10 15/32	15 3/4	23 9/16	39 1/4	47 1/8
34	1 5/32	5 29/32	11 27/32	17 13/16	26 23/32	44 17/32	53 3/8
36	1 7/32	6 1/4	12 17/32	18 7/8	28 7/32	47	56 17/32
42	1 7/16	7 5/16	14 5/8	22	32 31/32	54 31/32	65 15/16

ODD DEGREE LONG RADIUS ELBOWS							
NOM SIZE	\multicolumn{7}{c	}{INSIDE ARC}
	AA	BB	CC	DD	EE	FF	GG
2	1/32	5/32	5/16	15/32	23/32	1 3/16	1 7/16
2 1/2	3/64	3/16	13/32	19/32	29/32	1 1/2	1 13/16
3	3/64	1/4	1/2	23/32	1 3/32	1 13/16	2 5/32
3 1/2	1/16	9/32	9/16	27/32	1 9/32	2 1/8	2 9/16
4	1/16	5/16	21/32	31/32	1 15/32	2 7/16	2 15/16
5	5/64	13/32	13/16	1 1/4	1 27/32	3 3/32	3 23/32
6	3/32	1/2	1	1 1/2	2 7/32	3 23/32	4 15/32
8	1/8	11/16	1 11/32	2	3 1/2	5 1/2	6 1/2
10	5/32	27/32	1 11/16	2 17/32	3 25/32	6 5/16	7 9/16
12	7/32	1	2 1/2	3 3/16	4 9/16	7 19/32	9 1/8
14	1/4	1 7/32	2 7/16	3 21/32	5 1/2	9 5/32	11
16	9/32	1 13/32	2 13/16	4 3/16	6 9/32	10 15/32	12 5/8
18	5/16	1 9/16	3 1/8	4 23/32	7 1/16	11 25/32	14 1/8
20	11/32	1 3/4	3 1/2	5 1/4	7 27/32	13 9/32	15 11/16
22	3/8	1 29/32	3 27/32	5 3/4	8 5/8	14 3/8	17 9/32
24	13/32	2 3/32	4 3/16	6 9/32	9 9/16	15 11/16	18 27/32
26	15/32	2 1/32	4 17/32	6 13/16	10 7/32	17 1/32	20 13/32
30	17/32	2 5/8	5 1/4	7 7/8	11 25/32	19 5/8	23 9/16
34	19/32	2 31/32	5 29/32	8 29/32	13 3/8	22 9/32	26 1/16
36	5/8	2 13/16	6 1/4	9 7/16	14 1/8	23 5/8	28 1/4
42	23/32	3 21/32	7 5/16	10 19/32	16 1/2	26 3/8	32 31/32

To establish a rounded cut, locate point *D* by measuring the distance between points *C* and *D* which is equal to two times the thickness of the pipe wall. Connect point *D* with the lines extended from points *A* and *B* (freehand). The header is now ready to be cut.

To lay out the branch, divide its circumference into four equal parts. Locate and mark points *A* and *B* on the circumference of the branch (along its centerline), as shown in Fig. 7-3. Distances *E* and *F* are equal to one-half the diameter of the branch pipe. Locate point *C* on both sides of the pipe's circumference (along its centerline).

Using a wraparound, scribe the contour line from point *C* to point *A* and from point *C* to point *B* about the circumference of the pipe.

For a rounded cut, locate point *D* using two times the thickness of the pipe wall for the distance from point *C* to point *D*. The branch is now ready to be cut.

TABLE 7-2 CUTBACK AT 90 DEGREES FOR ID NOZZLE TO OD OF HEADER. STANDARD WEIGHT PIPE (*COURTESY TEXAS PIPE BENDING COMPANY.*)

Header \ Nozzle	3/4	1	1-1/2	2	2-1/2	3	3-1/2	4	5	6	8	10	12	14	16	18	20	22	24	30
3/4	0	1/2	7/8	1-1/8	1-3/8	1-11/16	1-15/16	2-3/16	2-3/4	3-5/16	4-5/16	5-3/8	6-3/8	7	8	9	10	11	12	15
1		0	13/16	1-1/16	1-5/16	1-11/16	1-15/16	2-3/16	2-3/4	3-1/4	4-1/4	5-3/8	6-3/8	7	8	9	10	11	12	15
1-1/2			0	7/8	1-3/8	1-9/16	1-13/16	2-1/8	2-11/16	3-3/16	4-1/4	5-5/16	6-5/16	6-15/16	7-15/16	8-15/16	9-15/16	11	12	15
2				0	1	1-7/16	1-11/16	2	2-9/16	3-1/8	4-3/16	5-1/4	6-5/16	6-15/16	7-15/16	8-15/16	9-15/16	10-15/16	11-15/16	14-15/16
2-1/2					0	1-1/4	1-9/16	1-7/8	2-1/2	3-1/16	4-1/8	5-1/4	6-1/4	6-7/8	7-7/8	8-15/16	9-15/16	10-15/16	11-15/16	14-15/16
3						0	1-5/16	1-5/8	2-5/16	2-15/16	4-1/16	5-1/8	6-3/16	6-13/16	7-7/8	8-7/8	9-7/8	10-7/8	11-7/8	14-15/16
3-1/2							0	1-3/8	2-1/8	2-13/16	3-15/16	5-1/16	6-1/8	6-3/4	7-13/16	8-13/16	9-13/16	10-7/8	11-7/8	14-7/8
4								0	1-15/16	2-5/8	3-13/16	5	6-1/16	6-11/16	7-3/4	8-3/4	9-13/16	10-13/16	11-13/16	14-7/8
5									0	2-1/8	3-1/2	4-3/4	5-7/8	6-1/2	7-9/16	8-5/8	9-11/16	10-11/16	11-3/4	14-13/16
6										0	3-1/16	4-7/16	5-5/8	6-5/16	7-3/8	8-1/2	9-1/2	10-9/16	11-5/8	14-11/16
8											0	3-9/16	4-15/16	5-3/4	6-15/16	8-1/16	9-3/16	10-1/4	11-5/8	14-7/8
10												0	3-15/16	4-7/8	6-1/4	7-1/2	8-11/16	9-13/16	10-15/16	14-1/8
12													0	3-5/8	5-5/16	6-11/16	8	9-1/2	10-3/8	13-3/4
14														0	4-1/2	6-1/16	7-1/2	8-3/16	10	13-7/16
16															0	4-3/4	6-1/2	7-15/16	9-1/4	12-15/16
18																0	5-1/16	6-13/16	8-5/8	12-1/4
20																	0	5-5/16	7-3/16	11-1/2
22																		0	5-9/16	10-9/16
24																			0	9-1/2
30																				0

TABLE 7-3 CUTBACK AT 90 DEGREES FOR ID OF NOZZLE TO OD OF HEADER. EXTRA HEAVY PIPE (*COURTESY TEXAS PIPE BENDING COMPANY.*)

	1/2	3/4	1	1-1/2	2	2-1/2	3	3-1/2	4	5	6	8	10	12	14	16	18	20	24	30
1/2	0	7/16	9/16	15/16	1-1/8	1-7/16	1-3/4	2	2-1/4	2-3/4	3-5/16	4-5/16	5-3/8	6-3/8	7	8	9	10	12	15
3/4		0	9/16	7/8	1-1/8	1-3/8	1-11/16	1-15/16	2-3/16	2-3/4	3-5/16	4-5/16	5-3/8	6-3/8	7	8	9	10	12	15
1			0	13/16	1-1/16	1-3/8	1-11/16	1-15/16	2-3/16	2-3/4	3-1/4	4-5/16	5-3/8	6-3/8	7	8	9	10	12	15
1-1/2				0	15/16	1-1/4	1-9/16	1-7/8	2-1/8	2-11/16	3-1/4	4-1/4	5-5/16	6-5/16	6-15/16	7-15/16	8-15/16	10	12	15
2					0	1-1/16	1-7/16	1-3/4	2	2-5/8	3-3/16	4-3/16	5-3/16	6-5/16	6-15/16	7-15/16	8-15/16	9-15/16	11-15/16	14-15/16
2-1/2						0	1-5/16	1-5/8	1-15/16	2-1/2	3-1/8	4-1/8	5-1/4	6-1/4	6-7/8	7-15/16	8-15/16	9-15/16	11-15/16	14-15/16
3							0	1-3/8	1-11/16	2-3/8	3	4-1/16	5-3/16	6-3/16	6-7/8	7-7/8	8-7/8	9-7/8	11-15/16	14-15/16
3-1/2								0	1-1/2	2-3/16	2-7/8	3-15/16	5-1/8	6-1/8	6-13/16	7-13/16	8-13/16	9-7/8	11-7/8	14-7/8
4									0	2	2-11/16	3-7/8	5	6-1/16	6-3/4	7-3/4	8-13/16	9-13/16	11-7/8	14-7/8
5										0	2-1/4	3-9/16	4-13/16	5-7/8	6-9/16	7-5/8	8-11/16	9-11/16	11-3/4	14-13/16
6											0	3-3/16	4-9/16	5-11/16	6-3/8	7-7/8	8-1/2	9-9/16	11-5/8	14-3/4
8												0	3-13/16	5-1/8	5-7/8	7-1/16	8-1/8	9-1/4	11-3/8	14-1/2
10													0	4-1/8	5	6-5/16	7-9/16	8-3/4	10-15/16	14-3/16
12														0	3-13/16	5-7/8	6-13/16	8-1/16	10-7/16	13-13/16
14															0	4-11/16	6-1/4	7-5/8	10-1/16	13-1/2
16																0	5	6-5/8	9-3/8	13
18																	0	5-1/4	8-1/2	12-3/8
20																		0	7-5/16	11-5/8
24																			0	9-5/8

PIPE SETUP

BRANCH AND HEADER LAYOUT FOR 45-DEGREE LATERALS

For the 45-degree full-size lateral, lay out the two centerlines of the pipes, as shown in Fig. 7-4. The lines on either side of the centerlines are then laid out at a distance equal to one-half the outside diameter of the pipes. These two lines will be drawn parallel to the centerlines of the pipes and will intersect at point A and B.

Draw a straight line from point A and point B to the intersection of the two centerlines located at point C. Line A–C and line B–C are the cut lines.

To lay out a header for a lateral, draw a line around the header at the point where the two centerlines intersect and mark point C. Divide the header's circumference into four equal parts using a wraparound.

Establish points A and B by measuring distances D and F along the top (upper) line of the pipe (Fig. 7-4). Scribe the contour cut line by using the wraparound, from point C to point A and point B along the upper half of the pipe (C to A to C and C to B to C). Cut the header along the cut line.

For the branch line, divide its circumference into four equal parts. Using distance F (from the header), locate point B. Find point A by subtracting distance D (from the header) from distance F (Fig. 7-4). Connect points C to A to C and points C to B to C using the wraparound. The branch is now ready to be cut.

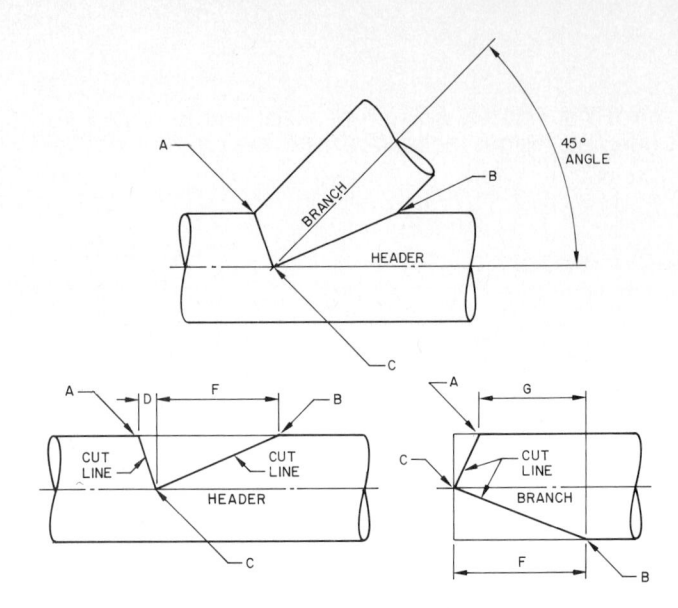

Fig. 7-4 Branch and header layout for 45-degree laterals (*Courtesy Texas Pipe Bending Company.*)

TABLE 7-4 ID OF NOZZLE TO OD OF LR ELL ON CENTERLINE. STANDARD WEIGHT PIPE (*COURTESY TEXAS PIPE BENDING COMPANY.*)

HEADER ELL

1/2	3/4	1	1-1/2	2	2-1/2	3	3-1/2	4	5	6	8	10	12	14	16	18	20	24	NOZZLE
13/16	5/16	5/16	5/16	7/16	1/2	1/2	5/8	11/16	13/16	15/16	1-5/16	1-9/16	1-15/16	2-13/16	3-3/16	3-9/16	3-7/8	4-5/8	1/2
	1/2	1/2	7/16	9/16	5/8	5/8	11/16	13/16	15/16	1-1/16	1-7/16	1-11/16	2	2-15/16	3-5/16	3-11/16	4	4-11/16	3/4
		9/16	11/16	3/4	13/16	13/16	7/8	15/16	1-1/16	1-3/16	1-9/16	1-13/16	2-3/16	3-1/8	3-7/16	3-13/16	4-3/16	4-7/8	1
			1-5/16	1-1/4	1-5/16	1-3/16	1-1/4	1-3/8	1-7/16	1-9/16	1-7/8	2-1/8	2-1/2	3-7/16	3-13/16	4-1/8	4-1/2	5-3/16	1-1/2
				1-7/8	1-3/4	1-9/16	1-5/8	1-11/16	1-3/4	1-7/8	2-3/16	2-7/16	2-3/4	3-3/4	4-1/16	4-7/16	4-3/4	5-7/16	2
				2-5/16	2	2	2-1/16	2-1/16	2-1/8	2-7/16	2-11/16	3-1/16	4	4-5/16	4-11/16	5	5-3/4	2-1/2
					2-7/8	2-11/16	2-5/8	2-9/16	2-5/8	2-7/8	3-1/16	3-7/16	4-3/8	4-11/16	5-1/16	5-3/8	6-1/16	3
						3-1/2	3-1/4	3-1/4	3-1/16	3-1/4	3-7/16	3-3/4	4-3/4	5-1/16	5-3/8	5-11/16	6-3/8	3-1/2
							4	3-9/16	3-7/16	3-5/8	3-3/4	4-1/16	5	5-3/8	5-11/16	6	6-11/16	4
								5-3/16	4-5/8	4-9/16	4-5/8	4-13/16	5-3/4	6-1/16	6-3/8	6-11/16	7-5/16	5
									6-3/8	5-11/16	5-1/2	5-11/16	6-5/8	6-7/8	7-1/8	7-7/16	8-1/16	6
										8-13/16	7-5/8	7-1/2	8-3/8	8-1/2	8-11/16	8-15/16	9-7/16	8
											11-1/8	9-15/16	10-5/8	10-1/2	10-1/2	10-5/8	11-1/16	10
												13-3/4	13-9/16	12-7/8	12-5/8	12-9/16	12-3/4	12
													16-7/16	14-3/4	14-1/8	13-15/16	13-15/16	14
														19-1/8	17-1/8	16-7/16	16	16
															21-13/16	19-5/8	18-5/16	18
																2 – 0-9/16	21-1/16	20
																	2 – 6	24

FORMULA:

$Z = R + \dfrac{O.D. \text{ of Ell}}{2}$

$S = 1/2 \text{ I.D. Nozzle}$

$Z^2 - (R + S)^2 = Y^2$

$R - Y = X$

TABLE 7-5 REINFORCING PADS (ARC OF NOZZLE DIAMETER ON CIRCUMFERENCE OF HEADER) (*COURTESY TEXAS PIPE BENDING COMPANY.*)

Header (in.) / Nozzle (in.)

Nozzle (in.)	3/4	1	1 1/2	2	2 1/2	3	3 1/2	4	5	6	8	10	12	14	16	18	20	24	
1/2	X	X	X	7/8	7/8	7/8	7/8	7/8	13/16	13/16	13/16	13/16	13/16	13/16	13/16	13/16	13/16	13/16	1/2
3/4		X	X	1 1/16	1 1/16	1 1/16	1 1/16	1 1/16	1 1/16	1 1/16	1 1/16	1 1/16	1 1/16	1 1/16	1 1/16	1 1/16	1 1/16	1 1/16	3/4
1			X	1 3/8	1 3/8	1 3/8	1 5/16	1 5/16	1 5/16	1 5/16	1 5/16	1 5/16	1 5/16	1 5/16	1 5/16	1 5/16	1 5/16	1 5/16	1
1 1/2				2 3/16	2 1/16	2	2	1 15/16	1 15/16	1 15/16	1 15/16	1 15/16	1 15/16	1 7/8	1 7/8	1 7/8	1 7/8	1 7/8	1 1/2
2					2 13/16	2 5/8	2 9/16	2 1/2	2 7/16	2 7/16	2 3/8	2 3/8	2 3/8	2 3/8	2 3/8	2 3/8	2 3/8	2 3/8	2
2 1/2						3 3/8	3 3/16	3 1/8	3	3	2 15/16	2 15/16	2 7/8	2 7/8	2 7/8	2 7/8	2 7/8	2 7/8	2 1/2
3							4 1/4	4	3 13/16	3 11/16	3 5/8	3 9/16	3 9/16	3 9/16	3 1/2	3 1/2	3 1/2	3 1/2	3
3 1/2								4 15/16	4 7/16	4 5/16	4 3/16	4 1/8	4 1/16	4 1/16	4 1/16	4 1/16	4	4	3 1/2
4									5 1/4	4 15/16	4 3/4	4 5/8	4 5/8	4 9/16	4 9/16	4 9/16	4 9/16	4 1/2	4
5										6 5/8	6 1/16	5 7/8	5 3/4	5 3/4	5 11/16	5 5/8	5 5/8	5 5/8	5
6											7 9/16	7 1/8	6 15/16	6 7/8	6 13/16	6 13/16	6 3/4	6 11/16	6
8												10	9 1/2	9 5/16	9 1/8	9	8 15/16	8 13/16	8
10													12 13/16	12 1/4	11 13/16	11 1/2	11 3/8	11 1/8	10
12														16	14 3/4	14 3/16	13 13/16	13 7/16	12
14															17 1/16	16 1/16	15 1/2	14 15/16	14
16																19 11/16	18 9/16	17 1/2	16
18																	22 3/8	20 3/8	18
20																		23 5/8	20
24																			

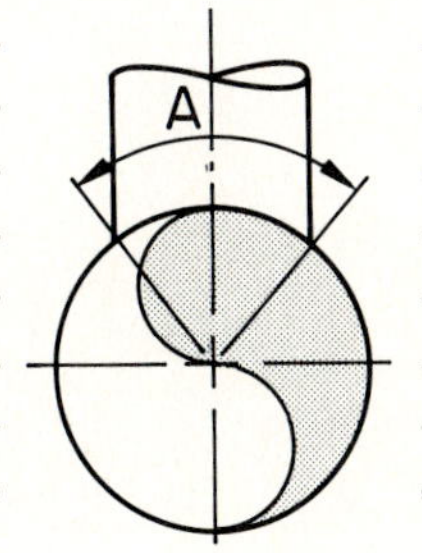

A = ARC OF NOZZLE DIAMETER ON CIRCUMFERENCE OF HEADER

☐ SEE SIZE ON SIZE CHART

TABLE 7-6 REINFORCING PADS (ARC FOR LARGE OD PIPE) (*COURTESY TEXAS PIPE BENDING COMPANY.*)

Header (in.) / Nozzle (in.)

Nozzle (in.)	30″	32″	34″	36″	38″	40″	42″	48″
6	6 3/4″	6 3/4″	6 3/4″	6 3/4″	6 5/8″	6 5/8″	6 5/8″	
8	8 3/4″	8 3/4″	8 3/4″	8 3/4″	8 3/4″	8 3/4″	8 3/4″	
10	11″	11″	11″	11″	10 7/8″	10 7/8″	10 7/8″	
12	13 1/4″	13 1/8″	13 1/8″	13 1/8″	13 1/8″	13″	13″	
14	14 5/8″	14 1/2″	14 1/2″	14 3/8″	14 1/4″	14 1/4″	14 1/4″	
16	16 7/8″	16 3/4″	16 5/8″	16 5/8″	16 5/8″	16 1/2″	16 3/8″	
18	19 3/8″	19 1/8″	19″	18 7/8″	18 7/8″	18 3/4″	18 5/8″	
20	21 7/8″	21 5/8″	21 3/8″	21 1/4″	21 1/8″	21″	20 7/8″	
24	2′ 3 7/8″	2′ 3 1/8″	2′ 2 5/8″	2′ 2 1/4″	2′ 2″	2′ 1 3/4″	2′ 1 1/2″	2′ 1 1/8″
26	2′ 7 1/2″	2′ 6 3/8″	2′ 5 5/8″	2′ 5 1/8″	2′ 4 5/8″	2′ 4 3/8″	2′ 4 1/8″	
28	3′ 0 1/8″	2′ 10 1/8″	2′ 8 7/8″	2′ 8 1/8″	2′ 7 1/2″	2′ 7″	2′ 6 5/8″	
30		3′ 2 7/8″	3′ 0 3/4″	2′ 11 1/2″	2′ 10 5/8″	2′ 10″	2′ 9 1/2″	
32			3′ 5 3/4″	3′ 3 1/2″	3′ 2 1/8″	3′ 1 1/8″	3′ 0 5/8″	
34				3′ 8 1/2″	3′ 6 1/8″	3′ 4 5/8″	3′ 3 5/8″	
36					3′ 11 3/8″	3′ 8 7/8″	3′ 7 1/4″	
38						4′ 2 1/8″	3′ 11 1/2″	
40							4′ 5″	
42								

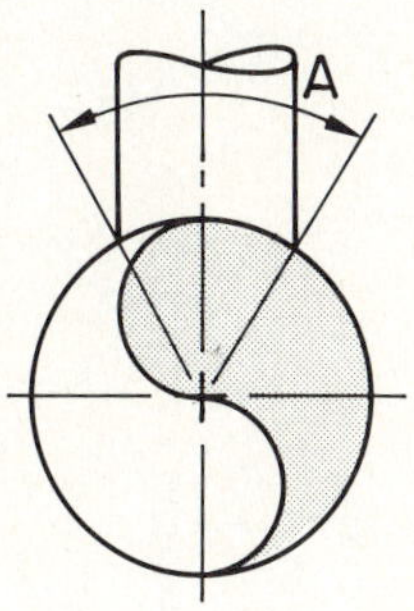

8
OFFSETS

Fig. 8-1 (*Courtesy Chicago Bridge and Iron Company.*)

CALCULATED PIPE OFFSETS*

An obstruction in the way of a run of pipe frequently results in the pipe being continued parallel to the original run, but no longer in the same line; that is, the pipe is offset. The pipe is commonly said to "break" over and/or up or down at this point. Frequently, these offset or break problems are solved by cut and try methods, while at other times application of mathematics is desirable to ensure a satisfactory result. The use of 90-degree elbows or bends makes pipe layout problems simple, but is frequently avoided in favor of lesser degree bends in order to cause less frictional loss of the fluid pressure, in order to use less pipe and fittings, and for a more streamlined appearance of the job.

The most common and satisfactory offset job is one employing 45-degree elbows. Since a World War II order discontinued the manufacture of screw elbows other than 45 degree and 90 degree offsets can be made at other angles only with pipe bends or fitting swings.

The pipe fitter naturally avoids written calculations, if with simple measurements or rule of thumb he can determine the cuts of pipe to be made. In the case of a 45-degree break on screw pipe the location of which is satisfactory within a range of several inches, the fitter installs the first 45-degree ell at the beginning of the break and screws into the 45-degree elbow hand tight a "space" piece of pipe longer than is needed. He may then hold the second 45-degree elbow at the desired location along the space piece, simply sighting by eye as to its desired location or possibly combining sighting and use of a rule, locating it from some object, ceiling, or by offset from the pipe run before the break, etc. Instead of sighting, he may tie a string in the desired location of the pipe run after the break and turn the first 45-degree ell until the space piece makes contact with the string. In any case, he marks off the cut point, making the usual allowances for depth of threads, etc.

However there are situations where the location of the break is critical, especially where larger size pipe 2 in. and over is involved and try and cut methods are risky. In this case, a bit of "figuring" is worthwhile. The pipe fitter, having soiled hands, often chalks or pencils his sketch and calculations on the nearest piece of board, carton, or wrapping, etc., in order to save time.

When the problem is a simple vertical or simple horizontal break in a line, the calculation is simple also, keeping in mind the properties of a 45-degree triangle (Fig. 8-2).

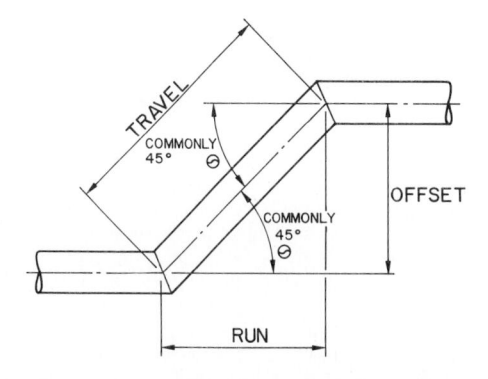

Fig. 8-2 45-degree common offset (*Courtesy G. K. Bachmann,* Pipefitter's and Plumber's Vest Pocket Reference Book, *Prentice-Hall, Inc., Englewood Cliffs, N.J.*)

$$\theta = 45°$$
$$A = \text{area}$$
$$A = \frac{a^2}{2}$$
$$b = 1.414a = 2\sqrt{A}$$
$$a = 0.707b = 1.414\sqrt{A}$$

Below is an example in a simple 45-degree offset using trigonometry data. See Table 8-1.

EXAMPLE 1:

45° offsets (angle $\theta = 45°$)

Let run = offset = 6'2'' = 6.166'

$$\text{Travel} = \frac{\text{Run}}{.7071} = 8.721' = 8'8\tfrac{3}{4}''$$

(sin 45° = .7071 = cos 45°)

$$\text{Travel} = \frac{\text{Offset}}{.7071} = 8.721' = 8'8\tfrac{5}{8}''$$

Run = .7071 × Travel = 6.166' = 6'2''

Offset = .7071 × Travel = 6.166' = 6'2''

The simple 45-degree offset problem is so common that the calculation for the 45-degree travel of pipe from center to center of 45-degree bends for various offsets is given in table 8-1.

If the problem is one of 45-degree offset with a vertical plus a horizontal break—that is, there is a twist or roll to the break, as shown in Fig. 8-3—the

---

*(Courtesy George K. Bachmann, Pipe Fitter's and Plumber's Vest Pocket Reference Book, *Prentice-Hall, Inc., Englewood Cliffs, N.J.*)

OFFSETS

TABLE 8-1 45-DEGREE PIPE BENDS. RUN, OFFSET, AND TRAVEL (RUN = OFFSET). TRAVEL IN TABLE FOR GIVEN RUN (OR OFFSET) BY 1/2-IN. INCREMENTS. TO OBTAIN TRAVEL FOR INTERMEDIATE OFFSET MEASUREMENTS, ADD 11/64 IN. TRAVEL FOR EACH 1/8 IN. ADDITIONAL OFFSET (*COURTESY G. K. BACHMANN*, PIPEFITTER'S AND PLUMBER'S VEST POCKET REFERENCE BOOK, *PRENTICE-HALL INC., ENGLEWOOD CLIFFS, N.J.*)

Offset	Travel	Offset	Travel	Offset	Travel	Offset	Travel	Offset	Travel	Offset	Travel
0'-2"	0'-2 53/64"	1'-6"	2'-1 29/64"	2'-10"	4'-5/64"	4'-2"	5'-10 45/64"	5'-6"	7'-9 21/64"	7'-6"	10'-7 17/64"
0'-2 1/2"	0'-3 17/32"	1'-6 1/2"	2'-2 1/16"	2'-10 1/2"	4'-25/32"	4'-2 1/2"	5'-11 13/32"	5'-6 1/2"	7'-10 1/32"	7'-7"	10'-8 45/64"
0'-3"	0'-4 15/64"	1'-7"	2'-2 55/64"	2'-11"	4'-1 31/64"	4'-3"	6'-7/64"	5'-7"	7'-10 47/64"	7'-8"	10'-9 3/32"
0'-3 1/2"	0'-4 61/64"	1'-7 1/2"	2'-3 37/64"	2'-11 1/2"	4'-2 13/64"	4'-3 1/2"	6'-53/64"	5'-7 1/2"	7'-11 7/16"	7'-9"	10'-11 1/2"
0'-4"	0'-5 21/32"	1'-8"	2'-4 9/32"	3'-0"	4'-2 29/32"	4'-4"	6'-1 17/32"	5'-8"	8'-5/32"	7'-10"	11'-59/64"
0'-4 1/2"	0'-6 23/64"	1'-8 1/2"	2'-4 63/64"	3'-1/2"	4'-3 39/64"	4'-4 1/2"	6'-2 15/64"	5'-8 1/2"	8'-55/64"	7'-11"	11'-2 21/64"
0'-5"	0'-7 1/16"	1'-9"	2'-5 11/16"	3'-1"	4'-4 5/16"	4'-5"	6'-2 15/16"	5'-9"	8'-1 9/16"	8'	11'-3 3/4"
0'-5 1/2"	0'-7 25/32"	1'-9 1/2"	2'-6 13/32"	3'-1 1/2"	4'-5 1/32"	4'-5 1/2"	6'-3 25/32"	5'-9 1/2"	8'-2 17/64"	8'-1"	11'-5 5/32"
0'-6"	0'-8 31/64"	1'-10"	2'-7 7/64"	3'-2"	4'-5 47/64"	4'-6"	6'-4 23/64"	5'-10"	8'-2 63/64"	8'-2"	11'-6 37/64"
0'-6 1/2"	0'-9 3/16"	1'-10 1/2"	2'-7 13/16"	3'-2 1/2"	4'-6 7/16"	4'-6 1/2"	6'-5 5/16"	5'-10 1/2"	8'-3 11/16"	8'-3"	11'-7 63/64"
0'-7"	0'-9 57/64"	1'-11"	2'-8 33/64"	3'-3"	4'-7 9/64"	4'-7"	6'-5 49/64"	5'-11"	8'-4 25/64"	8'-4"	11'-9 13/32"
0'-7 1/2"	0'-10 39/64"	1'-11 1/2"	2'-9 15/64"	3'-3 1/2"	4'-7 55/64"	4'-7 1/2"	6'-6 31/64"	5'-11 1/2"	8'-5 3/32"	8'-5"	11'-10 13/16"
0'-8"	0'-11 5/16"	2'-0"	2'-9 15/16"	3'-4"	4'-8 9/16"	4'-8"	6'-7 3/16"	6'-0"	8'-5 13/16"	8'-6"	12'-15/64"
0'-8 1/2"	1'-1/64"	2'-1/2"	2'-10 41/64"	3'-4 1/2"	4'-9 17/64"	4'-8 1/2"	6'-7 57/64"	6'-1/2"	8'-6 33/64"	8'-7"	12'-1 43/64"
0'-9"	1'-23/32"	2'-1"	2'-11 11/32"	3'-5"	4'-9 31/32"	4'-9"	6'-8 39/32"	6'-1"	8'-7 7/32"	8'-8"	12'-3 1/16"
0'-9 1/2"	1'-1 7/16"	2'-1 1/2"	3'-1 1/16"	3'-5 1/2"	4'-10 1/2"	4'-9 1/2"	6'-9 5/16"	6'-2"	8'-8 41/64"	8'-9"	12'-4 15/32"
0'-10"	1'-2 9/64"	2'-2"	3'-49/64"	3'-6"	4'-11 25/64"	4'-10"	6'-10 1/4"	6'-3"	8'-10 3/64"	8'-10"	12'-5 57/64"
0'-10 1/2"	1'-2 27/32"	2'-2 1/2"	3'-1 15/32"	3'-6 1/2"	5'-3/32"	4'-10 1/2"	6'-10 23/32"	6'-4"	8'-11 15/32"	8'-11"	12'-7 19/64"
0'-11"	1'-3 35/64"	2'-3"	3'-2 11/64"	3'-7"	5'-51/64"	4'-11"	6'-11 27/64"	6'-5"	9'-7/8"	9'	12'-8 23/32"
0'-11 1/2"	1'-4 17/64"	2'-3 1/2"	3'-2 57/64"	3'-7 1/2"	5'-1 33/64"	4'-11 1/2"	7'-9/64"	6'-6"	9'-2 19/64"	9'-1"	12'-10 1/8"
1'-0"	1'-4 31/32"	2'-4"	3'-3 19/32"	3'-8"	5'-2 7/32"	5'-0"	7'-27/32"	6'-7"	9'-3 45/64"	9'-2"	12'-11 35/64"
1'-1/2"	1'-5 43/64"	2'-4 1/2"	3'-4 19/64"	3'-8 1/2"	5'-2 59/32"	5'-1/2"	7'-1 15/32"	6'-8"	9'-5 1/8"	9'-3"	13'-61/64"
1'-1"	1'-6 3/8"	2'-5"	3'-5"	3'-9"	5'-3 5/8"	5'-1"	7'-2 1/4"	6'-9"	9'-6 17/32"	9'-4"	13'-2 3/8"
1'-1 1/2"	1'-7 3/32"	2'-5 1/2"	3'-5 23/32"	3'-9 1/2"	5'-4 11/32"	5'-1 1/2"	7'-2 31/32"	6'-10"	9'-7 61/64"	9'-5"	13'-3 25/32"
1'-2"	1'-7 51/64"	2'-6"	3'-6 27/64"	3'-10"	5'-5 3/64"	5'-2"	7'-3 43/64"	6'-11"	9'-9 23/64"	9'-6"	13'-5 13/64"
1'-2 1/2"	1'-8 1/2"	2'-6 1/2"	3'-7 1/8"	3'-10 1/2"	5'-5 3/4"	5'-2 1/2"	7'-4 3/8"	7'	9'-10 25/32"	9'-7"	13'-6 39/64"
1'-3"	1'-9 13/64"	2'-7"	3'-7 53/64"	3'-11"	5'-6 29/64"	5'-3"	7'-5 5/64"	7'-1"	10'-3/16"	9'-8"	13'-8 3/32"
1'-3 1/2"	1'-9 59/64"	2'-7 1/2"	3'-8 35/64"	3'-11 1/2"	5'-7 11/64"	5'-3 1/2"	7'-5 25/32"	7'-2"	10'-1 39/64"	9'-9"	13'-9 7/16"
1'-4"	1'-10 5/8"	2'-8"	3'-9 1/4"	4'-0"	5'-7 7/8"	5'-4"	7'-6 1/2"	7'-3"	10'-3 1/64"	9'-10"	13'-10 55/64"
1'-4 1/2"	1'-11 21/64"	2'-8 1/2"	3'-9 61/64"	4'-1/2"	5'-8 37/64"	5'-4 1/2"	7'-7 13/64"	7'-4"	10'-4 7/16"	9'-11"	14'-17/64"
1'-5"	2'-1/32"	2'-9"	3'-10 21/32"	4'-1"	5'-9 5/32"	5'-5"	7'-7 29/32"	7'-5"	10'-5 27/32"	10'	14'-1 11/16"
1'-5 1/2"	2'-3/4"	2'-9 1/2"	3'-11 3/8"	4'-1 1/2"	5'-10"	5'-5 1/2"	7'-8 39/64"				

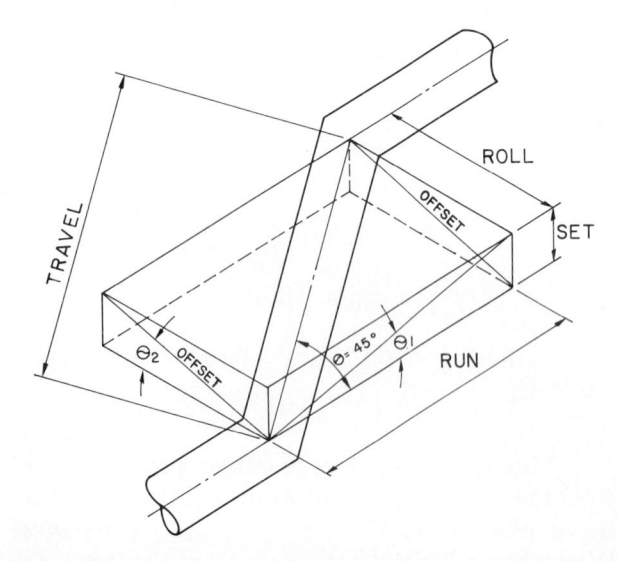

Fig. 8-3 45-degree rolling offset (*Courtesy G. K. Bachmann*, Pipefitter's and Plumber's Vest Pocket Reference Book, *Prentice-Hall Inc., Englewood Cliffs, N.J.*)

calculation involves square and square root. The roll reduces the effective angle of horizontal or vertical offset. A sample calculation is shown in Example 2, for a 45-degree roll ($\theta_2 = 45°$), intermediate between vertical and horizontal. Example 3 gives a calculation for angle of roll from given dimensions of set and run.

EXAMPLE 2: If there is to be a set of 2'2" on a 45-degree offset bend, that has a roll of 45 degrees from the vertical ($\theta_2 = 45°$), what will be the dimensions of the run and travel?

$$45° \text{ offset } (\theta = 45°) \text{ roll } \theta_2 = 45°$$

$$\text{Let set} = \text{roll} = 2'2''$$

$$\text{Offset} = \sqrt{2'2''^2 + 2'2''^2}$$

$$= \sqrt{2.166^2 + 2.166^2} =$$

$$\text{Offset} = \sqrt{4.69 + 4.69} = \sqrt{9.38} = 3.063$$

To Find Side	When You Know	Multiply Side	For 5⅛°	For 11¼°	For 22½°	For 30°	For 45°	For 60°
T	S	S	10.207	5.125	2.613	2.00	1.414	1.154
S	T	T	.096	.195	.382	.50	.707	.866
R	S	S	10.153	5.027	2.414	1.732	1.000	.577
S	R	R	.098	.198	.414	.577	1.000	1.732
T	R	R	1.004	1.019	1.062	1.154	1.141	2.000
R	T	T	.995	.960	.923	.866	.707	.500

Travel = $S \times$ factor or $R \times$ factor.
Set = $T \times$ factor or $R \times$ factor.
Run = $T \times$ factor or $S \times$ factor.

$$\text{Travel} = \frac{\text{Offset}}{\cos \theta} = \frac{3.063}{.707} = 4.33 = 4'4''$$

$$\text{Run} = \text{Offset} = 3.063' = 3'0\tfrac{3}{4}''$$

EXAMPLE 3: If there is to be a set of 2′2″ on a 45-degree offset bend ($\theta = 45°$) with a run or offset of 3′9″, what will be the travel, the roll, the angle of roll (θ_2), and the angle of rise along the run (θ_1)?

$$\text{Travel} = \frac{\text{Run}}{\sin \theta} = \frac{3.75}{.7071} = 5.303'$$

$$= 5'3\tfrac{5}{8}''$$

$$\text{Roll} = \sqrt{\text{Offset}^2 - \text{Set}^2} = \sqrt{3.75^2 - 2.166^2}$$

$$\sqrt{14.06 - 4.69} = \sqrt{9.37} = 3.061$$

θ_2 = angle whose tangent is

$$\left(\frac{2.166}{3.061} = .7076\right) = 35° \; 20'$$

θ_1 = angle whose tangent is

$$\left(\frac{2.166}{3.75} = .577\right) = 30°$$

SIMPLE OFFSETS

The offset shown in Fig. 8-4 can be solved for by labeling its three sides as shown. The three sides correspond to the sides of a right triangle. The set (S) equals the altitude (a), the run (R) equals the base (b), and the travel (T) equals the hypotenuse (c). See Chapters 17 and 18, on formulas and triangles, to solve for the unknowns using trigonometry or Pythagorean theorem.

In most cases, calculating an offset requires that the travel be established. The set and run or the set and the angle are normally given. The travel is the center-to-center distance between the two fittings (elbows).

All offsets are based on the right triangle, with the angle the fittings derive their name from equal to the offset angle. To solve for the travel of a particular offset, the set and angle are normally stated, for example, 22-in. set and 60-degree angle. Use of the factors provided in the chart will speed the calculation.

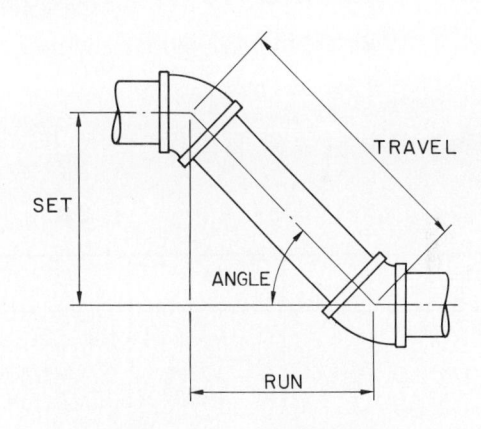

Fig. 8-4 Simple offsets

EXAMPLE 3:

Set S = 10 in., 30 degrees

Travel (T) = $S \times$ factor

Travel T = 100 × 2.000

T = 20 in.

ROLLING OFFSETS

To calculate the length of a piece of pipe between fittings where there is both a set, run and a roll

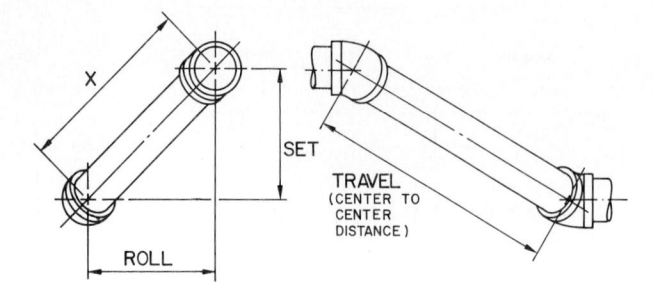

Fig. 8-5 Rolling offsets

(Fig. 8-5), the following formulas using factors can be used. Set = S, R = roll, and T equals travel. Travel (center-to-center distance between fittings) for a rolling offset equals the given factor times the square root of the sum of the set squared plus the roll squared. Travel can also be found by calculating the distance X and then multiplying it times the cosecant of the angle of the fitting.

EXAMPLES: Find the travel for a rolling offset using 45-degree elbows with a set of 12 in. and a roll of 8 in.

$$\text{Set} = 12 \text{ in.}$$
$$\text{Roll} = 8 \text{ in.}$$

$$\text{Travel} = \text{factor} \sqrt{\text{Set}^2 + \text{Roll}^2}$$
$$\text{Travel} = 1.414 \sqrt{12^2 + 8^2}$$
$$T = 1.414 \sqrt{208}$$
$$T = 1.414 \times 14.42$$
$$T = 20.38 \text{ in.}$$

To calculate rolling offsets for:

$5\frac{5}{8}°$ elbows $T = 10.207 \sqrt{S^2 + R^2}$

$11\frac{1}{4}°$ elbows $T = 5.126 \sqrt{S^2 + R^2}$

$22\frac{1}{2}°$ elbows $T = 2.613 \sqrt{S^2 + R^2}$

$30°$ elbows $T = 2.000 \sqrt{S^2 + R^2}$

$45°$ elbows $T = 1.414 \sqrt{S^2 + R^2}$

$60°$ elbows $T = 1.155 \sqrt{S^2 + R^2}$

Distance $X = \sqrt{\text{Set}^2 + \text{Roll}^2}$

Travel = $X \times$ cosecant of angle of fitting

Run = $X \times$ cotangent of angle of fitting

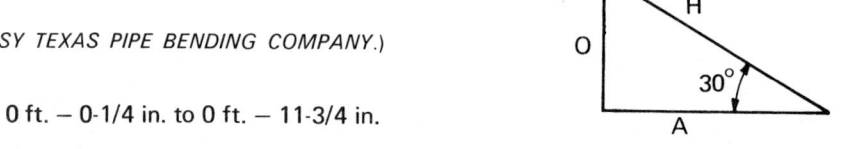

TABLE 8-2 30-DEGREE OFFSETS (*COURTESY TEXAS PIPE BENDING COMPANY.*)

0 ft. − 0-1/4 in. to 0 ft. − 11-3/4 in.

O	H	A	O	H	A	O	H	A
			0 − 4	0 − 8	0 − 6-15/16	0 − 8	1 − 4	1 − 1-15/16
0 − 0-1/4	0 − 0-1/2	0 − 0-7/16	0 − 4-1/4	0 − 8-1/2	0 − 7-3/8	0 − 8-1/4	1 − 4-1/2	1 − 2-5/16
0 − 0-1/2	0 − 1	0 − 0-7/8	0 − 4-1/2	0 − 9	0 − 7-13/16	0 − 8-1/2	1 − 5	1 − 2-3/4
0 − 0-3/4	0 − 1-1/2	0 − 1-5/16	0 − 4-3/4	0 − 9-1/2	0 − 8-1/4	0 − 8-3/4	1 − 5-1/2	1 − 3-1/8
0 − 1	0 − 2	0 − 1-3/4	0 − 5	0 − 10	0 − 8-11/16	0 − 9	1 − 6	1 − 3-9/16
0 − 1-1/4	0 − 2-1/2	0 − 2-3/16	0 − 5-1/4	0 − 10-1/2	0 − 9-1/16	0 − 9-1/4	1 − 6-1/2	1 − 4
0 − 1-1/2	0 − 3	0 − 2-5/8	0 − 5-1/2	0 − 11	0 − 9-1/2	0 − 9-1/2	1 − 7	1 − 4-7/16
0 − 1-3/4	0 − 3-1/2	0 − 3	0 − 5-3/4	0 − 11-1/2	0 − 9-15/16	0 − 9-3/4	1 − 7-1/2	1 − 4-7/8
0 − 2	0 − 4	0 − 3-7/16	0 − 6	1 − 0	0 − 10-3/8	0 − 10	1 − 8	1 − 5-5/16
0 − 2-1/4	0 − 4-1/2	0 − 3-7/8	0 − 6-1/4	1 − 0-1/2	0 − 10-13/16	0 − 10-1/4	1 − 8-1/2	1 − 5-3/4
0 − 2-1/2	0 − 5	0 − 4-5/16	0 − 6-1/2	1 − 1	0 − 11-1/4	0 − 10-1/2	1 − 9	1 − 6-3/16
0 − 2-3/4	0 − 5-1/2	0 − 4-3/4	0 − 6-3/4	1 − 1-1/2	0 − 11-11/16	0 − 10-3/4	1 − 9-1/2	1 − 6-5/8
0 − 3	0 − 6	0 − 5-3/16	0 − 7	1 − 2	1 − 0-1/8	0 − 11	1 − 10	1 − 7-1/16
0 − 3-1/4	0 − 6-1/2	0 − 5-5/8	0 − 7-1/4	1 − 2-1/2	1 − 0-9/16	0 − 11-1/4	1 − 10-1/2	1 − 7-1/2
0 − 3-1/2	0 − 7	0 − 6-1/16	0 − 7-1/2	1 − 3	1 − 1	0 − 11-1/2	1 − 11	1 − 7-15/16
0 − 3-3/4	0 − 7-1/2	0 − 6-1/2	0 − 7-3/4	1 − 3-1/2	1 − 1-7/16	0 − 11-3/4	1 − 11-1/2	1 − 8-3/8

TABLE 8-2 *(CONTD.)*

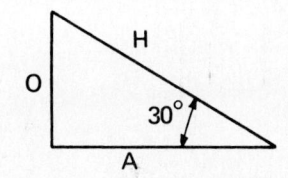

1 ft. − 0 in. to 1 ft. − 11-3/4 in.

O	H	A	O	H	A	O	H	A
1 − 0	2 − 0	1 − 8-13/16	1 − 4	2 − 8	2 − 3-11/16	1 − 8	3 − 4	2 − 10-5/8
1 − 0-1/4	2 − 0-1/2	1 − 9-3/16	1 − 4-1/4	2 − 8-1/2	2 − 4-1/8	1 − 8-1/4	3 − 4-1/2	2 − 11-1/16
1 − 0-1/2	2 − 1	1 − 9-5/8	1 − 4-1/2	2 − 9	2 − 4-9/16	1 − 8-1/2	3 − 5	2 − 11-1/2
1 − 0-3/4	2 − 1-1/2	1 − 10-1/16	1 − 4-3/4	2 − 9-1/2	2 − 5	1 − 8-3/4	3 − 5-1/2	2 − 11-15/16
1 − 1	2 − 2	1 − 10-1/2	1 − 5	2 − 10	2 − 5-7/16	1 − 9	3 − 6	3 − 0-3/8
1 − 1-1/4	x2 − 2-1/2	1 − 10-15/16	1 − 5-1/4	2 − 10-1/2	2 − 5-7/8	1 − 9-1/4	3 − 6-1/2	3 − 0-13/16
1 − 1-1/2	2 − 3	1 − 11-3/8	1 − 5-1/2	2 − 11	2 − 6-5/16	1 − 9-1/2	3 − 7	3 − 1-1/4
1 − 1-3/4	2 − 3-1/2	1 − 11-13/16	1 − 5-3/4	2 − 11-1/2	2 − 6-3/4	1 − 9-3/4	3 − 7-1/2	3 − 1-11/16
1 − 2	2 − 4	2 − 0-1/4	1 − 6	3 − 0	3 − 7-3/16	1 − 10	3 − 8	3 − 2-1/8
1 − 2-1/4	2 − 4-1/2	2 − 0-11/16	1 − 6-1/4	3 − 0-1/2	2 − 7-5/8	1 − 10-1/4	3 − 8-1/2	3 − 2-9/16
1 − 2-1/2	2 − 5	2 − 1-1/8	1 − 6-1/2	3 − 1	2 − 8-1/16	1 − 10-1/2	3 − 9	3 − 3
1 − 2-3/4	2 − 5-1/2	2 − 1-9/16	1 − 6-3/4	3 − 1-1/2	2 − 8-1/2	1 − 10-3/4	3 − 9-1/2	3 − 3-3/8
1 − 3	2 − 6	2 − 2	1 − 7	3 − 2	2 − 8-15/16	1 − 11	3 − 10	3 − 3-13/16
1 − 3-1/4	2 − 6-1/2	2 − 2-7/16	1 − 7-1/4	2 − 2-1/2	2 − 9-5/16	1 − 11-1/4	3 − 10-1/2	3 − 4-1/4
1 − 3-1/2	2 − 7	2 − 2-7/8	1 − 7-1/2	3 − 3	2 − 9-3/4	1 − 11-1/2	3 − 11	3 − 4-11/16
1 − 3-3/4	2 − 7-1/2	2 − 3-1/4	1 − 7-3/4	3 − 3-1/2	2 − 10-3/16	1 − 11-3/4	3 − 11-1/2	3 − 5-1/8

2′ − 0″ to 2′ − 11-3/4″

O	H	A	O	H	A	O	H	A
2 − 0	4 − 0	3 − 5-9/16	2 − 4	4 − 8	4 − 0-1/2	2 − 8	5 − 4	4 − 7-7/16
2 − 0-1/4	4 − 0-1/2	3 − 6	2 − 4-1/4	4 − 8-1/2	4 − 0-15/16	2 − 8-1/4	5 − 4-1/2	4 − 7-7/8
2 − 0-1/2	4 − 1	3 − 6-7/16	2 − 4-1/2	4 − 9	4 − 1-3/8	2 − 8-1/2	5 − 5	4 − 8-5/16
2 − 0-3/4	4 − 1-1/2	3 − 6-7/8	2 − 4-3/4	4 − 9-1/2	4 − 1-13/16	2 − 8-3/4	5 − 5-1/2	4 − 8-3/4
2 − 1	4 − 2	3 − 7-5/16	2 − 5	4 − 10	4 − 2-1/4	2 − 9	5 − 6	4 − 9-3/16
2 − 1-1/4	4 − 2-1/2	3 − 7-3/4	2 − 5-1/4	4 − 10-1/2	4 − 2-11/16	2 − 9-1/4	5 − 6-1/2	4 − 9-9/16
2 − 1-1/2	4 − 3	3 − 8-3/16	2 − 5-1/2	4 − 11	4 − 3-1/8	2 − 9-1/2	5 − 7	4 − 10
2 − 1-3/4	4 − 3-1/2	3 − 8-5/8	2 − 5-3/4	4 − 11-1/2	4 − 3-1/2	2 − 9-3/4	5 − 7-1/2	4 − 10-7/16
2 − 2	4 − 4	3 − 9-1/16	2 − 6	5 − 0	4 − 3-15/16	2 − 10	5 − 8	4 − 10-7/8
2 − 2-1/4	4 − 4-1/2	3 − 9-1/2	2 − 6-1/4	5 − 0-1/2	4 − 4-3/8	2 − 10-1/4	5 − 8-1/2	4 − 11-5/16
2 − 2-1/2	4 − 5	3 − 9-7/8	2 − 6-1/2	5 − 1	4 − 4-13/16	2 − 10-1/2	5 − 9	4 − 11-3/4
2 − 2-3/4	4 − 5-1/2	3 − 10-5/16	2 − 6-3/4	5 − 1-1/2	4 − 5-1/4	2 − 10-3/4	5 − 9-1/2	5 − 0-3/16
2 − 3	4 − 6	3 − 10-3/4	2 − 7	5 − 2	4 − 5-11/16	2 − 11	5 − 10	5 − 0-5/8
2 − 3-1/4	4 − 6-1/2	3 − 11-3/16	2 − 7-1/4	5 − 2-1/2	4 − 6-1/8	2 − 11-1/4	5 − 10-1/2	5 − 1-1/16
2 − 3-1/2	4 − 7	3 − 11-5/8	2 − 7-1/2	5 − 3	4 − 6-9/16	2 − 11-1/2	5 − 11	5 − 1-1/2
2 − 3-3/4	4 − 7-1/2	4 − 0-1/16	2 − 7-3/4	5 − 3-1/2	4 − 7	2 − 11-3/4	5 − 11-1/2	5 − 1-13/16

OFFSETS

TABLE 8-2 *(CONTD.)*

3′ − 0″ to 3′ − 11-3/4″

O	H	A	O	H	A	O	H	A
3 − 0	6 − 0	5 − 2-3/8	3 − 4	6 − 8	5 − 9-5/16	3 − 8	7 − 4	6 − 4-3/16
3 − 0-1/4	6 − 0-1/2	5 − 2-13/16	3 − 4-1/4	6 − 8-1/2	5 − 9-11/16	3 − 8-1/4	7 − 4-1/2	6 − 4-5/8
3 − 0-1/2	6 − 1	5 − 3-1/4	3 − 4-1/2	6 − 9	5 − 10-1/8	3 − 8-1/2	7 − 5	6 − 5-1/16
3 − 0-3/4	6 − 1-1/2	5 − 3-5/8	3 − 4-3/4	6 − 9-1/2	5 − 10-9/16	3 − 8-3/4	7 − 5-1/2	6 − 6-1/2
3 − 1	6 − 2	5 − 4-1/16	3 − 5	6 − 10	5 − 11	3 − 9	7 − 6	6 − 5-15/16
3 − 1-1/4	6 − 2-1/2	5 − 4-1/2	3 − 5-1/4	6 − 10-1/2	5 − 11-7/16	3 − 9-1/4	7 − 6-1/2	6 − 6-3/8
3 − 1-1/2	6 − 3	5 − 4-15/16	3 − 5-1/2	6 − 11	5 − 11-7/8	3 − 9-1/2	7 − 7	6 − 6-13/16
3 − 1-3/4	6 − 3-1/2	5 − 5-3/8	3 − 5-3/4	6 − 11-1/2	6 − 0-5/16	3 − 9-3/4	7 − 7-1/2	6 − 7-1/4
3 − 2	6 − 4	5 − 5-13/16	3 − 6	7 − 0	6 − 0-3/4	3 − 10	7 − 8	6 − 7-11/16
3 − 2-1/4	6 − 4-1/2	5 − 6-1/4	3 − 6-1/4	7 − 0-1/2	6 − 1-3/16	3 − 10-1/4	7 − 8-1/2	6 − 8-1/8
3 − 2-1/2	6 − 5	5 − 6-11/16	3 − 6-1/2	7 − 1	6 − 1-5/8	3 − 10-1/2	7 − 9	6 − 8-9/16
3 − 2-3/4	6 − 5-1/2	5 − 7-1/8	3 − 6-3/4	7 − 1-1/2	6 − 2-1/16	3 − 10-3/4	7 − 9-1/2	6 − 9
3 − 3	6 − 6	5 − 7-9/16	3 − 7	7 − 2	6 − 2-1/2	3 − 11	7 − 10	6 − 9-7/16
3 − 3-1/4	6 − 6-1/2	5 − 8	3 − 7-1/4	7 − 2-1/2	6 − 2-15/16	3 − 11-1/4	7 − 10-1/2	6 − 9-13/16
3 − 3-1/2	6 − 7	5 − 8-7/16	3 − 7-1/2	7 − 3	6 − 3-3/8	3 − 11-1/2	7 − 11	6 − 10-1/4
3 − 3-3/4	6 − 7-1/2	5 − 8-7/8	3 − 7-3/4	7 − 3-1/2	6 − 3-3/4	3 − 11-3/4	7 − 11-1/2	6 − 10-11/16

TABLE 8-3 45-DEGREE OFFSETS *(COURTESY TEXAS PIPE BENDING COMPANY.)*

	0	1	2	3	4	5	6	7	8	9	10	11	
0 1/16	0 1/16	1-7/16 1-1/2	2-13/16 2-7/8	4-1/4 4-5/16	5-11/16 5-3/4	7-1/16 7-3/16	8-1/2 8-9/16	9-7/8 10	11-5/16 11-3/8	12-3/4 12-13/16	14-1/8 14-1/4	15-9/16 15-5/8	0 1/16
1/8 3/16	3/16 1/4	1-9/16 1-11/16	3 3-1/16	4-7/16 4-1/2	5-13/16 5-7/8	7-1/4 7-5/16	8-11/16 8-3/4	10-1/16 10-3/16	11-1/2 11-9/16	12-7/8 13	14-5/16 14-7/16	15-3/4 15-13/16	1/8 3/16
1/4 5/16	3/8 7/16	1-3/4 1-7/8	3-3/16 3-1/4	4-5/8 4-11/16	6 6-0/8	7-7/16 7-1/2	8-13/16 8-15/16	10-1/4 10-3/8	11-11/16 11-3/4	13-1/16 13-3/16	14-1/2 14-9/16	15-15/16 16	1/4 5/16
3/8 7/16	1/2 5/8	1-15/16 2-1/16	3-3/8 3-7/16	4-3/4 4-7/8	6-3/16 6-1/4	7-5/8 7-11/16	9 9-1/8	10-7/16 10-1/2	11-7/8 11-15/16	13-1/4 13-3/8	14-11/16 14-3/4	16-1/16 16-3/16	3/8 7/16
1/2 9/16	11/16 13/16	2-1/8 2-3/16	3-9/16 3-5/8	4-15/16 5-1/16	6-3/8 6-7/16	7-3/4 7-7/8	9-3/16 9-1/4	10-5/8 10-11/16	12 12-1/8	13-7/16 13-9/16	14-7/8 14-15/16	16-1/4 16-3/8	1/2 9/16
5/8 11/16	7/8 1	2-5/16 2-3/8	3-11/16 3-13/16	5-1/8 5-3/16	6-9/16 6-5/8	7-15/16 8-1/16	9-3/8 9-7/16	10-13/16 10-7/8	12-3/16 12-5/16	13-5/8 13-11/16	15 15-1/8	16-7/16 16-1/2	5/8 11/16
3/4 13/16	1-1/16 1-1/8	2-1/2 2-9/16	3-7/8 4	5-5/16 5-3/8	6-11/16 6-13/16	8-1/8 8-1/4	9-9/16 9-5/8	11 11-1/16	12-3/8 12-7/16	13-13/16 13-7/8	15-3/16 15-5/16	16-5/8 16-11/16	3/4 13/16
7/8 15/16	1-1/4 1-5/16	2-5/8 2-3/4	4-1/16 4-1/8	5-1/2 5-9/16	6-7/8 7	8-5/16 8-3/8	9-11/16 9-13/16	11-1/8 11-1/4	12-9/16 12-5/8	13-15/16 14-1/16	15-3/8 15-7/16	16-13/16 16-7/8	7/8 15/16

TABLE 8-4 45-DEGREE TRIANGLES—BASE TO HYPOTENUSE (COURTESY TEXAS PIPE BENDING COMPANY.)

	2 – 4	2 – 1	2 – 2	2 – 3	2 – 4	2 – 5	2 – 6	2 – 7	2 – 8	2 – 9	2 – 10	2 – 11	
0	2 – 9-15/16	2 – 11-3/8	3 – 0-3/4	3 – 2-3/16	3 – 3-5/8	3 – 5	3 – 6-7/16	3 – 7-13/16	3 – 9-1/4	3 – 10-11/16	4 – 0-1/16	4 – 1-1/2	0
1/16	2 – 10	2 – 11-7/16	3 – 0-7/8	3 – 2-1/4	3 – 3-11/16	3 – 5-1/8	3 – 6-1/2	3 – 7-15/16	3 – 9-5/16	3 – 10-3/4	4 – 0-3/16	4 – 1-9/16	1/16
1/8	2 – 10-1/8	2 – 11-9/16	3 – 0-15/16	3 – 2-3/8	3 – 3-3/4	3 – 5-3/16	3 – 6-5/8	3 – 8	3 – 9-7/16	3 – 10-7/8	4 – 0-1/4	4 – 4-11/16	1/8
3/16	2 – 10-3/16	2 – 11-5/8	3 – 1-1/16	3 – 2-7/16	3 – 3-7/8	3 – 5-1/4	3 – 6-11/16	3 – 8-1/8	3 – 9-1/2	3 – 10-15/16	4 – 0-3/8	4 – 1-3/4	3/16
1/4	2 – 10-5/16	2 – 11-11/16	3 – 1-1/8	3 – 2-9/16	3 – 3-15/16	3 – 5-3/8	3 – 6-3/4	3 – 8-3/16	3 – 9-5/8	3 – 11	4 – 0-7/16	4 – 1-7/8	1/4
3/16	2 – 10-3/8	2 – 11-13/16	3 – 1-3/16	3 – 2-5/8	3 – 4-1/16	3 – 5-7/16	3 – 6-7/8	3 – 8-5/16	3 – 9-11/16	3 – 11-1/8	4 – 0-1/2	4 – 1-15/16	3/16
3/8	2 – 10-1/2	2 – 11-7/8	3 – 1-5/16	3 – 2-11/16	3 – 4-1/8	3 – 5-9/16	3 – 6-15/16	3 – 8-3/8	3 – 9-3/4	3 – 11-3/16	4 – 0-5/8	4 – 2	3/8
7/16	2 – 10-9/16	3 – 0	3 – 1-3/8	3 – 2-13/16	3 – 4-3/16	3 – 5-5/8	3 – 7-1/16	3 – 8-7/16	3 – 9-7/8	3 – 11-5/16	4 – 0-11/16	4 – 2-1/16	7/16
1/2	2 – 10-5/8	3 – 0-1/16	3 – 1-1/2	3 – 2-7/8	3 – 4-5/16	3 – 5-3/4	3 – 7-1/8	3 – 8-9/16	3 – 9-15/16	3 – 11-3/8	4 – 0-13/16	4 – 2-3/16	1/2
9/16	2 – 10-3/4	3 – 0-1/8	3 – 1-9/16	3 – 3	3 – 4-3/8	3 – 5-13/16	3 – 7-1/4	3 – 8-5/8	3 – 10-1/16	3 – 11-7/16	4 – 0-7/8	4 – 2-5/16	9/16
3/8	2 – 10-13/16	3 – 0-1/4	3 – 1-11/16	3 – 3-1/16	3 – 4-1/2	3 – 5-7/8	3 – 7-5/16	3 – 8-3/4	3 – 10-1/8	3 – 11-9/16	4 – 0-15/16	4 – 2-3/8	3/8
11/16	2 – 10-15/16	3 – 0-5/16	3 – 1-3/4	3 – 3-1/8	3 – 4-9/16	3 – 6	3 – 7-3/8	3 – 8-13/16	3 – 10-1/4	3 – 11-5/8	4 – 1-1/16	4 – 2-1/2	11/16
3/4	2 – 11	3 – 0-7/16	3 – 1-13/16	3 – 3-1/4	3 – 4-11/16	3 – 6-1/16	3 – 7-1/2	3 – 8-15/16	3 – 10-5/16	3 – 11-11/16	4 – 1-1/8	4 – 2-9/16	3/4
13/16	2 – 11-1/16	3 – 0-1/2	3 – 1-15/16	3 – 3-5/16	3 – 4-3/4	3 – 6-3/16	3 – 7-9/16	3 – 9	3 – 10-3/8	3 – 11-13/16	4 – 1-1/4	4 – 2-5/8	13/16
7/8	2 – 11-3/16	3 – 0-9/16	3 – 2	3 – 3-7/16	3 – 4-13/16	3 – 6-1/4	3 – 7-11/16	3 – 9-1/16	3 – 10-1/2	3 – 11-13/16	4 – 1-3/16	4 – 2-3/4	7/8
15/16	2 – 11-1/4	3 – 0-11/16	3 – 2-1/8	3 – 3-1/2	3 – 4-15/16	3 – 6-5/16	3 – 7-3/4	3 – 9-3/16	3 – 10-9/16	4 – 0	4 – 1-7/16	4 – 2-13/16	15/16

	3 – 0	3 – 1	3 – 2	3 – 3	3 – 4	3 – 5	3 – 6	3 – 7	3 – 8	3 – 9	3 – 10	3 – 11	
0	4 – 2-13/16	4 – 4-3/16	4 – 5-3/4	4 – 7-1/8	4 – 8-9/16	4 – 10	4 – 11-3/8	5 – 0-13/16	5 – 2-1/4	5 – 3-5/8	5 – 5-1/16	5 – 6-7/16	0
1/16	4 – 3	4 – 4-7/16	4 – 5-13/16	4 – 7-1/4	4 – 8-5/8	4 – 10-1/16	4 – 11-1/2	5 – 0-7/8	5 – 2-5/16	5 – 3-3/4	5 – 5-1/8	5 – 6-9/16	1/16
1/8	4 – 3-1/16	4 – 4-1/2	4 – 5-15/16	4 – 7-5/16	4 – 8-3/4	4 – 10-3/16	4 – 11-9/16	5 – 1	5 – 2-3/8	5 – 3-13/16	5 – 5-3/16	5 – 6-5/8	1/8
3/16	4 – 3-3/16	4 – 4-9/16	4 – 6	4 – 7-7/16	4 – 8-13/16	4 – 10-1/4	4 – 11-11/16	5 – 1-1/16	5 – 2-1/2	5 – 3-7/8	5 – 5-5/16	5 – 6-3/4	3/16
1/4	4 – 3-1/4	4 – 4-11/16	4 – 6-1/8	4 – 7-1/2	4 – 8-13/16	4 – 10-5/16	4 – 11-3/4	5 – 1-3/16	5 – 2-9/16	5 – 4	5 – 5-7/16	5 – 6-13/16	1/4
5/16	4 – 3-3/8	4 – 4-3/4	4 – 6-3/8	4 – 7-5/8	4 – 9	4 – 10-7/16	4 – 11-13/16	5 – 1-1/4	5 – 2-11/16	5 – 4-1/16	5 – 5-1/2	5 – 6-15/16	5/16
3/8	4 – 3-7/16	4 – 4-7/8	4 – 6-1/4	4 – 7-11/16	4 – 9-1/8	4 – 10-1/2	4 – 11-15/16	5 – 1-5/16	5 – 2-3/4	5 – 4-3/16	5 – 5-9/16	5 – 7	3/8
7/16	4 – 3-1/2	4 – 4-13/16	4 – 6-3/8	4 – 7-3/4	4 – 9-3/16	4 – 10-5/8	5 – 0	5 – 1-7/16	5 – 2-7/8	5 – 4-1/4	5 – 5-11/16	5 – 7-1/16	7/16
1/2	4 – 3-5/8	4 – 5-1/16	4 – 6-7/16	4 – 7-7/8	4 – 9-1/4	4 – 10-11/16	5 – 0-1/8	5 – 1-1/2	5 – 3	5 – 4-3/8	5 – 5-3/4	5 – 7-3/16	1/2
9/16	4 – 3-11/16	4 – 5-1/8	4 – 6-9/16	4 – 7-15/16	4 – 9-3/8	4 – 10-3/4	5 – 0-3/16	5 – 1-5/8	5 – 3	5 – 4-7/16	5 – 5-7/8	5 – 7-1/4	9/16
5/8	4 – 3-13/16	4 – 5-1/16	4 – 6-3/8	4 – 8-1/16	4 – 9-7/16	4 – 10-7/8	5 – 0-1/4	5 – 1-11/16	5 – 3-1/8	5 – 4-1/2	5 – 5-15/16	5 – 7-3/8	5/8
11/16	4 – 3-7/8	4 – 5-5/16	4 – 6-11/16	4 – 8-1/8	4 – 9-9/16	4 – 10-15/16	5 – 0-3/8	5 – 1-13/16	5 – 3-3/16	5 – 4-5/8	5 – 6	5 – 7-7/16	11/16
3/4	4 – 4	4 – 5-3/8	4 – 6-13/16	4 – 8-3/8	4 – 9-5/8	4 – 11-1/16	5 – 0-7/16	5 – 1-7/8	5 – 3-5/16	5 – 4-11/16	5 – 6-1/8	5 – 7-1/2	3/4
13/16	4 – 4-1/16	4 – 5-1/2	4 – 6-7/8	4 – 8-3/16	4 – 9-11/16	4 – 11-1/8	5 – 0-9/16	5 – 1-15/16	5 – 3-3/8	5 – 4-13/16	5 – 6-3/16	5 – 7-5/16	13/16
7/8	4 – 4-1/8	4 – 5-9/16	4 – 7	4 – 8-3/8	4 – 9-13/16	4 – 11-1/4	5 – 0-5/8	5 – 2-1/16	5 – 3-7/16	5 – 4-7/8	5 – 6-5/16	5 – 7-11/16	7/8
15/16	4 – 4-1/4	4 – 5-5/8	4 – 7-1/16	4 – 8-1/2	4 – 9-7/8	4 – 11-5/16	5 – 0-3/4	5 – 2-1/8	5 – 3-9/16	5 – 4-15/16	5 – 6-3/8	5 – 7-13/16	15/16

95

TABLE 8-4 (*CONTD.*)

	12	13	14	15	16	17	18	19	20	21	22	23	
0 1/16	17 17-1/16	18-3/8 18-1/2	19-13/16 19-7/8	21-3/16 21-5/16	22-5/8 22-11/16	2 – 0-1/16 2 – 0-1/8	2 – 1-7/16 2 – 1-9/16	2 – 2-7/8 2 – 2-15/16	2 – 4-5/16 2 – 4-3/8	2 – 5-11/16 2 – 5-13/16	2 – 7-1/8 2 – 7-3/16	2 – 8-1/2 2 – 8-5/8	0 1/16
1/8 3/16	17-1/8 17-1/4	18-9/16 18-5/8	20 20-1/16	21-3/8 21-1/2	22-13/16 22-7/8	2 – 0-1/4 2 – 0-5/16	2 – 1-5/8 2 – 1-3/4	2 – 3-1/16 2 – 3-1/8	2 – 4-1/2 2 – 4-9/16	2 – 5-7/8 2 – 5-15/16	2 – 7-5/16 2 – 7-3/8	2 – 8-11/16 2 – 8-13/16	1/8 3/16
1/4 5/16	17-5/16 17-7/8	18-3/4 18-13/16	20-1/8 20-1/4	21-9/16 21-5/8	23 23-1/16	2 – 0-3/8 2 – 0-1/2	2 – 1-13/16 2 – 1-7/8	2 – 3-1/4 2 – 3-5/16	2 – 4-5/8 2 – 4-3/4	2 – 6-1/16 2 – 6-1/8	2 – 7-7/16 2 – 7-9/16	2 – 8-7/8 2 – 9	1/4 5/16
3/8 7/16	17-1/2 17-9/16	18-7/8 19	20-5/16 20-7/16	21-3/4 21-13/16	23-3/16 23-1/4	2 – 0-9/16 2 – 0-11/16	2 – 2 2 – 2-1/16	2 – 3-3/8 2 – 3-1/2	2 – 4-13/16 2 – 4-7/8	2 – 6-1/4 2 – 6-5/16	2 – 7-5/8 2 – 7-3/4	2 – 9-1/16 2 – 9-1/8	3/8 7/16
1/2 9/16	17-11/16 17-3/4	19-1/8 19-3/16	20-1/2 20-5/8	21-15/16 22	23-5/16 23-7/16	2 – 0-3/4 2 – 0-13/16	2 – 2-3/16 2 – 2-1/4	2 – 3-9/16 2 – 3-11/16	2 – 5 2 – 5-1/16	2 – 6-3/8 2 – 6-7/16	2 – 7-13/16 2 – 7-15/16	2 – 9-1/4 2 – 9-5/16	1/2 9/16
5/8 11/16	17-7/8 17-15/16	19-1/4 19-3/8	20-11/16 20-3/4	22-1/8 22-3/16	23-1/2 23-5/8	2 – 0-15/16 2 – 1	2 – 2-5/16 2 – 2-7/16	2 – 3-3/4 2 – 3-13/16	2 – 5-3/16 2 – 5-1/4	2 – 6-9/16 2 – 6-11/16	2 – 8 2 – 8-1/16	2 – 9-7/16 2 – 9-1/2	5/8 11/16
3/4 13/16	18-1/16 18-1/8	19-7/16 19-9/16	20-7/8 20-15/16	22-1/4 22-3/8	23-11/16 22-3/4	2 – 1-1/16 2 – 1-3/16	2 – 2-1/2 2 – 2-5/8	2 – 3-15/16 2 – 4	2 – 5-3/8 2 – 5-7/16	2 – 6-3/4 2 – 6-7/8	2 – 8-3/16 2 – 8-1/4	2 – 9-9/16 2 – 9-11/16	3/4 13/16
7/8 15/16	18-3/16 18-5/16	19-5/8 19-11/16	21-1/16 21-1/8	22-7/16 22-9/16	23-7/8 23-15/16	2 – 1-1/4 2 – 1-3/8	2 – 2-11/16 2 – 2-13/16	2 – 4-1/8 2 – 4-3/16	2 – 5-1/2 2 – 5-5/8	2 – 6-15/16 2 – 7	2 – 8-3/8 2 – 8-7/16	2 – 9-3/4 2 – 9-7/8	7/8 15/16

9
BENDS

Fig. 9-1

DEVELOPED LENGTH = 1.571 R
QUARTER BEND-90°

DEVELOPED LENGTH = 0.785 R
45° BEND

ARC = Rα FACTOR-SEE
OFFSET BEND

DEVELOPED LENGTH = 3.142 R
U-BEND-180°

DEVELOPED LENGTH = 2 × DEVELOPED LENGTH OF OFFSET BEND
CROSSOVER BEND

DEVELOPED LENGTH = 3.14 R
SINGLE OFFSET QUARTER BEND

DEVELOPED LENGTH = 6.127 R
SINGLE OFFSET U BEND

DEVELOPED LENGTH = 6.283 R
DOUBLE OFFSET U BEND

DEVELOPED LENGTH = 6.283 R + 2X
EXPANSION U BEND
(WHEN X IS 2 FEET OR LESS)

DEVELOPED LENGTH = 9.425 R
DOUBLE OFFSET EXPANSION BEND

DEVELOPED LENGTH = 6.283 R
CIRCLE BEND

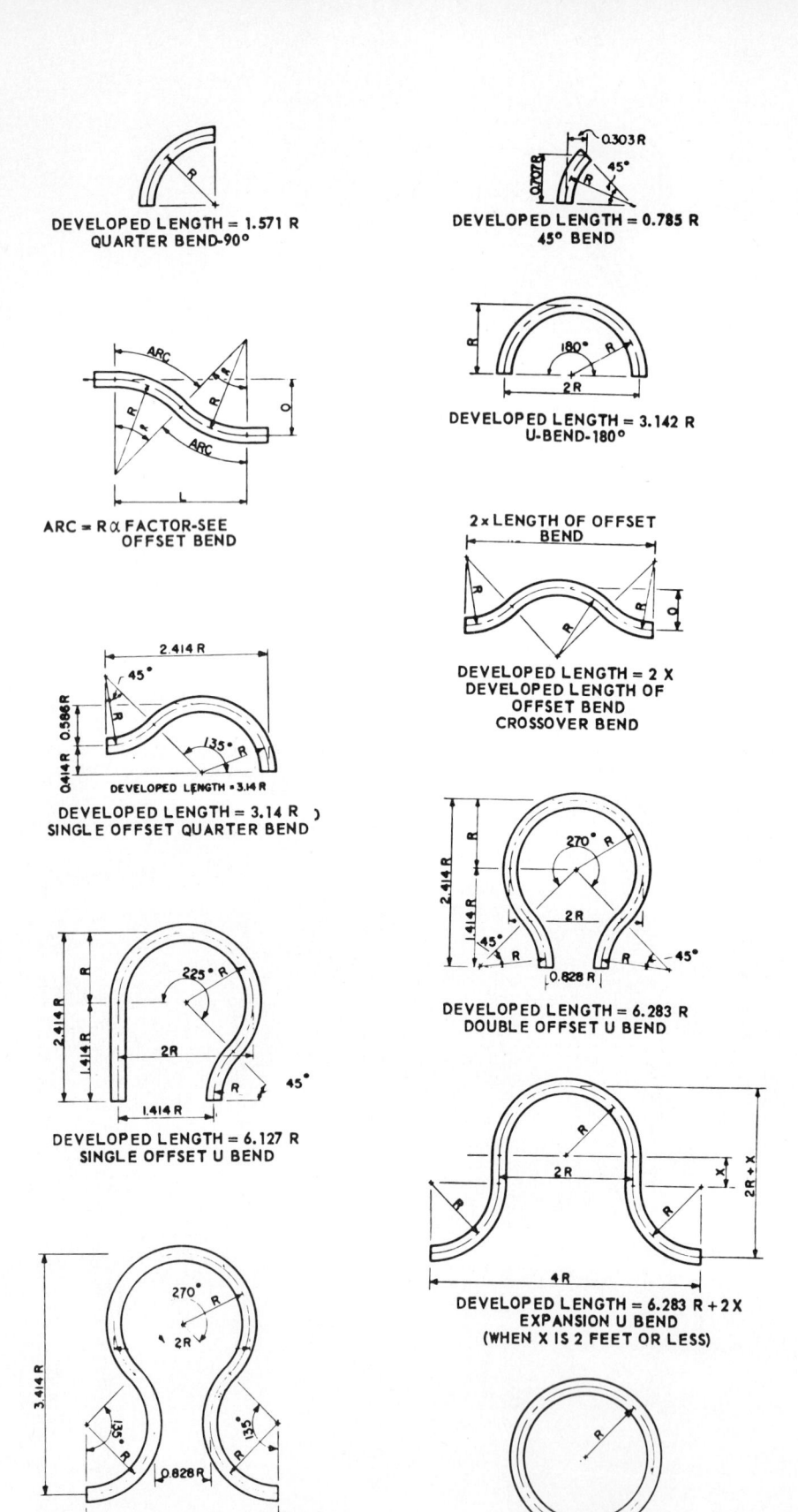

TABLE 9-1 LENGTHS OF ARCS FOR RADIUS 1 IN FIG. 9-2 (*COURTESY ITT GRINNELL CORPORATION.*)

Degrees		Degrees	
0°	0.000000	0°	0.000000
1	0.017453	31	0.541052
2	0.034907	32	0.558505
3	0.052360	33	0.575959
4	0.069813	34	0.593412
5	0.087266	35	0.610865
6	0.104720	36	0.628319
7	0.122173	37	0.645772
8	0.139626	38	0.663225
9	0.157080	39	0.680678
10	0.174533	40	0.698132
11	0.191986	41	0.715585
12	0.209440	42	0.733038
13	0.226893	43	0.750492
14	0.244346	44	0.767945
15	0.261799	45	0.785398
16	0.279253	46	0.802851
17	0.296706	47	0.820305
18	0.314159	48	0.837758
19	0.331613	49	0.855211
20	0.349066	50	0.872665
21	0.366519	51	0.890118
22	0.383972	52	0.907571
23	0.401426	53	0.925025
24	0.418879	54	0.942478
25	0.436332	55	0.959931
26	0.453786	56	0.977384
27	0.471239	57	0.994838
28	0.488692	58	1.012291
29	0.506145	59	1.029744
30	0.523599	60	1.047198

Fig. 9-2 Standard pipe bends (*Courtesy ITT Grinnell Corporation.*)

TABLE 9-2 LENGTH OF PIPE IN BENDS (*COURTESY CRANE COMPANY.*)

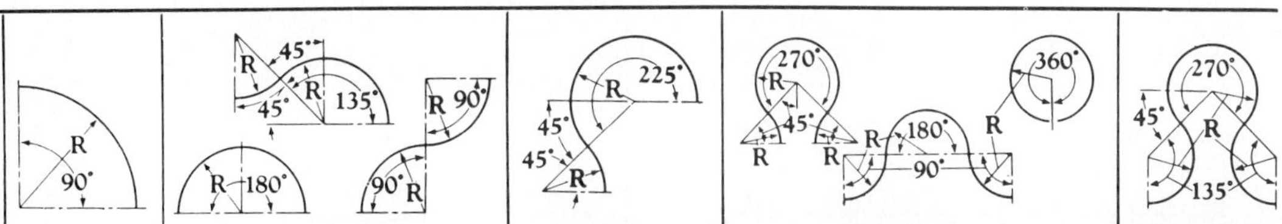

Radius of Pipe Bends (Inches)	Feet	90° Bends	180° Bends	270° Bends	360° Bends	540° Bends
1		1½"	3"	4¾"	6¼"	9½"
2		3	6¼	9½	12½	18¾
3	¼	4¾	9½	14¼	18¾	28¼
4		6¼	12½	18¾	25¼	37¾
5		7¾	15¾	23½	31½	47¼
6	½	9½	18¾	28¼	37¾	56½
7		11	22	33	44	66
8		12½	25¼	37¾	50¼	75½
9	¾	14¼	28¼	42½	56½	84¾
10		15¾	31½	47¼	62¾	94¼
11		17¼	34½	51¾	69	103¾
12	1	18¾	37¾	56½	75½	113
12.5		19¾	39¼	59	78½	117¾
14		22	44	66	88	132
15	1¼	23½	47	70¾	94¼	141¼
16		25¼	50¼	75½	100½	150¾
17.5		27½	55	82½	110	165
18	1½	28¼	56½	84¾	113	169¾
20		31½	62¾	94¼	125¾	188½
21	1¾	33	66	99	132	198
24	2	37¾	75½	113	150¾	226¼
25		39¼	78½	117¾	157	235½
30	2½	47¼	94¼	141¼	188½	282¾
32		50¼	100½	150¾	201	301½
36	3	56½	113	169½	226¼	339¼
40		62¾	125¾	188½	251¼	377
48	4	75½	150¾	226¼	301½	452½
50		78½	157	235½	314¼	471¼
56		88	176	264	351¾	527¾
60	5	94¼	188½	282¾	377	565½
64		100½	201	301½	402	603¼
70		110	220	329¾	439¾	659¾
72	6	113	226¼	339¼	452½	678½
80		125¾	251¼	377	502¾	754
84	7	132	263¾	395¾	527¾	791½
90	7½	141¼	282¾	424	565½	848¼
96	8	150¾	301½	452½	603	904¾
100		157	314¼	471¼	628¼	942½
108	9	169½	339¼	509	678½	1017¾
120	10	188½	377	565½	754	1131
132	11	207¼	414¾	622	829½	1244
144	12	226¼	452½	678½	904¾	1357¼
156	13	245	490	735¼	980¼	1470¼
168	14	263¾	527¾	791½	1055½	1583½
180	15	282¾	565½	848¼	1131	1696½
192	16	301½	603	904¾	1206¼	1809½
204	17	320½	640¾	961¼	1281¾	1922½
216	18	339¼	678½	1017¾	1357¼	2035¾
228	19	358	716¼	1074½	1432½	2148¾
240	20	377	754	1131	1508	2262

To find the length of pipe in a bend having a radius not given above, add together the length of pipe in bends whose combined radii equal the required radius.

Example: Find length of pipe in 90° bend of 5' 9" radius.
Length of pipe in 90° bend of 5' radius = 94¼"
Length of pipe in 90° bend of 9" radius = 14¼"
Then, length of pipe in 90° bend of 5' 9" radius = 108½"

BENDS

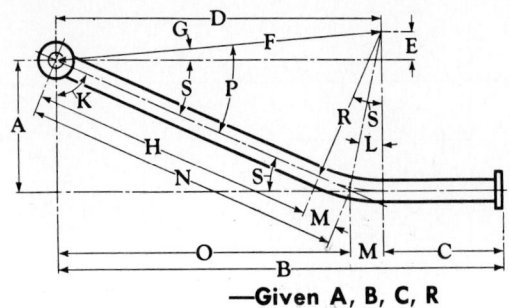

Diagram 1 — Given A, B, C, R

$D = B - C$
$E = R - A$
$F = \sqrt{D^2 + E^2}$
$\dfrac{E}{F} = \sin \angle G$
$H = \sqrt{F^2 - R^2}$

$\dfrac{R}{F} = \sin \angle P$
$\angle S = \angle P - \angle G$
$\angle K = 90° - \angle S$
$\angle L = \tfrac{1}{2} \angle S$
$M = \tan \angle L \times R$
$N = H + M$
$O = B - C - M$

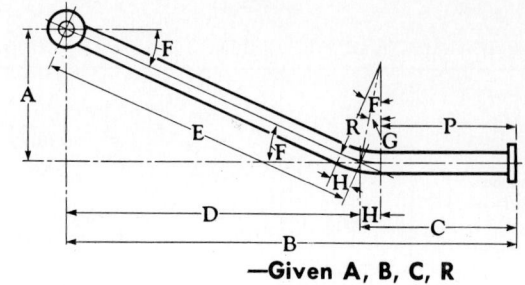

Diagram 2 — Given A, B, C, R

$D = B - C$
$E = \sqrt{A^2 + D^2}$
$\dfrac{A}{E} = \sin \angle F$
$\angle G = \tfrac{1}{2} \angle F$
$H = \tan \angle G \times R$
$P = C - H$

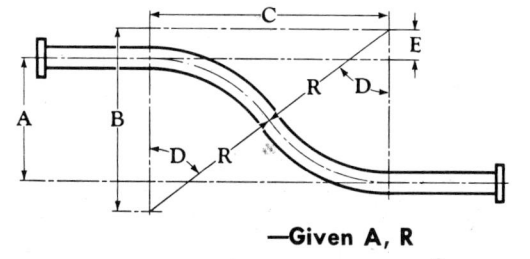

Diagram 3 — Given A, R

$B = 2R - A \qquad C = \sqrt{(2R)^2 - B^2} \qquad \dfrac{C}{2R} = \sin \angle D$

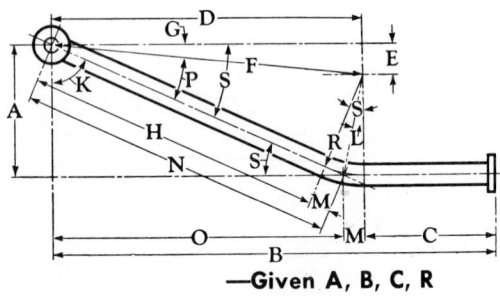

Diagram 4 — Given A, B, C, R

$D = B - C$
$E = A - R$
$F = \sqrt{D^2 + E^2}$
$\dfrac{E}{F} = \sin \angle G$
$H = \sqrt{F^2 - R^2}$

$\dfrac{R}{F} = \sin \angle P$
$\angle S = \angle P + \angle G$
$\angle K = 90° - \angle S$
$\angle L = \tfrac{1}{2} \angle S$
$M = \tan \angle L \times R$
$N = H + M$
$O = B - C - M$

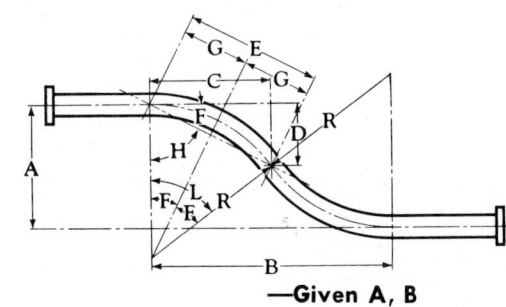

Diagram 5 — Given A, B

$C = \tfrac{1}{2} B$
$D = \tfrac{1}{2} A$
$E = \sqrt{C^2 + D^2}$

$\dfrac{D}{E} = \sin \angle F$
$G = \tfrac{1}{2} E$

$\angle H = 90° - \angle F$
$R = \dfrac{A^2 + B^2}{4A}$
$L = 2F$

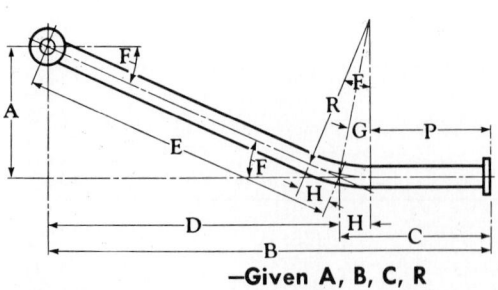

Diagram 6 — Given A, B, C, R

$D = B - C$
$E = \sqrt{A^2 + D^2}$
$\dfrac{A}{E} = \sin \angle F$

$\angle G = \tfrac{1}{2} \angle F$
$H = \tan \angle G \times R$
$P = C - H$

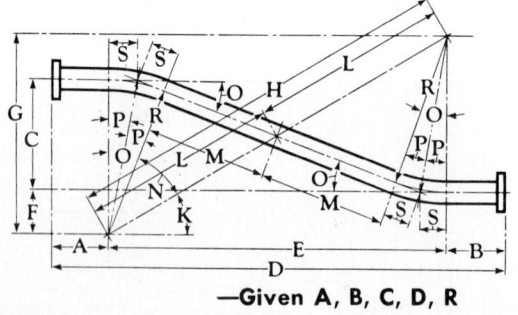

Diagram 7 — Given A, B, C, D, R

$E = D - A - B$
$F = R - C$
$G = R + F$
$H = \sqrt{E^2 + G^2}$

$G/H = \sin \angle K$
$L = \tfrac{1}{2} H$
$M = \sqrt{L^2 - R^2}$

$M/L = \sin \angle N$
$\angle O = 90° - \angle K - \angle N$
$\angle P = \tfrac{1}{2} \angle O$
$S = \tan \angle P \times R$

Fig. 9-3 Calculation of pipe bends (*Courtesy Crane Company.*)

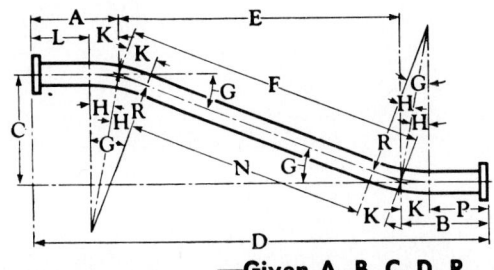

—Given A, B, C, D, R

$E = D - A - B$
$F = \sqrt{E^2 + C^2}$
$\dfrac{C}{F} = \sin \angle G$

$\angle H = \tfrac{1}{2} \angle G$
$K = \tan \angle H \times R$
$L = A - K$
$P = B - K$
$N = F - 2K$

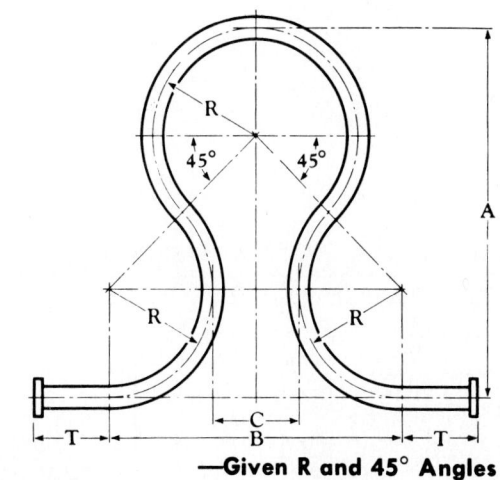

—Given R and 45° Angles

$A = 3.414 \times R$
$B = 2.828 \times R$
$C = 0.828 \times R$

T = Tangent
Length of pipe in bend =
$9.425 \times R + 2T$

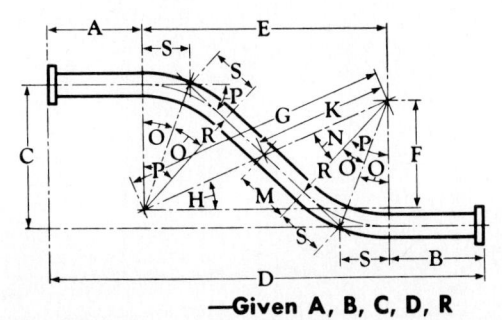

—Given A, B, C, D, R

$E = D - A - B$
$F = 2R - C$
$G = \sqrt{E^2 + F^2}$
$\dfrac{F}{E} = \tan \angle H$
$K = \tfrac{1}{2} G$

$M = \sqrt{K^2 - R^2}$
$\dfrac{M}{K} = \sin \angle N$
$\angle P = 90° - \angle H - \angle N$
$\angle O = \tfrac{1}{2} \angle P$
$S = \tan \angle O \times R$

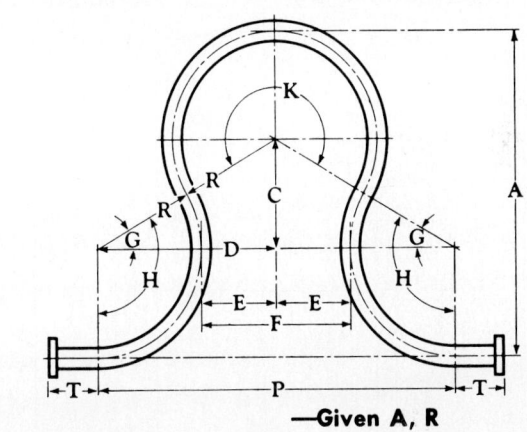

—Given A, R

$C = A - 2R$
$D = \sqrt{(2R)^2 - C^2}$
$E = D - R$
$F = 2E$

$P = 2D$
$C/2R = \sin \angle G$
$\angle H = 90° + \angle G$
$\angle K = 180° + 2 \angle G$

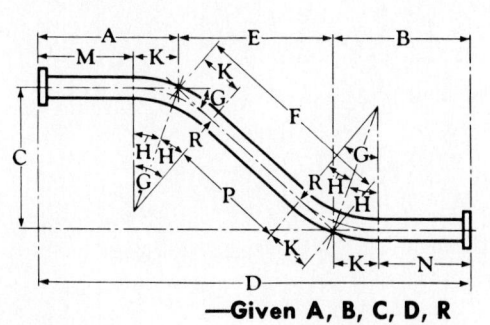

—Given A, B, C, D, R

$E = D - A - B$
$F = \sqrt{C^2 + E^2}$
$\dfrac{C}{F} = \sin \angle G$

$\angle H = \tfrac{1}{2} \angle G$
$K = \tan \angle H \times R$
$M = A - K$
$N = B - K$
$P = F - 2K$

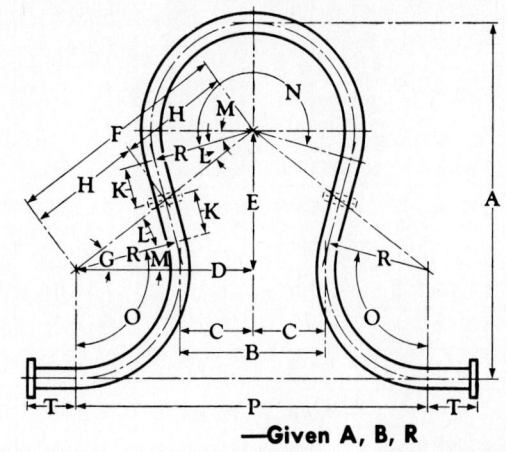

—Given A, B, R

$C = \tfrac{1}{2} B$
$D = R + C$
$E = A - 2R$
$F = \sqrt{D^2 + E^2}$

$E/F = \sin \angle G$
$H = \tfrac{1}{2} F$
$K = \sqrt{H^2 - R^2}$
$K/H = \sin \angle L$

$\angle M = \angle G - \angle L$
$\angle N = 180° + 2 \angle M$
$\angle O = 90° + \angle M$
$P = 2D$

Fig. 9-3 (contd.)

BENDS

101

TABLE 9-3 CENTER-TO-END (CE), BACK CENTER-TO-END (B/CE), AND ARC LENGTH FOR 30-DEGREE BENDS OF VARYING RADII AND PIPE SIZES (*COURTESY TEXAS PIPE BENDING COMPANY*.)

Rad.	Feet — Inches Arc	(CE)	(B/CE) Nominal Pipe Sizes (inches) 2	3	4	5	6	8	10	12
6	3-1/8	1-5/8	1-15/16							
7	3-11/16	1-7/8	2-3/16							
8	4-3/16	2-1/8	2-7/16							
9	4-11/16	2-7/16	2-3/4	2-15/16						
10	5-1/4	2-11/16	3	3-3/16						
11	5-3/4	2-15/16	3-1/4	3-7/16						
1 — 0	6-5/16	3-3/16	3-1/2	3-11/16	3-13/16					
1 — 1	6-13/16	3-1/2	3-13/16	4	4-1/8					
1 — 2	7-5/16	3-3/4	4-1/16	4-1/4	4-3/8					
1 — 3	7-7/8	4	4-5/16	4-1/2	4-5/8	4-3/4				
1 — 4	8-3/8	4-5/16	4-5/8	4-13/16	4-15/16	5-1/16				
1 — 5	8-7/8	4-9/16	4-7/8	5-1/16	5-3/16	5-5/16				
1 — 6	9-7/16	4-13/16	5-1/8	5-5/16	5-7/16	5-9/16	5-11/16			
1 — 7	9-15/16	5-1/16	5-3/8	5-9/16	5-11/16	5-13/16	5-15/16			
1 — 8	10-1/2	5-3/8	5-11/16	5-7/8	6	6-1/8	6-1/4			
1 — 9	11	5-5/8	5-15/16	6-1/8	6-1/4	6-3/8	6-1/2			
1 — 10	11-1/2	5-7/8	6-3/16	6-3/8	6-1/2	6-5/8	6-3/4			
1 — 11	1 — 0-1/16	6-3/16	6-1/2	6-11/16	6-13/16	6-15/16	7-1/16			
2 — 0	1 — 0-9/16	6-7/16	6-3/4	6-15/16	7-1/16	7-3/16	7-5/16	7-9/16		
2 — 1	1 — 1-1/16	6-11/16	7	7-3/16	7-5/16	7-7/16	7-9/16	7-13/16		
2 — 2	1 — 1-5/8	6-15/16	7-1/4	7-7/16	7-9/16	7-11/16	7-13/16	8-1/16		
2 — 3	1 — 2-1/8	7-1/4	7-9/16	7-3/4	7-7/8	8	8-1/8	8-3/8		
2 — 4	1 — 2-5/8	7-1/2	7-13/16	8	8-1/8	8-1/4	8-3/8	8-5/8		
2 — 5	1 — 3-3/16	7-3/4	8-1/16	8-1/4	8-3/8	8-1/2	8-5/8	8-7/8		
2 — 6	1 — 3-11/16	8-1/16	8-3/8	8-9/16	8-11/16	8-13/16	8-15/16	9-3/16	9-1/2	
2 — 7	1 — 4-1/4	8-5/16	8-5/8	8-13/16	8-15/16	9-1/16	9-3/16	9-7/16	9-3/4	
2 — 8	1 — 4-3/4	8-9/16	8-7/8	9-1/16	9-3/16	9-5/16	9-7/16	9-11/16	10	
2 — 9	1 — 5-1/4	8-13/16	9-1/8	9-5/16	9-7/16	9-9/16	9-11/16	9-15/16	10-1/4	
2 — 10	1 — 5-13/16	9-1/8	9-7/16	9-5/8	9-3/4	9-7/8	10	10-1/4	10-9/16	
2 — 11	1 — 6-5/16	9-3/8	9-11/16	9-7/8	10	10-1/8	10-1/4	10-1/2	10-13/16	
3 — 0	1 — 6-7/8	9-5/8	9-15/16	10-1/8	10-1/4	10-3/8	10-1/2	10-3/4	11-1/16	11-5/16
3 — 1	1 — 7-3/8	9-15/16	10-1/4	10-7/16	10-9/16	10-11/16	10-13/16	11-1/16	11-3/8	11-5/8
3 — 2	1 — 7-7/8	10-3/16	10-1/2	10-11/16	10-13/16	10-15/16	11-1/16	11-5/16	11-5/8	11-7/8
3 — 3	1 — 8-7/16	10-7/16	10-3/4	10-15/16	11-1/16	11-3/16	11-5/16	11-9/16	11-7/8	1 — 0-1/8

TABLE 9-3 (*CONTD.*)

Feet – Inches			(B/CE) Nominal Pipe Sizes							
Rad.	Arc	(CE)	2	3	4	5	6	8	10	12
3 – 4	1 – 8-15/16	10-11/16	11	11-3/16	11-5/16	11-7/16	11-9/16	11-13/16	1 – 0-1/8	1 – 0-3/8
3 – 5	1 – 9-1/2	11	11-5/16	11-1/2	11-5/8	11-3/4	11-7/8	1 – 0-1/8	1 – 0-7/16	1 – 0-11/16
3 – 6	1 – 10	11-1/4	11-9/16	11-3/4	11-7/8	1 – 0	1 – 0-1/8	1 – 0-3/8	1 – 0-11/16	1 – 0-15/16
3 – 7	1 – 10-1/2	11-1/2	11-13/16	1 – 0	1 – 0-1/8	1 – 0-1/4	1 – 0-3/8	1 – 0-5/8	1 – 0-15/16	1 – 1-3/16
3 – 8	1 – 11-1/16	11-13/16	1 – 0-1/8	1 – 0-5/16	1 – 0-7/16	1 – 0-9/16	1 – 0-11/16	1 – 0-15/16	1 – 1-1/4	1 – 1-1/2
3 – 9	1 – 11-9/16	1 – 0-1/16	1 – 0-3/8	1 – 0-9/16	1 – 0-11/16	1 – 0-13/16	1 – 0-15/16	1 – 1-3/16	1 – 1-1/2	1 – 1-3/4
3 – 10	2 – 0-1/8	1 – 0-5/16	1 – 0-5/8	1 – 0-13/16	1 – 0-15/16	1 – 1-1/16	1 – 1-3/16	1 – 1-7/16	1 – 1-3/4	1 – 2
3 – 11	2 – 0-5/8	1 – 0-5/8	1 – 0-15/16	1 – 1-1/8	1 – 1-1/4	1 – 1-3/8	1 – 1-1/2	1 – 1-3/4	1 – 2-1/16	1 – 2-5/16
4 – 0	2 – 1-1/8	1 – 0-7/8	1 – 1-3/16	1 – 1-3/8	1 – 1-1/2	1 – 1-5/8	1 – 1-3/4	1 – 2	1 – 2-5/16	1 – 2-9/16
4 – 1	2 – 1-11/16	1 – 1-1/8	1 – 1-7/16	1 – 1-5/8	1 – 1-3/4	1 – 1-7/8	1 – 2	1 – 2-1/4	1 – 2-9/16	1 – 2-13/16
4 – 2	2 – 2-3/16	1 – 1-3/8	1 – 1-11/16	1 – 1-7/8	1 – 2	1 – 2-1/8	1 – 2-1/4	2 – 2-1/2	1 – 2-13/16	1 – 3-1/16
4 – 3	2 – 2-11/16	1 – 1-11/16	1 – 2	1 – 2-3/16	1 – 2-5/16	1 – 2-7/16	1 – 2-9/16	1 – 2-13/16	1 – 3-1/8	1 – 3-3/8
4 – 4	2 – 3-1/4	1 – 1-15/16	1 – 2-1/4	1 – 2-7/16	1 – 2-9/16	1 – 2-11/16	1 – 2-13/16	1 – 3-1/16	1 – 3-3/8	1 – 3-5/8
4 – 5	2 – 3-3/4	1 – 2-3/16	1 – 2-1/2	1 – 2-11/16	1 – 2-13/16	1 – 2-15/16	1 – 3-1/16	1 – 3-5/16	1 – 3-5/8	1 – 3-7/8
4 – 6	2 – 4-1/4	1 – 2-1/2	1 – 2-13/16	1 – 3	1 – 3-1/8	1 – 3-1/4	1 – 3-3/8	1 – 3-5/8	1 – 3-15/16	1 – 4-3/16
4 – 7	2 – 4-13/16	1 – 2-3/4	1 – 3-1/16	1 – 3-1/4	1 – 3-3/8	1 – 3-1/2	1 – 3-5/8	1 – 3-7/8	1 – 4-3/16	1 – 4-7/16
4 – 8	2 – 5-5/16	1 – 3	1 – 3-5/16	1 – 3-1/2	1 – 3-5/8	1 – 3-3/4	1 – 3-7/8	1 – 4-1/8	1 – 4-7/16	1 – 4-11/16
4 – 9	2 – 5-7/8	1 – 3-1/4	1 – 3-9/16	1 – 3-3/4	1 – 3-7/8	1 – 4	1 – 4-1/8	1 – 4-3/8	1 – 4-11/16	1 – 4-15/16
4 – 10	2 – 6-3/8	1 – 3-9/16	1 – 3-7/8	1 – 4-1/16	1 – 4-3/16	1 – 4-1/4	1 – 4-3/8	1 – 4-5/8	1 – 4-15/16	1 – 5-3/16
4 – 11	2 – 6-7/8	1 – 3-13/16	1 – 4-1/8	1 – 4-5/16	1 – 4-7/16	1 – 4-9/16	1 – 4-11/16	1 – 4-15/16	1 – 5-1/4	1 – 5-1/2
5 – 0	2 – 7-7/16	1 – 4-1/16	1 – 4-3/8	1 – 4-9/16	1 – 4-11/16	1 – 4-13/16	1 – 4-15/16	1 – 5-3/16	1 – 5-1/2	1 – 5-3/4
5 – 1	2 – 7-13/16	1 – 4-3/8	1 – 4-11/16	1 – 4-7/8	1 – 5	1 – 5-1/8	1 – 5-1/4	1 – 5-1/2	1 – 5-13/16	1 – 6-1/16
5 – 2	2 – 8-7/16	1 – 4-5/8	1 – 4-15/16	1 – 5-1/8	1 – 5-1/4	1 – 5-3/8	1 – 5-1/2	1 – 5-3/4	1 – 6-1/16	1 – 6-5/16
5 – 3	2 – 9	1 – 4-7/8	1 – 5-3/16	1 – 5-3/8	1 – 5-1/2	1 – 5-5/8	1 – 5-3/4	1 – 6	1 – 6-5/16	1 – 6-9/16
5 – 4	2 – 9-1/2	1 – 5-1/8	1 – 5-7/16	1 – 5-5/8	1 – 5-3/4	1 – 5-7/8	1 – 6	1 – 6-1/4	1 – 6-9/16	1 – 6-13/16
5 – 5	2 – 10-1/16	1 – 5-7/16	1 – 5-3/4	1 – 5-15/16	1 – 6-1/16	1 – 6-3/16	1 – 6-5/16	1 – 6-9/16	1 – 6-7/8	1 – 7-1/8
5 – 6	2 – 10-9/16	1 – 5-11/16	1 – 6	1 – 6-3/16	1 – 6-5/16	1 – 6-7/16	1 – 6-9/16	1 – 6-13/16	1 – 7-1/8	1 – 7-3/8
5 – 7	2 – 11-1/16	1 – 5-15/16	1 – 6-1/4	1 – 6-7/16	1 – 6-9/16	1 – 6-11/16	1 – 6-13/16	1 – 7-1/16	1 – 7-3/8	1 – 7-5/8
5 – 8	2 – 11-5/8	1 – 6-1/4	1 – 6-9/16	1 – 6-3/4	1 – 6-7/8	1 – 7	1 – 7-1/8	1 – 7-3/8	1 – 7-11/16	1 – 7-15/16
5 – 9	3 – 0-1/8	1 – 6-1/2	1 – 6-13/16	1 – 7	1 – 7-1/8	1 – 7-1/4	1 – 7-3/8	1 – 7-5/8	1 – 7-15/16	1 – 8-3/16
5 – 10	3 – 0-5/8	1 – 6-3/4	1 – 7-1/16	1 – 7-1/4	1 – 7-3/8	1 – 7-1/2	1 – 7-5/8	1 – 7-7/8	1 – 8-3/16	1 – 8-7/16
5 – 11	3 – 1-3/16	1 – 7	1 – 7-5/16	1 – 7-1/2	1 – 7-5/8	1 – 7-3/4	1 – 7-7/8	1 – 8-1/8	1 – 8-7/16	1 – 8-11/16
6 – 0	3 – 1-11/16	1 – 7-5/16	1 – 7-5/8	1 – 7-13/16	1 – 7-15/16	1 – 8-1/16	1 – 8-3/16	1 – 8-7/16	1 – 8-3/4	1 – 9

BENDS

TABLE 9-4 CENTER-TO-END (CE), BACK CENTER-TO-END (B/CE), AND ARC LENGTH FOR 45-DEGREE BENDS OF VARYING RADII AND PIPE SIZES *(COURTESY TEXAS PIPE BENDING COMPANY.)*

Rad.	Feet — Inches Arc	CE	2	3	4	5	6	8	10	12
6	4-11/16	2-1/2	3							
7	5-1/2	2-7/8	3-3/8							
8	6-5/16	3-5/16	3-13/16							
9	7-1/16	3-3/4	4-1/4	4-1/2						
10	7-7/8	4-1/8	4-5/8	4-7/8						
11	8-5/8	4-9/16	5-1/16	5-5/16	5-15/16					
1—0	9-7/16	5	5-1/2	5-3/4	6-5/16					
1—1	10-3/16	5-3/8	5-7/8	6-1/8	6-3/4					
1—2	11	5-13/16	6-5/16	6-9/16	7-1/8	7-5/16				
1—3	11-3/4	6-3/16	6-11/16	6-15/16	7-3/8	7-3/4				
1—4	1—0-9/16	6-5/8	7-1/8	7-3/8	7-9/16	8-3/16				
1—5	1—1-3/8	7-1/16	7-9/16	7-13/16	8	8-9/16	8-13/16			
1—6	1—2-1/8	7-7/16	7-15/16	8-3/16	8-3/8	9	9-1/4			
1—7	1—2-15/16	7-7/8	8-3/8	8-5/8	8-13/16	9-7/16	9-11/16			
1—8	1—3-11/16	8-5/16	8-13/16	9-1/16	9-1/4	9-13/16	10-1/16			
1—9	1—4-1/2	8-11/16	9-3/16	9-7/16	9-5/8	10-1/4	10-1/2			
1—10	1—5-1/4	9-1/8	9-5/8	9-7/8	10-1/16	10-11/16	10-15/16	11-3/4		
1—11	1—6-1/16	9-1/2	10	10-1/4	10-7/16	11-1/16	11-5/16	11-3/4		
2—0	1—6-7/8	9-15/16	10-7/16	10-11/16	10-7/8	11-1/2	11-3/4	1—0-3/16		
2—1	1—7-5/8	10-3/8	10-7/8	11-1/8	11-5/16	11-7/8	1—0-1/8	1—0-9/16		
2—2	1—8-7/16	10-3/4	11-1/4	11-1/2	11-11/16	1—0-5/16	1—0-9/16	1—1		
2—3	1—9-3/16	11-3/16	11-11/16	11-15/16	1—0-1/8	1—0-3/4	1—1	1—1-7/16		
2—4	1—10	11-5/8	1—0-1/8	1—0-3/8	1—0-9/16	1—1-1/8	1—1-3/8	1—1-13/16		
2—5	1—10-3/4	1—0	1—0-1/2	1—0-3/4	1—0-15/16	1—1-9/16	1—1-13/16	1—2-1/4	1—2-11/16	
2—6	1—11-9/16	1—0-7/16	1—0-15/16	1—1-3/16	1—1-3/8	1—1-15/16	1—2-3/16	1—2-5/8	1—3-1/16	
2—7	2—0-3/8	1—0-13/16	1—1-5/16	1—1-9/16	1—1-3/4	1—2-3/8	1—2-5/8	1—3-1/16	1—3-1/2	
2—8	2—1-1/8	1—1-1/4	1—1-3/4	1—2	1—2-3/16	1—2-13/16	1—3-1/16	1—3-1/2	1—3-15/16	
2—9	2—1-15/16	1—1-11/16	1—2-3/16	1—2-7/16	1—2-5/8	1—3-3/16	1—3-7/16	1—3-7/8	1—4-5/16	
2—10	2—2-11/16	1—2-1/16	1—2-9/16	1—2-13/16	1—3	1—3-5/8	1—3-7/8	1—4-5/16	1—4-3/4	
2—11	2—3-1/2	1—2-1/2	1—3	1—3-1/4	1—3-7/16	1—4-1/16	1—4-5/16	1—4-3/4	1—5-3/16	
3—0	2—4-1/4	1—2-15/16	1—3-7/16	1—3-11/16	1—3-7/8	1—4-7/16	1—4-11/16	1—5-1/8	1—5-9/16	1—5-9/16
3—1	2—5-1/16	1—3-5/16	1—3-13/16	1—4-1/16	1—4-1/4	1—4-7/8	1—5-1/8	1—5-9/16	1—6	1—5-15/16
3—2	2—5-7/8	1—3-3/4	1—4-1/4	1—4-1/2	1—4-11/16	1—5-1/4	1—5-1/2	1—5-15/16	1—6-3/8	1—6-3/8
3—3	2—6-5/8	1—4-1/8	1—4-5/8	1—4-7/8	1—5-1/16					1—6-3/4

104

TABLE 9-4 (CONTD.)

| Rad. | Feet – Inches Arc | (CE) | \multicolumn{8}{c}{(B/CE) Nominal Pipe Sizes} |
			2	3	4	5	6	8	10	12
3 – 4	2 – 7-7/16	1 – 4-9/16	1 – 5-1/16	1 – 5-5/16	1 – 5-1/2	1 – 5-11/16	1 – 5-15/16	1 – 6-3/8	1 – 6-13/16	1 – 7-3/16
3 – 5	2 – 8-3/16	1 – 5	1 – 5-1/2	1 – 5-3/4	1 – 5-15/16	1 – 6-1/8	1 – 6-3/8	1 – 6-13/16	1 – 7-1/4	1 – 7-5/8
3 – 6	2 – 9	1 – 5-3/8	1 – 5-7/8	1 – 6-1/8	1 – 6-5/16	1 – 6-1/2	1 – 6-3/4	1 – 7-3/16	1 – 7-5/8	1 – 8
3 – 7	2 – 9-3/4	1 – 5-13/16	1 – 6-5/16	1 – 6-9/16	1 – 6-3/4	1 – 6-15/16	1 – 7-3/16	1 – 7-5/8	1 – 8-1/16	1 – 8-7/16
3 – 8	2 – 10-9/16	1 – 6-1/4	1 – 6-3/4	1 – 7	1 – 7-3/16	1 – 7-3/8	1 – 7-5/8	1 – 8-1/16	1 – 8-1/2	1 – 8-7/8
3 – 9	2 – 11-5/16	1 – 6-5/8	1 – 7-1/8	1 – 7-3/8	1 – 7-9/16	1 – 7-3/4	1 – 8	1 – 8-7/8	1 – 8-7/8	1 – 9-1/4
3 – 10	3 – 0-1/8	1 – 7-1/16	1 – 7-9/16	1 – 7-13/16	1 – 8	1 – 8-3/16	1 – 8-7/16	1 – 8-7/8	1 – 9-5/16	1 – 9-11/16
3 – 11	3 – 0-15/16	1 – 7-7/16	1 – 7-15/16	1 – 8-3/16	1 – 8-3/8	1 – 8-9/16	1 – 8-13/16	1 – 9-1/4	1 – 9-11/16	1 – 10-1/16
4 – 0	3 – 1-11/16	1 – 7-7/8	1 – 8-3/8	1 – 8-5/8	1 – 8-13/16	1 – 9	1 – 9-1/4	1 – 9-11/16	1 – 10-1/8	1 – 10-1/2
4 – 1	3 – 2-1/2	1 – 8-5/16	1 – 8-13/16	1 – 9-1/16	1 – 9-1/4	1 – 9-7/16	1 – 9-11/16	1 – 10-1/8	1 – 10-9/16	1 – 10-15/16
4 – 2	3 – 3-1/4	1 – 8-11/16	1 – 9-3/16	1 – 9-7/16	1 – 9-5/8	1 – 9-13/16	1 – 10-1/16	1 – 10-1/2	1 – 1-015/16	1 – 11-5/16
4 – 3	3 – 4-1/16	1 – 9-1/8	1 – 9-5/8	1 – 9-7/8	1 – 10-1/16	1 – 10-1/4	1 – 10-1/2	1 – 10-15/16	1 – 11-3/8	1 – 11-3/4
4 – 4	3 – 4-13/16	1 – 9-9/16	1 – 10-1/16	1 – 10-5/16	1 – 10-1/2	1 – 10-11/16	1 – 10-15/16	1 – 11-3/8	1 – 11-13/16	2 – 0-3/16
4 – 5	3 – 5-5/8	1 – 9-15/16	1 – 10-7/16	1 – 10-11/16	1 – 10-7/8	1 – 11-1/16	1 – 11-5/16	1 – 11-3/4	2 – 0-3/16	2 – 0-9/16
4 – 6	3 – 6-7/16	1 – 10-3/8	1 – 10-7/8	1 – 11-1/8	1 – 11-5/16	1 – 11-1/2	1 – 11-3/4	2 – 0-3/16	2 – 0-5/8	2 – 1
4 – 7	3 – 7-3/16	1 – 10-13/16	1 – 11-5/16	1 – 11-9/16	1 – 11-3/4	1 – 11-15/16	2 – 0-3/16	2 – 0-5/8	2 – 1-1/16	2 – 1-7/16
4 – 8	3 – 8	1 – 11-3/16	1 – 11-11/16	1 – 11-15/16	2 – 0-1/8	2 – 0-5/16	2 – 0-9/16	2 – 1	2 – 1-7/16	2 – 1-13/16
4 – 9	3 – 8-3/4	1 – 11-5/8	2 – 0-1/8	2 – 0-3/8	2 – 0-9/16	2 – 0-3/4	2 – 1	2 – 1-7/16	2 – 1-7/8	2 – 2-1/2
4 – 10	3 – 9-9/16	2 – 0	2 – 0-1/2	2 – 0-3/4	2 – 0-15/16	2 – 1-1/8	2 – 1-3/8	2 – 1-13/16	2 – 2-1/4	2 – 2-5/8
4 – 11	3 – 10-5/16	2 – 0-7/16	2 – 0-15/16	2 – 1-3/16	2 – 1-3/8	2 – 1-9/16	2 – 1-13/16	2 – 2-1/4	2 – 2-11/16	2 – 3-1/16
5 – 0	3 – 11-1/8	2 – 0-7/8	2 – 1-3/8	2 – 1-5/8	2 – 1-13/16	2 – 2	2 – 2-1/4	2 – 2-11/16	2 – 3-1/8	2 – 3-1/2
5 – 1	3 – 11-15/16	2 – 1-1/4	2 – 1-3/4	2 – 2	2 – 2-3/16	2 – 2-3/8	2 – 2-5/8	2 – 3-1/16	2 – 3-1/2	2 – 3-7/8
5 – 2	4 – 0-11/16	2 – 1-11/16	2 – 2-3/16	2 – 2-7/16	2 – 2-5/8	2 – 2-13/16	2 – 3-1/16	2 – 3-1/2	2 – 3-15/16	2 – 4-5/16
5 – 3	4 – 1-1/2	2 – 2-1/8	2 – 2-5/8	2 – 2-7/8	2 – 3-1/16	2 – 3-1/4	2 – 3-1/2	2 – 3-15/16	2 – 4-3/8	2 – 4-3/4
5 – 4	4 – 2-1/4	2 – 2-1/2	2 – 3	2 – 3-1/4	2 – 3-7/16	2 – 3-5/8	2 – 3-7/8	2 – 4-5/16	2 – 4-3/4	2 – 5-1/8
5 – 5	4 – 3-1/16	2 – 2-15/16	2 – 3-7/16	2 – 3-11/16	2 – 3-7/8	2 – 4-1/16	2 – 4-5/16	2 – 4-3/4	2 – 5-3/16	2 – 5-9/16
5 – 6	4 – 3-13/16	2 – 3-5/16	2 – 3-13/16	2 – 4-1/16	2 – 4-1/4	2 – 4-7/16	2 – 4-11/16	2 – 5-1/8	2 – 5-9/16	2 – 6-15/16
5 – 7	4 – 4-5/8	2 – 3-3/4	2 – 4-1/4	2 – 4-1/2	2 – 4-11/16	2 – 4-7/8	2 – 5-1/8	2 – 5-9/16	2 – 6	2 – 6-3/8
5 – 8	4 – 5-7/16	2 – 4-3/16	2 – 4-11/16	2 – 4-15/16	2 – 5-1/8	2 – 5-5/16	2 – 5-9/16	2 – 6	2 – 6-7/16	2 – 6-13/16
5 – 9	4 – 6-3/16	2 – 4-9/16	2 – 5-1/16	2 – 5-5/16	2 – 5-1/2	2 – 5-11/16	2 – 5-15/16	2 – 6-3/8	2 – 6-13/16	2 – 7-3/16
5 – 10	4 – 7	2 – 5	2 – 5-1/2	2 – 5-3/4	2 – 5-15/16	2 – 6-1/8	2 – 6-3/8	2 – 6-13/16	2 – 7-1/4	2 – 7-5/8
5 – 11	4 – 7-3/4	2 – 5-3/8	2 – 5-7/8	2 – 6-1/8	2 – 6-5/16	2 – 6-1/2	2 – 6-3/4	2 – 7-3/16	2 – 7-5/8	2 – 8
6 – 0	4 – 8-9/16	2 – 5-13/16	2 – 6-5/16	2 – 6-9/16	2 – 6-3/4	2 – 6-15/16	2 – 7-3/16	2 – 7-5/8	2 – 8-1/16	2 – 8-7/16

TABLE 9-5 CENTER-TO-END (CE), BACK CENTER-TO-END (B/CE), AND ARC LENGTH FOR 60-DEGREE BENDS OF VARYING RADII AND PIPE SIZES *(COURTESY TEXAS PIPE BENDING COMPANY.)*

Rad.	Feet – Inches Arc	(CE)	2	3	(B/CE) Nominal Pipe Sizes 4	5	6	8	10	12
6	6-5/16	3-7/16	4-1/8							
7	7-5/16	4-1/16	4-3/4							
8	8-3/8	4-5/8	5-5/16							
9	9-7/16	5-3/16	5-7/8	6-3/16						
10	10-1/2	5-3/4	6-7/16	6-3/4						
11	11-1/2	6-3/8	7-1/16	7-3/8						
1 – 0	1 – 0-9/16	6-15/16	7-5/8	7-15/16	8-1/4					
1 – 1	1 – 1-5/8	7-1/2	8-3/16	8-1/2	8-13/16					
1 – 2	1 – 2-5/8	8-1/16	8-3/4	9-1/16	9-3/8					
1 – 3	1 – 3-11/16	8-11/16	9-3/8	9-11/16	10					
1 – 4	1 – 4-3/4	9-1/4	9-15/16	10-1/4	10-9/16	10-5/16				
1 – 5	1 – 5-13/16	9-13/16	10-1/2	10-13/16	11-1/8	10-7/8				
1 – 6	1 – 6-7/8	10-3/8	11-1/16	11-3/8	11-11/16	11-7/16				
1 – 7	1 – 7-7/8	11	11-11/16	1 – 0	1 – 0-5/16	1 – 0	1 – 0-5/16			
1 – 8	1 – 8-15/16	11-9/16	1 – 0-1/4	1 – 0-9/16	1 – 0-7/8	1 – 0-5/8	1 – 0-15/16			
1 – 9	1 – 10	1 – 0-1/8	1 – 0-13/16	1 – 1-1/8	1 – 1-7/16	1 – 1-3/16	1 – 1-1/2			
1 – 10	1 – 11-1/16	1 – 0-11/16	1 – 1-3/8	1 – 1-11/16	1 – 2	1 – 1-3/4	1 – 2-1/16			
1 – 11	2 – 0-1/8	1 – 1-1/4	1 – 1-15/16	1 – 2-1/4	1 – 2-9/16	1 – 2-5/16	1 – 2-5/8			
2 – 0	2 – 1-1/8	1 – 1-7/8	1 – 2-9/16	1 – 2-7/8	1 – 3-3/16	1 – 2-7/8	1 – 3-3/16			
2 – 1	2 – 2-3/16	1 – 2-7/16	1 – 3-1/16	1 – 3-7/16	1 – 3-3/4	1 – 3-1/2	1 – 3-13/16	1 – 4-3/8		
2 – 2	2 – 3-1/4	1 – 3	1 – 3-11/16	1 – 4	1 – 4-5/16	1 – 4-1/16	1 – 4-3/8	1 – 4-15/16		
2 – 3	2 – 4-1/4	1 – 3-9/16	1 – 4-1/4	1 – 4-9/16	1 – 4-7/8	1 – 4-5/8	1 – 4-15/16	1 – 5-1/2		
2 – 4	2 – 5-5/16	1 – 4-3/16	1 – 4-7/8	1 – 5-3/16	1 – 5-1/2	1 – 5-3/16	1 – 5-1/2	1 – 6-1/16		
2 – 5	2 – 6-3/8	1 – 4-3/4	1 – 5-7/16	1 – 5-3/4	1 – 6-1/16	1 – 5-13/16	1 – 6-1/8	1 – 6-11/16		
2 – 6	2 – 7-7/16	1 – 5-5/16	1 – 6	1 – 6-5/16	1 – 6-5/8	1 – 6-3/8	1 – 6-11/16	1 – 7-1/4		
2 – 7	2 – 8-7/16	1 – 5-7/8	1 – 6-9/16	1 – 6-7/8	1 – 7-3/16	1 – 6-15/16	1 – 7-1/4	1 – 7-13/16	1 – 8-7/16	
2 – 8	2 – 9-1/2	1 – 6-1/2	1 – 7-3/16	1 – 7-1/2	1 – 7-13/16	1 – 7-1/2	1 – 7-13/16	1 – 8-3/8	1 – 9	
2 – 9	2 – 10-9/16	1 – 7-1/16	1 – 7-3/4	1 – 8-1/16	1 – 8-3/8	1 – 8-1/8	1 – 8-7/16	1 – 9	1 – 9-5/8	
2 – 10	2 – 11-5/8	1 – 7-5/8	1 – 8-5/16	1 – 8-5/8	1 – 8-15/16	1 – 8-11/16	1 – 9	1 – 9-9/16	1 – 10-3/16	
2 – 11	3 – 0-5/8	1 – 8-3/16	1 – 8-7/8	1 – 9-3/16	1 – 9-1/2	1 – 9-1/4	1 – 9-9/16	1 – 10-1/8	1 – 10-3/4	
3 – 0	3 – 1-11/16	1 – 8-13/16	1 – 9-1/2	1 – 9-13/16	1 – 10-1/8	1 – 9-13/16	1 – 10-1/8	1 – 10-11/16	1 – 11-5/16	2 – 0-1/2
3 – 1	3 – 2-3/4	1 – 9-3/8	1 – 10-1/16	1 – 10-3/8	1 – 10-11/16	1 – 10-7/8	1 – 10-3/4	1 – 11-5/16	1 – 11-15/16	2 – 1-1/16
3 – 2	3 – 3-13/16	1 – 9-15/16	1 – 10-5/8	1 – 10-15/16	1 – 11-1/4	1 – 11	1 – 11-5/16	1 – 11-7/8	2 – 0-1/2	2 – 1-5/8
3 – 3	3 – 4-13/16	1 – 10-1/2	1 – 11-3/16	1 – 11-1/2	1 – 11-13/16	1 – 11-9/16	1 – 11-7/8	2 – 0-7/16	2 – 1-1/16	2 – 2-3/16

TABLE 9-5 (CONTD).

Rad.	Feet — Inches Arc	(CE)	2	3	(B/CE) Nominal Pipe Sizes 4	5	6	8	10	12
3 – 4	3 – 5-7/8	1 – 11-1/8	1 – 11-13/16	2 – 0-1/8	2 – 0-7/16	2 – 0-3/4	2 – 1-1/16	2 – 1-5/8	1 – 2-1/4	2 – 2-13/16
3 – 5	3 – 6-15/16	1 – 11-11/16	2 – 0-3/8	2 – 0-11/16	2 – 1	2 – 1-5/16	2 – 1-5/8	2 – 2-3/16	2 – 2-13/16	2 – 3-3/8
3 – 6	3 – 8	2 – 0-1/4	2 – 0-15/16	2 – 1-1/4	2 – 1-9/16	2 – 1-7/8	2 – 2-3/16	2 – 2-3/4	2 – 3-3/8	2 – 3-15/16
3 – 7	3 – 9	2 – 0-13/16	2 – 1-1/2	2 – 1-13/16	2 – 2-1/8	2 – 2-7/16	2 – 2-3/4	2 – 3-5/16	2 – 3-15/16	2 – 4-1/2
3 – 8	3 – 10-1/16	2 – 1-3/8	2 – 2-1/16	2 – 2-3/8	2 – 2-11/16	2 – 3	2 – 3-5/16	2 – 3-7/8	2 – 4-1/2	2 – 5-1/16
3 – 9	3 – 11-1/8	2 – 2	2 – 2-11/16	2 – 3	2 – 3-5/16	2 – 3-5/8	2 – 3-15/16	2 – 4-1/2	2 – 5-1/8	2 – 5-11/16
3 – 10	4 – 0-3/16	2 – 2-9/16	2 – 3-1/4	2 – 3-9/16	2 – 3-7/8	2 – 4-3/16	2 – 4-1/2	2 – 5-1/16	2 – 5-11/16	2 – 6-1/4
3 – 11	4 – 1-3/16	2 – 3-1/8	2 – 3-13/16	2 – 4-1/8	2 – 4-7/16	2 – 4-3/4	2 – 5-1/16	2 – 5-5/8	2 – 6-1/4	2 – 6-13/16
4 – 0	4 – 2-1/4	2 – 3-11/16	2 – 4-3/8	2 – 4-11/16	2 – 5	2 – 5-5/16	2 – 5-5/8	2 – 6-3/16	2 – 6-13/16	2 – 7-3/8
4 – 1	4 – 3-5/16	2 – 4-5/16	2 – 5	2 – 5-5/16	2 – 5-5/8	2 – 5-15/16	2 – 6-1/4	2 – 6-13/16	2 – 7-7/16	2 – 8
4 – 2	4 – 4-3/8	2 – 4-7/8	2 – 5-9/16	2 – 5-7/8	2 – 6-3/16	2 – 6-1/2	2 – 6-13/16	2 – 7-3/8	2 – 8	2 – 8-9/16
4 – 3	4 – 5-7/16	2 – 5-7/16	2 – 6-1/8	2 – 6-7/16	2 – 6-3/4	2 – 7-1/16	2 – 7-3/8	2 – 7-15/16	2 – 8-9/16	2 – 9-1/8
4 – 4	4 – 6-7/16	2 – 6	2 – 6-11/16	2 – 7	2 – 7-5/16	2 – 7-5/8	2 – 7-15/16	2 – 8-1/2	2 – 9-1/8	2 – 9-11/16
4 – 5	4 – 7-1/2	2 – 6-5/8	2 – 7-5/16	2 – 7-5/8	2 – 7-15/16	2 – 8-1/4	2 – 8-9/16	2 – 9-1/8	2 – 9-3/4	2 – 10-5/16
4 – 6	4 – 8-9/16	2 – 7-3/16	2 – 7-7/8	2 – 8-3/16	2 – 8-1/2	2 – 8-13/16	2 – 9-1/8	2 – 9-11/16	2 – 10-5/16	2 – 10-7/8
4 – 7	4 – 9-5/8	2 – 7-3/4	2 – 8-7/16	2 – 8-3/4	2 – 9-1/16	2 – 9-3/8	2 – 9-11/16	2 – 10-1/4	2 – 10-7/8	2 – 11-7/16
4 – 8	4 – 10-5/8	2 – 8-5/16	2 – 9	2 – 9-5/16	2 – 9-5/8	2 – 9-15/16	2 – 10-1/4	2 – 10-13/16	2 – 11-7/16	3 – 0
4 – 9	4 – 11-11/16	2 – 8-15/16	2 – 9-5/8	2 – 9-15/16	2 – 10-1/4	2 – 10-9/16	2 – 10-7/8	2 – 11-7/16	2 – 0-1/16	3 – 0-5/8
4 – 10	5 – 0-3/4	2 – 9-1/2	2 – 10-3/16	2 – 10-1/2	2 – 10-13/16	2 – 11-1/8	2 – 11-7/16	3 – 0	3 – 0-5/8	3 – 1-3/16
4 – 11	5 – 1-13/16	2 – 10-1/16	2 – 10-3/4	2 – 11-1/16	2 – 11-3/8	2 – 11-11/16	3 – 0	3 – 0-9/16	3 – 1-3/16	3 – 1-3/4
5 – 0	5 – 2-13/16	2 – 10-5/8	2 – 11-5/16	2 – 11-5/8	2 – 11-15/16	3 – 0-1/4	3 – 0-9/16	3 – 1-1/8	3 – 1-3/4	3 – 2-5/16
5 – 1	5 – 3-7/8	2 – 11-1/4	2 – 11-15/16	3 – 0-1/4	3 – 0-9/16	3 – 0-7/8	3 – 1-3/16	3 – 1-3/4	3 – 2-3/8	3 – 2-15/16
5 – 2	5 – 4-15/16	2 – 11-13/16	3 – 0-1/2	3 – 0-13/16	3 – 1-1/8	3 – 1-7/16	3 – 1-3/4	3 – 2-5/16	3 – 2-15/16	3 – 3-1/2
5 – 3	5 – 6	3 – 0-3/8	3 – 1-1/16	3 – 1-3/8	3 – 1-11/16	3 – 2	3 – 2-5/16	3 – 2-7/8	3 – 3-1/2	3 – 4-1/16
5 – 4	5 – 7	3 – 0-15/16	3 – 1-5/8	3 – 1-15/16	3 – 2-1/4	3 – 2-9/16	3 – 2-7/8	3 – 3-7/16	3 – 4-1/16	3 – 4-5/8
5 – 5	5 – 8-1/16	3 – 1-1/2	3 – 2-3/16	3 – 2-1/2	3 – 2-13/16	3 – 3-1/8	3 – 3-7/16	3 – 4	3 – 4-5/8	3 – 5-3/16
5 – 6	5 – 9-1/8	3 – 2-1/8	3 – 2-13/16	3 – 3-1/8	3 – 3-7/16	3 – 3-3/4	3 – 4-1/16	3 – 4-5/8	3 – 5-1/4	3 – 5-13/16
5 – 7	5 – 10-3/16	3 – 2-11/16	3 – 3-3/8	3 – 3-11/16	3 – 4	3 – 4-5/16	3 – 4-5/8	3 – 5-3/16	3 – 5-13/16	3 – 6-3/8
5 – 8	5 – 11-3/16	3 – 3-1/4	3 – 3-15/16	3 – 4-1/4	3 – 4-9/16	3 – 4-7/8	3 – 5-3/16	3 – 5-3/4	3 – 6-3/8	3 – 6-15/16
5 – 9	6 – 0-1/4	3 – 3-13/16	3 – 4-1/2	3 – 4-13/16	3 – 5-1/8	3 – 5-7/16	3 – 5-3/4	3 – 6-5/16	3 – 6-15/16	3 – 7-1/2
5 – 10	6 – 1-5/16	3 – 4-7/16	3 – 5-1/8	3 – 5-7/16	3 – 5-3/4	3 – 6-1/16	3 – 6-3/8	3 – 6-15/16	3 – 7-9/16	3 – 8-1/8
5 – 11	6 – 2-3/8	3 – 5	3 – 5-11/16	3 – 6	3 – 6-5/16	3 – 6-5/8	3 – 6-15/16	3 – 7-1/2	3 – 8-1/8	3 – 8-11/16
6 – 0	6 – 3-3/8	3 – 5-9/16	3 – 6-1/4	3 – 6-9/16	3 – 6-7/8	3 – 7-3/16	3 – 7-1/2	3 – 8-1/16	3 – 8-11/16	3 – 9-1/4

107

TABLE 9-6 ARC FOR 90-DEGREE BENDS *(COURTESY TEXAS PIPE BENDING COMPANY.)*

Radius (inches)

Radius (feet)

	0 – 0	1 – 0	2 – 0	3 – 0	4 – 0	5 – 0	6 – 0	7 – 0	8 – 0		Rad.	Arc
0		1 – 6-7/8	3 – 1-11/16	4 – 8-9/16	6 – 3-3/8	7 – 10-1/4	9 – 5-1/8	10 – 11-15/16	12 – 6-13/16	0	1-1/4	1-15/16
1	1-9/16	1 – 8-7/16	3 – 3-1/4	4 – 10-1/8	6 – 5	7 – 11-13/16	9 – 6-11/16	11 – 1-1/2	12 – 8-3/8	1	1-1/2	2-3/8
2	3-1/8	1 – 10	3 – 4-13/16	4 – 11-11/16	6 – 6-9/16	8 – 1-3/8	9 – 8-1/4	11 – 3-1/16	12 – 9-15/16	2	1-3/4	2-3/4
3	4-11/16	1 – 11-9/16	3 – 6-7/16	5 – 1-1/4	6 – 8-1/8	8 – 2-15/16	9 – 9-13/16	11 – 4-11/16	12 – 11-1/2	3	2-1/4	3-9/16
4	6-5/16	2 – 1-1/8	3 – 8	5 – 2-13/16	6 – 9-11/16	8 – 4-1/2	9 – 11-3/8	11 – 6-1/4	13 – 1-1/16	4	2-1/2	3-15/16
5	7-7/8	2 – 2-11/16	3 – 9-9/16	5 – 4-3/8	6 – 11-1/4	8 – 6-1/8	10 – 0-15/16	11 – 7-13/16	13 – 2-5/8	5	2-3/4	4-5/16
6	9-7/16	2 – 4-1/4	3 – 11-1/8	5 – 6	7 – 0-13/16	8 – 7-11/16	10 – 2-1/2	11 – 9-3/8	13 – 4-1/8	6	3-1/4	5-1/8
7	11	2 – 5-7/8	4 – 0-11/16	5 – 7-9/16	7 – 2-3/8	8 – 9-1/4	10 – 4-1/16	11 – 10-15/16	13 – 5-13/16	7	3-1/2	5-1/2
8	1 – 0-9/16	2 – 7-7/16	4 – 2-1/4	5 – 9-1/8	7 – 3-15/16	8 – 10-13/16	10 – 5-11/16	12 – 0-1/2	13 – 7-3/8	8	3-3/4	5-7/8
9	1 – 2-1/8	2 – 9	4 – 3-13/16	5 – 10-11/16	7 – 5-9/16	9 – 0-3/8	10 – 7-1/4	12 – 2-1/16	13 – 8-15/16	9	4-1/4	6-11/16
10	1 – 3-11/16	2 – 10-9/16	4 – 5-7/16	6 – 0-1/4	7 – 7-1/8	9 – 1-15/16	10 – 8-13/16	12 – 3-5/8	13 – 10-1/2	10	4-1/2	7-1/16
11	1 – 5-1/4	3 – 0-1/8	4 – 7	6 – 1-13/16	7 – 8-11/16	9 – 3-1/2	10 – 10-3/8	12 – 5-1/4	14 – 0-1/16	11	4-3/4	7-7/16

	9 – 0	10 – 0	11 – 0	12 – 0	13 – 0	14 – 0	15 – 0	16 – 0	17 – 0		Rad.	Arc
0	14 – 1-5/8	15 – 8-1/2	17 – 3-3/8	18 – 10-3/16	20 – 5-1/16	21 – 11-7/8	23 – 6-3/4	25 – 1-5/8	26 – 8-7/16	0	5-1/4	8-1/4
1	14 – 3-3/16	15 – 10-1/16	17 – 4-15/16	18 – 11-3/4	20 – 6-5/8	22 – 1-7/16	23 – 8-5/16	25 – 3-3/16	26 – 10	1	5-1/2	8-5/8
2	14 – 4-13/16	15 – 11-5/8	17 – 6-1/2	19 – 1-5/16	20 – 8-3/16	22 – 3-1/16	23 – 9-7/8	25 – 4-3/4	26 – 11-9/16	2	5-3/4	9-1/16
3	14 – 6-3/8	16 – 1-3/16	17 – 8-1/16	19 – 2-15/16	20 – 9-3/4	22 – 4-5/8	23 – 11-7/16	25 – 6-5/16	27 – 1-1/8	3	6-1/2	10-3/16
4	14 – 7-15/16	16 – 2-3/4	17 – 9-5/8	19 – 4-1/2	20 – 11-5/16	22 – 6-3/16	24 – 1	25 – 7-7/8	27 – 2-3/4	4	7-1/2	11-3/4
5	14 – 9-1/2	16 – 4-3/8	17 – 11-3/16	19 – 6-1/16	21 – 0-7/8	22 – 7-3/4	24 – 2-5/8	25 – 9-7/16	27 – 4-5/16	5	8-1/2	13-3/8
6	14 – 11-1/16	16 – 5-15/16	18 – 0-3/4	19 – 7-5/8	21 – 2-1/2	22 – 9-5/16	24 – 4-3/16	25 – 11	27 – 5-7/8	6	9-1/2	14-15/16
7	15 – 0-5/8	16 – 7-1/2	18 – 2-5/16	19 – 9-3/16	21 – 4-1/16	22 – 10-7/8	24 – 5-3/4	26 – 0-9/16	27 – 7-7/16	7	10-1/2	16-1/2
8	15 – 2-3/16	16 – 9-1/16	18 – 3-15/16	19 – 10-3/4	21 – 5-5/8	23 – 0-7/16	24 – 7-5/16	26 – 2-3/16	27 – 9	8	11-1/2	18-1/16
9	15 – 3-13/16	16 – 10-5/8	18 – 5-1/2	20 – 0-5/16	21 – 7-3/16	23 – 2-1/16	24 – 8-7/8	26 – 3-3/4	27 – 10-9/16	9	12-1/2	19-5/8
10	15 – 5-3/8	17 – 0-3/16	18 – 7-1/16	20 – 1-7/8	21 – 8-3/4	23 – 3-5/8	24 – 10-7/16	26 – 5-5/16	28 – 0-1/8	10	13-1/2	21-3/16
11	15 – 6-15/16	17 – 1-3/4	18 – 8-5/8	20 – 3-1/2	21 – 10-5/16	23 – 5-3/16	25 – 0	26 – 6-7/8	28 – 1-3/4	11	14-1/2	22-3/4
											15-1/2	24-3/8

Arc (inches)

TABLE 9-7 TRUE ANGLE OF BENDS IN TWO PLANES (*COURTESY TEXAS PIPE BENDING COMPANY.*)

A\B	0°	5°	10°	15°	20°	25°	30°	35°	40°	45°	50°	55°	60°	65°	70°	75°	80°	85°	90°
0°	90°	90	90	90	90	90	90	90	90	90	90	90	90	90	90	90	90	90	90
5°	90	89-33	89-08	88-42	88-17	87-53	87-30	87-08	86-47	86-28	86-10	85-54	85-40	85-28	85-18	85-10	85-05	85-01	85-00
10°	90	89-08	88-46	87-25	86-36	85-47	85-01	84-17	83-36	82-57	82-21	81-50	81-21	80-56	80-37	80-21	80-09	80-02	80-00
15°	90	88-42	87-25	86-10	84-55	83-43	82-34	81-28	80-25	79-27	78-34	77-45	77-03	76-26	75-55	75-31	75-14	75-03	75-00
20°	90	88-17	86-36	81-55	83-17	81-41	80-09	78-41	77-18	76-00	74-49	73-44	72-46	71-56	71-15	70-42	70-19	70-05	70-00
25°	90	87-53	85-47	83-43	81-41	79-43	77-48	75-58	74-14	72-37	71-07	67-45	68-32	67-29	66-36	65-54	65-24	65-06	65-00
30°	90	87-30	85-01	82-34	80-09	77-48	75-31	73-20	71-15	69-18	67-29	65-49	61-20	63-03	61-59	61-07	60-30	60-08	60-00
35°	90	87-08	84-17	81-28	78-41	75-58	73-20	70-48	68-22	66-04	63-56	61-59	60-13	58-41	57-23	56-21	55-36	55-09	55-00
40°	90	86-47	83-36	80-25	77-18	74-14	71-15	68-22	65-36	52-58	60-30	58-14	56-10	54-22	52-50	51-37	50-44	50-11	50-00
45°	90	86-28	82-57	79-27	76-00	72-37	69-18	66-04	62-55	60-00	57-12	54-36	52-14	50-09	48-21	46-55	45-52	45-13	45-00
50°	90	86-10	82-21	78-34	74-49	71-07	67-29	63-56	60-30	57-12	54-04	51-08	48-26	46-02	43-58	42-16	41-01	40-16	40-00
55°	90	85-54	81-50	77-45	73-44	69-45	65-49	61-59	58-14	54-36	51-08	47-51	44-49	42-04	39-40	37-42	36-13	35-18	35-00
60°	90	85-40	81-21	77-03	72-46	68-32	64-20	60-13	56-10	52-14	48-26	44-49	41-25	38-17	35-32	33-14	31-29	30-23	30-00
65°	90	85-28	80-56	76-26	71-56	67-29	63-03	58-41	54-22	50-09	46-02	42-04	38-17	34-47	31-36	28-54	26-48	25-28	25-00
70°	90	85-18	80-37	75-55	71-15	66-36	61-59	57-23	52-50	48-21	43-58	39-40	35-32	31-36	27-52	24-48	22-16	20-35	20-00
75°	90	85-10	80-21	75-31	70-42	65-54	61-07	56-21	51-37	46-55	42-16	37-42	33-14	28-54	24-48	21-06	17-58	15-48	15-00
80°	90	85-05	80-09	75-14	70-19	65-24	60-30	55-36	50-44	45-52	41-01	36-13	31-29	26-48	22-16	17-58	14-66	11-11	10-00
85°	90	85-01	80-02	75-03	70-05	65-06	60-08	55-09	50-11	45-13	40-16	35-18	30-23	25-28	20-35	15-48	11-11	7-05	5-00
90°	90	85-00	80-00	75-00	70-00	65-00	60-00	55-00	50-00	45-00	40-00	35-00	30-00	25-00	20-00	15-00	10-00	5-00	—

(*Courtesy Texas Pipe Bending Company.*)

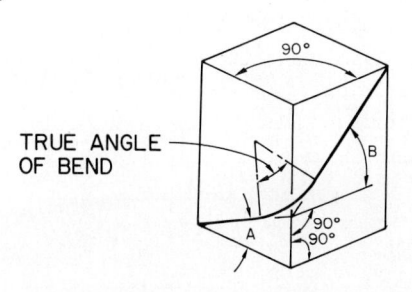

COS of TRUE ANGLE = SIN·A × SIN·B

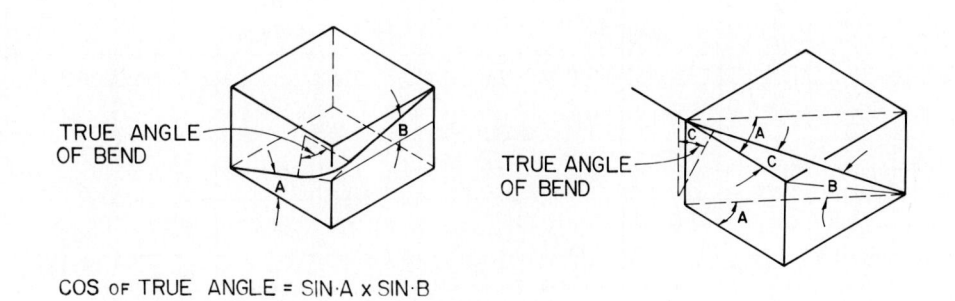

COS of TRUE ANGLE = SIN·A × SIN·B

COS·C = COS·A × COS·B

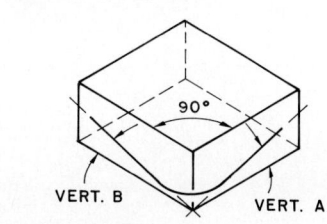

SIN·Q = SIN·A × SIN·B
TRUE ANGLE OF BEND = 90° + Q

BENDS

TABLE 9-8 DIMENSIONS FOR 30-DEGREE AND 45-DEGREE COMPOUND BENDS—BENDING AND ELBOWS (*COURTESY TEXAS PIPE BENDING COMPANY.*)

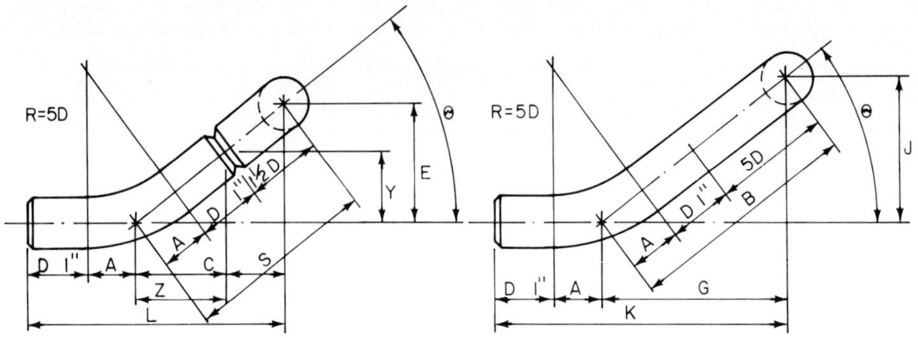

ELBOWS AND BENDS

$\phi = 30°$

Diameter	$1\frac{1}{2}$"	2"	3"	4"	6"	8"	10"	12"
A	2	$2\frac{11}{16}$	4	$5\frac{3}{8}$	$8\frac{1}{16}$	$10\frac{11}{16}$	$13\frac{3}{8}$	$16\frac{1}{16}$
Z	$3\frac{7}{8}$	$4\frac{15}{16}$	$6\frac{15}{16}$	9	$13\frac{1}{16}$	$17\frac{1}{16}$	$21\frac{1}{8}$	$25\frac{3}{16}$
Y	$2\frac{1}{4}$	$2\frac{13}{16}$	4	$5\frac{3}{16}$	$7\frac{1}{2}$	$9\frac{5}{16}$	$12\frac{3}{16}$	$14\frac{1}{2}$
S	$6\frac{3}{4}$	$8\frac{11}{16}$	$12\frac{1}{2}$	$16\frac{3}{8}$	$24\frac{11}{16}$	$31\frac{11}{16}$	$39\frac{3}{8}$	$47\frac{1}{16}$
C	$5\frac{7}{8}$	$7\frac{1}{2}$	$10\frac{13}{16}$	$14\frac{3}{16}$	$20\frac{13}{16}$	$27\frac{7}{16}$	$34\frac{1}{8}$	$40\frac{3}{4}$
E	$3\frac{3}{8}$	$4\frac{5}{16}$	$6\frac{1}{4}$	$8\frac{3}{16}$	12	$15\frac{13}{16}$	$19\frac{11}{16}$	$23\frac{1}{2}$
L	$10\frac{3}{8}$	$13\frac{3}{16}$	$18\frac{13}{16}$	$24\frac{3}{16}$	$35\frac{7}{8}$	$47\frac{1}{8}$	$58\frac{1}{2}$	$69\frac{13}{16}$
B	12	$15\frac{11}{16}$	23	$30\frac{3}{8}$	$45\frac{1}{16}$	$59\frac{11}{16}$	$74\frac{3}{8}$	$89\frac{1}{16}$
G	$10\frac{3}{8}$	$13\frac{9}{16}$	$19\frac{15}{16}$	$26\frac{5}{16}$	39	$51\frac{11}{16}$	$64\frac{7}{16}$	$78\frac{15}{16}$
J	6	$7\frac{13}{16}$	$11\frac{1}{2}$	$15\frac{3}{16}$	$22\frac{1}{2}$	$29\frac{13}{16}$	$37\frac{3}{16}$	$44\frac{1}{2}$
K	$14\frac{7}{8}$	$19\frac{1}{4}$	$27\frac{15}{16}$	$36\frac{11}{16}$	$54\frac{1}{16}$	$71\frac{3}{8}$	$88\frac{13}{16}$	108

$\phi = 45°$

D	$1\frac{1}{2}$"	2"	3"	4"	6"	8"	10"	12"
A	$3\frac{1}{8}$	$4\frac{1}{8}$	$6\frac{3}{16}$	$8\frac{5}{16}$	$12\frac{7}{16}$	$16\frac{9}{16}$	$20\frac{11}{16}$	$24\frac{7}{8}$
Z / Y	4	$5\frac{1}{16}$	$7\frac{3}{16}$	$9\frac{7}{16}$	$13\frac{3}{4}$	$18\frac{1}{16}$	$22\frac{7}{16}$	$26\frac{7}{16}$
S	$7\frac{7}{8}$	$10\frac{1}{8}$	$14\frac{11}{16}$	$19\frac{5}{16}$	$28\frac{7}{16}$	$37\frac{9}{16}$	$46\frac{11}{16}$	$55\frac{7}{8}$
C / E	$5\frac{9}{16}$	$7\frac{3}{16}$	$10\frac{3}{8}$	$13\frac{5}{8}$	$20\frac{1}{8}$	$26\frac{9}{16}$	33	$39\frac{1}{2}$
L	$11\frac{3}{16}$	$14\frac{5}{16}$	$20\frac{9}{16}$	$26\frac{15}{16}$	$39\frac{9}{16}$	$52\frac{1}{8}$	$64\frac{11}{16}$	$77\frac{3}{8}$
B	$13\frac{1}{8}$	$17\frac{1}{8}$	$25\frac{3}{16}$	$33\frac{5}{16}$	$49\frac{7}{16}$	$65\frac{9}{16}$	$81\frac{11}{16}$	$97\frac{7}{8}$
G / J	$9\frac{1}{4}$	$12\frac{1}{8}$	$17\frac{13}{16}$	$23\frac{9}{16}$	$34\frac{15}{16}$	$46\frac{3}{8}$	$57\frac{3}{4}$	$69\frac{3}{16}$
K	$14\frac{15}{16}$	$19\frac{1}{4}$	28	$36\frac{7}{8}$	$54\frac{3}{8}$	$71\frac{15}{16}$	$89\frac{7}{16}$	$107\frac{1}{16}$

10
MITERS

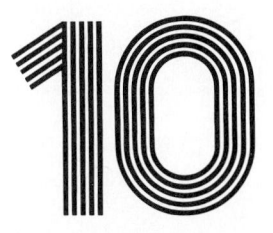

Fig. 10-1

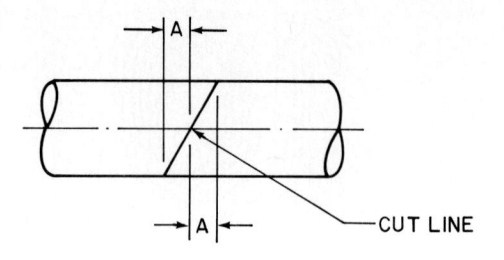

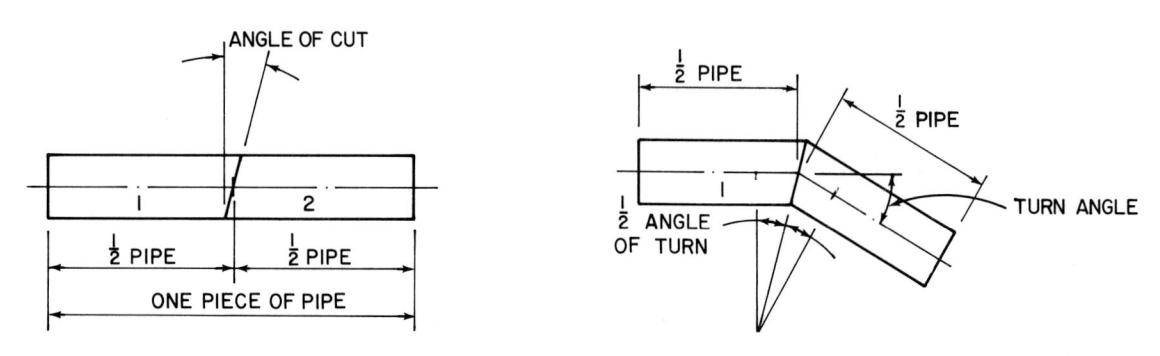

Fig. 10-2 One weld, two-piece miter. Angle of cut equals 1/2 angle of turn. Angle of cut equals angle of turn/divided by 2. See Table 10-1 for dimension A (cut back) (*Courtesy Texas Pipe Bending Company.*)

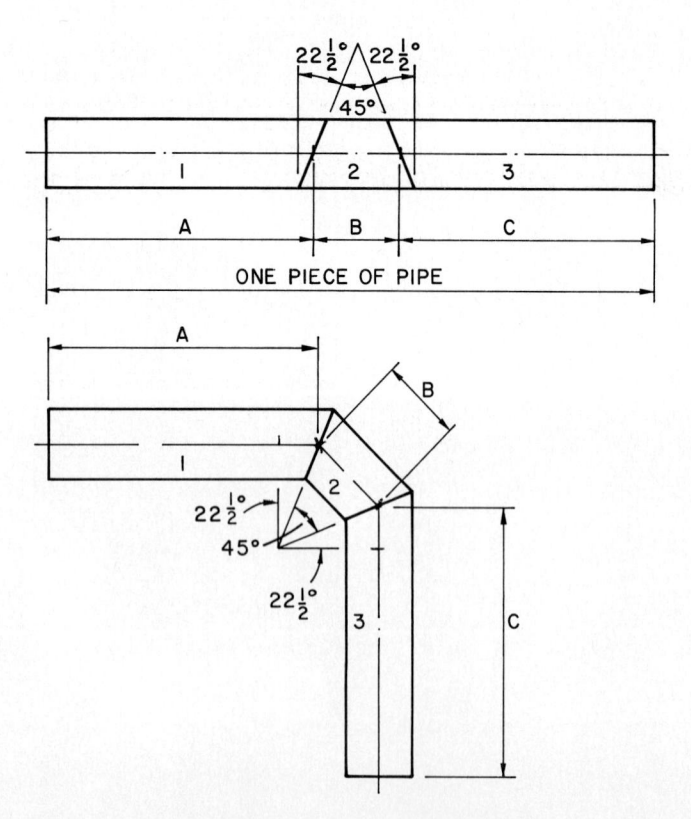

Fig. 10-3 90-degree miter using two welds. See Table 10-1 for setback and dimensions (*Courtesy Texas Pipe Bending Company.*)

CHAPTER 10

TABLE 10-1 MITER WELDING DIMENSIONS FOR ONE-MITER AND TWO-MITER BEND (*COURTESY TEXAS PIPE BENDING COMPANY.*)

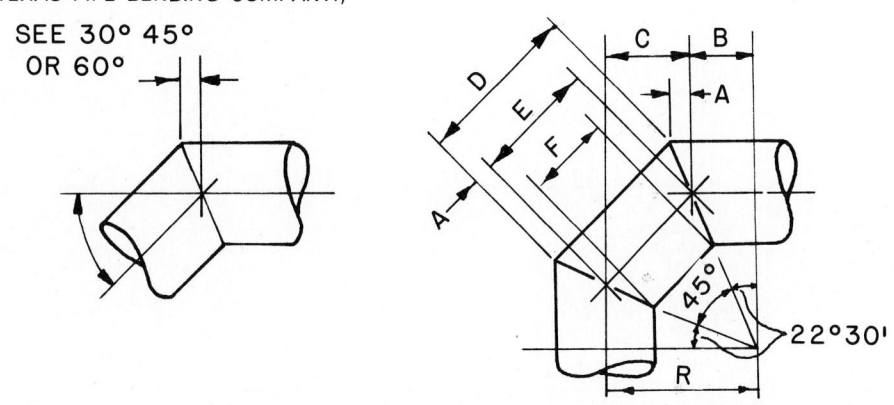

Size	30°	45°	60°	R	A	B	C	D	E	F
3	1/2	3/4	1	4-1/2	3/4	1-7/8	2-5/8	5-1/4	3-3/4	2-1/4
4	5/8	15/16	1-5/16	6	15/16	2-1/2	3-1/2	6-7/8	5	3-1/8
6	7/8	1-3/8	1-15/16	9	1-3/8	3-3/4	5-1/4	10-3/16	7-7/16	4-11/16
8	1-1/8	1-13/16	2-1/2	1 – 0	1-13/16	5	7	1 – 1-9/16	9-15/16	6-5/16
10	1-7/16	2-1/4	3-1/8	1 – 3	2-1/4	6-3/16	8-13/16	1 – 4-15/16	1 – 0-7/16	7-15/16
12	1-11/16	2-5/8	3-11/16	1 – 6	2-5/8	7-7/16	10-9/16	1 – 8-3/16	1 – 2-15/16	9-11/16
14	1-7/8	2-7/8	4-1/16	1 – 9	2-7/8	8-11/16	1 – 0-5/16	1 – 11-1/8	1 – 5-3/8	11-5/8
16	2-1/8	3-5/16	4-5/8	2 – 0	3-5/16	9-15/16	1 – 2-1/16	2 – 2-1/2	1 – 7-7/8	1 – 1-1/4
18	2-7/16	3-3/4	5-3/16	2 – 3	3-3/4	11-3/16	1 – 3-13/16	2 – 5-7/8	1 – 10-3/8	1 – 2-7/8
20	2-11/16	4-1/8	5-3/4	2 – 6	4-1/8	1 – 0-7/16	1 – 5-9/16	2 – 9-1/8	2 – 0-7/8	1 – 4-5/8
22	2-15/16	4-9/16	6-3/8	2 – 9	4-9/16	1 – 1-11/16	1 – 7-5/16	3 – 0-7/16	2 – 3-5/16	1 – 6-3/16
24	3-3/16	5	6-15/16	3 – 0	5	1 – 2-15/16	1 – 9-1/16	3 – 3-13/16	2 – 5-13/16	1 – 7-13/16
26	3-1/2	5-3/8	7-1/2	3 – 3	5-3/8	1 – 4-1/8	1 – 10-7/8	3 – 7-1/16	2 – 8-5/16	1 – 9-9/16
28	3-3/4	5-13/16	8-1/16	3 – 6	5-13/16	1 – 5-3/8	2 – 0-5/8	3 – 10-7/16	2 – 10-13/16	1 – 11-3/16
30	4	6-3/16	8-5/8	3 – 9	6-3/16	1 – 6-5/8	2 – 2-3/8	4 – 1-5/8	3 – 1-1/4	2 – 0-7/8
32	4-5/16	6-5/8	9-1/4	4 – 0	6-5/8	1 – 7-7/8	2 – 4-1/8	4 – 5	3 – 3-3/4	2 – 2-1/2
34	4-9/16	7-1/16	9-13/16	4 – 3	7-1/16	1 – 9-1/8	2 – 5-7/8	4 – 8-3/8	3 – 6-1/4	2 – 4-1/8
36	4-13/16	7-7/16	10-3/8	4 – 6	7-7/16	1 – 10-3/8	2 – 7-5/8	4 – 11-5/8	3 – 8-3/4	2 – 5-7/8
38	5-1/16	7-7/8	11	4 – 9	7-7/8	1 – 11-5/8	2 – 9-3/8	5 – 3	3 – 11-1/4	2 – 7-1/2
40	5-3/8	8-5/16	11-9/16	5 – 0	8-5/16	2 – 0-7/8	2 – 11-1/8	5 – 6-5/16	4 – 1-11/16	2 – 9-1/16
42	5-5/8	8-11/16	1 – 0-1/8	5 – 3	8-11/16	2 – 2-1/8	3 – 0-7/8	5 – 9-9/16	4 – 4-3/16	2 – 10-13/16
48	6-7/16	9-15/16	1 – 1-7/8	6 – 0	9-15/16	2 – 5-13/16	3 – 6-3/16	6 – 7-1/2	4 – 11-5/8	3 – 3-3/4
54	7-1/4	11-3/16	1 – 3-9/16	6 – 9	11-3/16	2 – 9-9/16	3 – 11-7/8	7 – 5-1/2	5 – 7-1/8	3 – 8-3/4
60	8-1/16	1 – 0-7/16	1 – 5-5/16	7 – 6	1 – 0-7/16	3 – 1-1/4	4 – 4-3/4	8 – 3-7/16	6 – 2-9/16	4 – 1-11/16
72	9-5/8	1 – 2-15/16	1 – 8-13/16	9 – 0	1 – 2-15/16	3 – 8-3/4	5 – 3-1/4	9 – 11-3/8	7 – 5-1/2	4 – 11-5/8

MITERS

TABLE 10-2 MITER WELDING DIMENSIONS FOR THREE- AND FOUR-MITER BENDS
(*COURTESY TEXAS PIPE BENDING COMPANY.*)

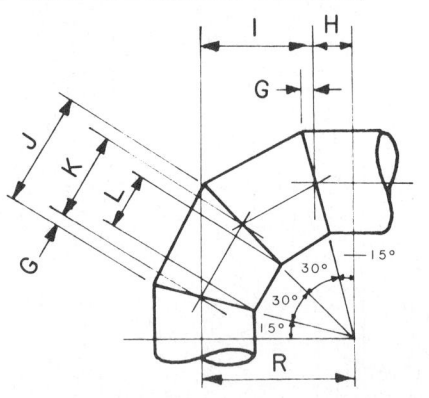

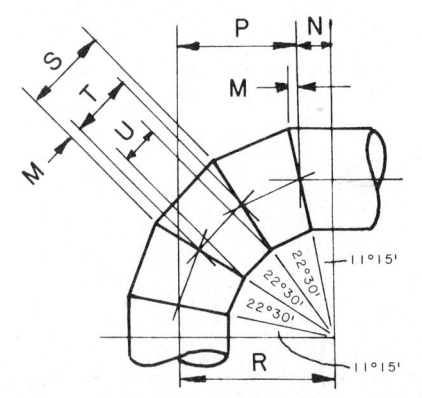

G	H	I	J	K	L	M	N	P	S	T	U
1/2	1-3/16	3-5/16	3-7/16	2-7/16	1-7/16	3/8	7/8	3-5/8	2-9/16	1-13/16	1-1/16
5/8	1-5/8	4-3/8	4-7/16	3-3/8	1-15/16	7/16	1-3/16	4-13/16	3-1/4	2-3/8	1-1/2
7/8	2-7/16	6-9/16	6-9/16	4-13/16	3-1/16	11/16	1-13/16	7-3/16	4-15/16	3-9/16	2-3/16
1-1/8	3-3/16	8-13/16	8-11/16	6-7/16	4-3/16	7/8	2-3/8	9-5/8	6-1/2	4-3/4	3
1-7/16	4	11	10-15/16	8-1/16	5-3/16	1-1/16	3	1 – 0	8-1/16	5-15/16	3-13/16
1-11/16	4-13/16	1 – 1-3/16	1 – 1	9-5/8	6-1/4	1-1/4	3-9/16	1 – 2-7/16	9-11/16	7-3/16	4-11/16
1-7/8	5-5/8	1 – 3-3/8	1 – 3	11-1/4	7-1/2	1-3/8	4-3/16	1 – 4-13/16	11-1/8	8-3/8	5-5/8
2-1/8	6-7/16	1 – 5-9/16	1 – 5-1/8	1 – 0-7/8	8-5/8	1-9/16	4-3/4	1 – 7-1/4	1 – 0-11/16	9-9/16	6-7/16
2-7/16	7-1/4	1 – 7-3/4	1 – 7-5/16	1 – 2-7/16	9-9/16	1-13/16	5-3/8	1 – 9-5/8	1 – 2-3/8	10-3/4	7-1/8
2-11/16	8-1/16	1 – 9-15/16	1 – 9-7/16	1 – 4-1/16	10-11/16	2	5-15/16	2 – 0-1/16	1 – 3-15/16	11-15/16	7-15/16
2-15/16	8-13/16	2 – 0-3/16	1 – 11-9/16	1 – 5-11/16	11-13/16	2-3/16	6-9/16	2 – 2-7/16	1 – 5-1/2	1 – 1-1/8	8-3/4
3-3/16	9-5/8	2 – 2-3/8	2 – 1-11/16	1 – 7-5/16	1 – 0-15/16	2-3/8	7-3/16	2 – 4-13/16	1 – 7-1/16	1 – 2-5/16	9-9/16
3-1/2	10-7/16	2 – 4-9/16	2 – 3-7/8	1 – 8-7/8	1 – 1-7/8	2-9/16	7-3/4	2 – 7-1/4	1 – 8-5/8	1 – 3-1/2	10-3/8
3-3/4	11-1/4	2 – 6-3/4	2 – 6	1 – 10-1/2	1 – 3	2-13/16	8-3/8	2 – 9-5/8	1 – 10-5/16	1 – 4-11/16	11-1/16
4	1 – 0-1/16	2 – 8-15/16	2 – 8-1/8	2 – 0-1/8	1 – 4-1/8	3	8-15/16	3 – 0-1/16	1 – 11-7/8	1 – 5-7/8	11-7/8
4-5/16	1 – 0-7/8	2 – 11-1/8	2 – 10-5/16	2 – 1-11/16	1 – 5-1/16	3-3/16	9-9/16	3 – 2-7/16	2 – 1-1/2	1 – 7-1/8	1 – 0-3/4
4-9/16	1 – 1-11/16	3 – 1-5/16	3 – 0-7/16	2 – 3-5/16	1 – 6-3/16	3-3/8	10-1/8	3 – 4-7/8	2 – 3-1/16	1 – 8-5/16	1 – 1-9/16
4-13/16	1 – 2-7/16	3 – 3-9/16	3 – 2-9/16	2 – 4-15/16	1 – 7-5/16	3-9/16	10-3/4	3 – 7-1/4	2 – 4-5/8	1 – 9-1/2	1 – 2-3/8
5-1/16	1 – 3-1/4	3 – 5-3/4	3 – 4-11/16	2 – 6-9/16	1 – 8-7/16	3-3/4	11-5/16	3 – 9-11/16	2 – 6-3/16	1 – 10-11/16	1 – 3-3/16
5-3/8	1 – 4-1/16	3 – 7-15/16	3 – 6-7/8	2 – 8-1/8	1 – 9-3/8	4	11-15/16	4 – 0-1/16	2 – 7-7/8	1 – 11-7/8	1 – 3-7/8
5-5/8	1 – 4-7/8	3 – 10-1/8	3 – 9	2 – 9-3/4	1 – 10-1/2	4-3/16	1 – 0-1/2	4 – 2-1/2	2 – 9-7/16	2 – 1-1/16	1 – 4-11/16
6-7/16	1 – 7-5/16	4 – 4-11/16	4 – 3-7/16	3 – 2-9/16	2 – 1-11/16	4-3/4	1 – 2-5/16	4 – 9-11/16	3 – 2-1/8	2 – 4-5/8	1 – 7-1/8
7-1/4	1 – 9-11/16	4 – 11-5/16	4 – 9-7/8	3 – 7-3/8	2 – 4-7/8	5-3/8	1 – 4-1/8	5 – 4-7/8	3 – 7	2 – 8-1/4	1 – 9-1/2
8-1/16	2 – 0-1/8	5 – 5-7/8	5 – 4-3/8	4 – 0-1/4	2 – 8-1/8	5-15/16	1 – 5-7/8	6 – 0-1/8	3 – 11-5/8	2 – 11-3/4	1 – 11-7/8
9-5/8	2 – 5	6 – 7	6 – 5-1/4	4 – 10	3 – 2-3/4	7-3/16	1 – 9-1/2	7 – 2-1/2	4 – 9-3/8	3 – 7	2 – 4-5/8

11
DEVELOPMENTS AND PATTERNS

Fig. 11-1

USE OF TEMPLATE LAYOUTS*

The fact that a length of pipe with square ends can be fabricated by wrapping a rectangular section of plate into a cylindrical form makes available a method of developing pipe surfaces, and hence developing the lines of intersection between pipe walls, known as parallel forms. Based on this principle, wraparound templates can be made for marking all manner of pipe fittings for cutting preparatory to welding, (Fig. 11-2).

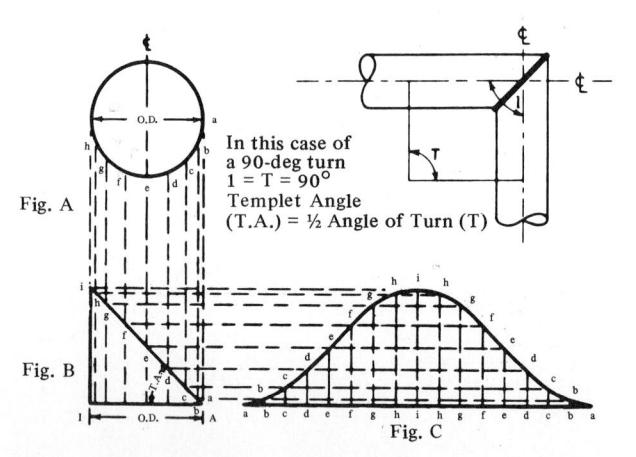

Fig. 11-2 Template layout (*Courtesy G. K. Bachmann, Pipefitter's and Plumber's Vest Pocket Reference Book, Prentice-Hall, Inc., Englewood Cliffs, N.J.*)

The development of a template is done in practice by dividing the circumference (in the end view) of the pipe into a certain number of equal sections, which are then projected onto the side view of the desired pipe section. The lengths of the various segments which make up the pipe wall may then be laid out, evenly spaced, on a base line which is in effect the unwrapped circumference. If the template, developed in Fig. C, is wrapped around the pipe with the base line square with the pipe, the curved line a-b-c-d-e-f-etc. will locate the position for cutting to make a 90-degree, two-piece turn. Draw a circle (Fig. A) equal to the OD of the pipe and divide half of it into equal sections. The more sections, the more accurate will be the final result. Perpendicular to the centerline and bisected by it, draw AI equal to OD (Fig. B) and to this line construct the template angle (TA) equal to one-half of the angle of turn—in this case, 45 degrees. Draw lines parallel to the centerline from points a,b,c, etc., on the circle and mark the points where these lines intersect line AI with corresponding letters. As an extension of AI, but a little distance from it, draw a straight line equal to the pipe circumference, or that of the circle in Fig. A. This line (Fig. C) should then be divided into twice as many equal spaces as the semicircle a-b-c-etc. and lettered as shown. Perpendiculars should then be erected from these points. Their intersections with lines drawn from the points on AI in Fig. B parallel to the base line in Fig. C determine the curve of the template.

MULTIPIECE WELDED TURNS

It is usually preferable to use more than two pieces when the desired angle of turn exceeds 45 degrees; the most popular design for a 90-degree turn is the fourpiece turn illustrated in Fig. 11-3.

In general, the layout of a multipiece turn is governed by the total angle of turn, the number of pieces, and the radius of bend. After these factors have been determined, the next operation consists of deciding upon the length of the shortest element of the individual sections of the turn. This may be determined from the following formula:

where

$$L = (2R\text{-}D) \tan TA$$

L = length of shortest element in inches

R = radius of turn in inches

D = OD of pipe in inches

TA = template angle

$= \dfrac{\text{total angle of turn}}{2N}$

N = number of welds

= number of pieces in turn less 1

The templates for such a multipiece turn are laid out as for a two-piece turn, except that the template angle is determined by dividing the total angle of turn by twice the number of welds.

---

*(Courtesy George K. Bachmann, *Pipefitter's and Plumber's Vest Pocket Reference Book*, Prentice-Hall, Inc., Englewood Cliffs, N.J.)

Fig. 11-3 Multipiece turn (*Courtesy G. K. Bachmann*, Pipefitter's and Plumber's Vest Pocket Reference Book, *Prentice-Hall, Inc., Englewood Cliffs, N.J.*)

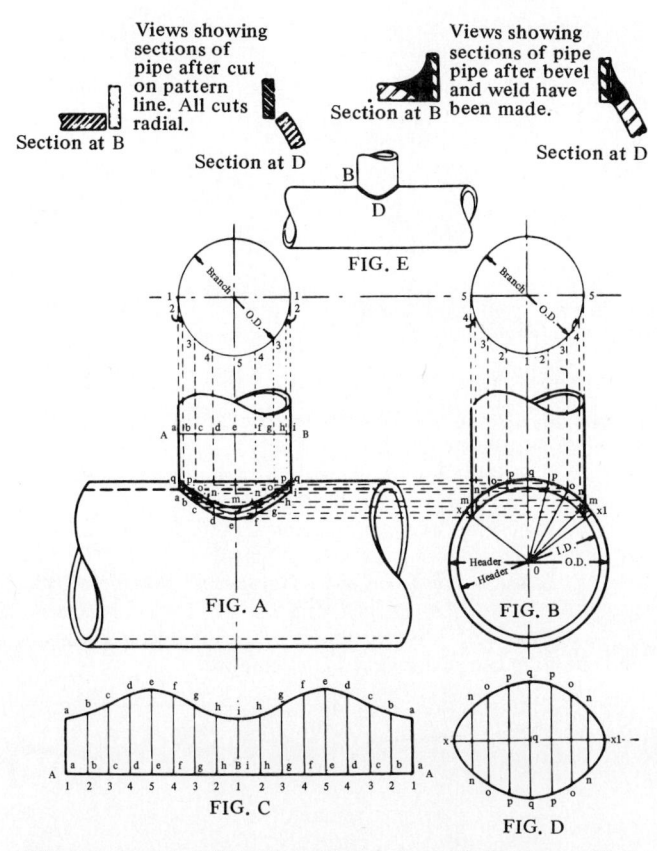

Fig. 11-4 Template for 90-degree branch (*Courtesy G. K. Bachmann*, Pipefitter's and Plumber's Vest Pocket Reference Book, *Prentice-Hall, Inc., Englewood Cliffs, N.J.*)

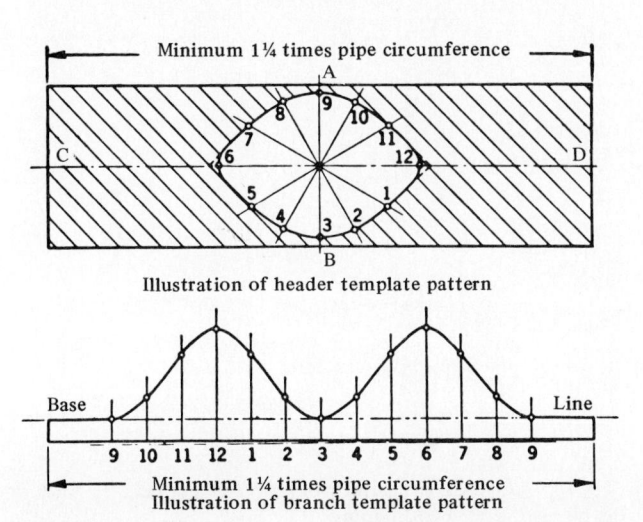

Fig. 11-5 Pattern for wraparound template (*Courtesy G. K. Bachmann*, Pipefitter's and Plumber's Vest Pocket Reference Book, *Prentice-Hall, Inc., Englewood Cliffs, N.J.*)

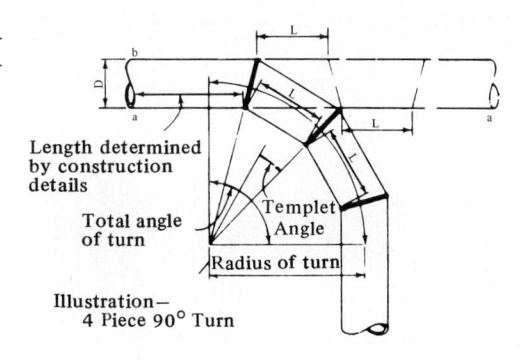

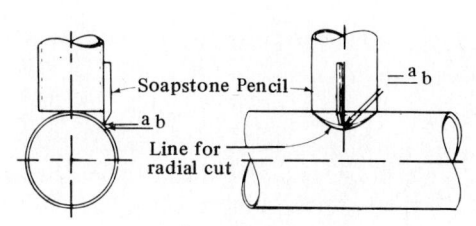

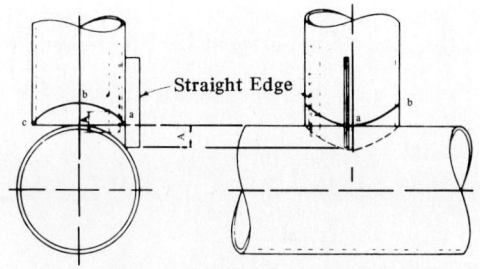

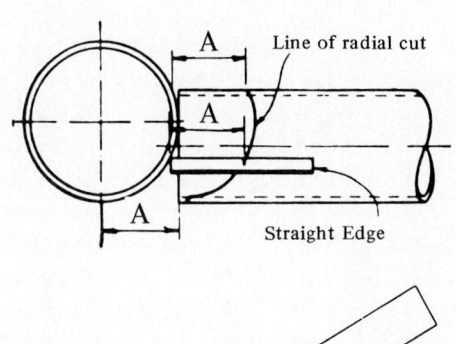

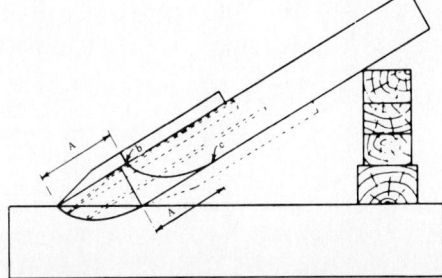

Fig. 11-6 Marking cuts without templates (*Courtesy G. K. Bachmann*, Pipefitter's and Plumber's Vest Pocket Reference Book, *Prentice-Hall, Inc., Englewood Cliffs, N.J.*)

DEVELOPMENTS AND PATTERNS

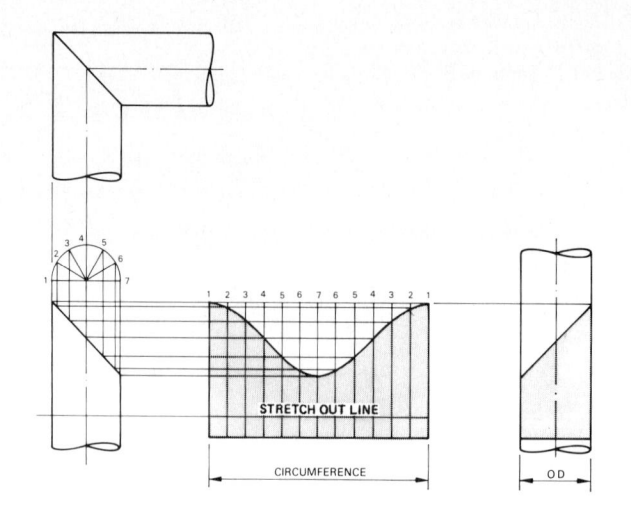

Fig. 11-7 90-degree, two-piece turn. Development and pattern

1. Draw a half circle and divide into equal parts as shown. The half section corresponds to the end view (right section) of the pipe. Label the intersection of the division lines from 1 to 7.
2. Project points 1 through 7 to the front view where they intersect the miter line.
3. Extend a stretch-out line perpendicular to the front view of the pipe (axis line) and lay off the length of the development using the calculated circumference (or set off the chord distances, 1-2, 2-3, etc.)
4. Divide the circumference into 12 equal parts along the stretch out line and label as shown.
5. Project the height dimension of each element from the front view to the development.
6. Connect points on the development with a smooth curve.
7. The development can now be transferred to a pattern and cutout to use as a wraparound template.

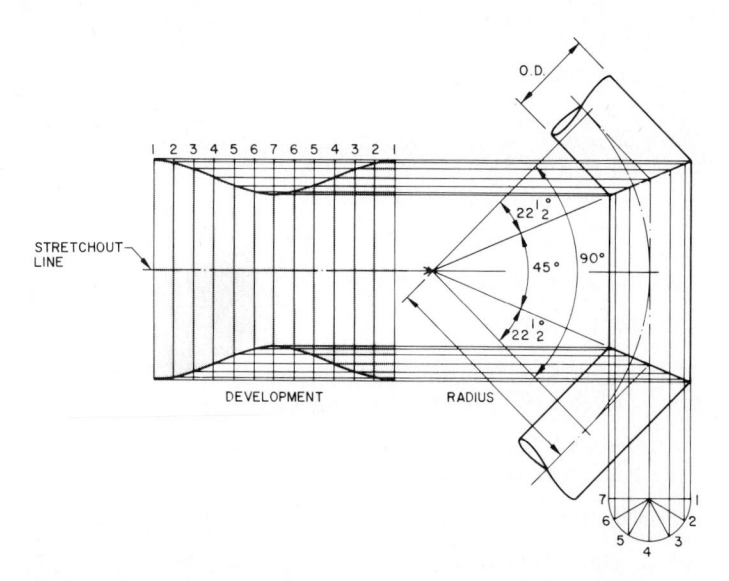

Fig. 11-8 90-degree, three-piece turn. Development and pattern

1. Draw the mitered sections as shown.
2. Since each piece is the same only one development will be needed. Divide the end view into equal parts as in Fig. 11-8.
3. Project the points to the front view where they intersect the two mitered edges.
4. Draw a stretch-out line perpendicular to the front view of the pipe.
5. Calculate the length of the development by finding the circumference of the pipe.
6. Divide the circumference into 12 equal parts and project the miter points from the front view to the development.
7. Connect the points with a smooth curve.

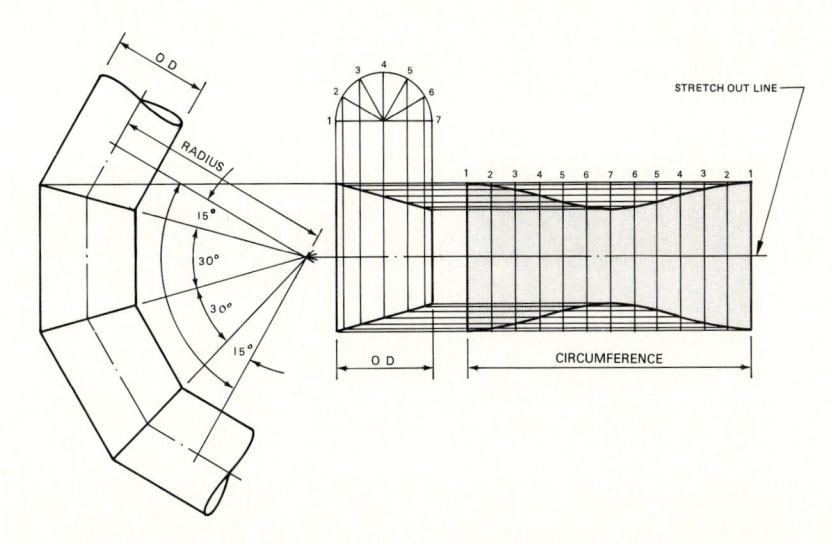

Fig. 11-9 90-degree, multipiece welded turn. Development and pattern

1. Same as Fig. 11-8.

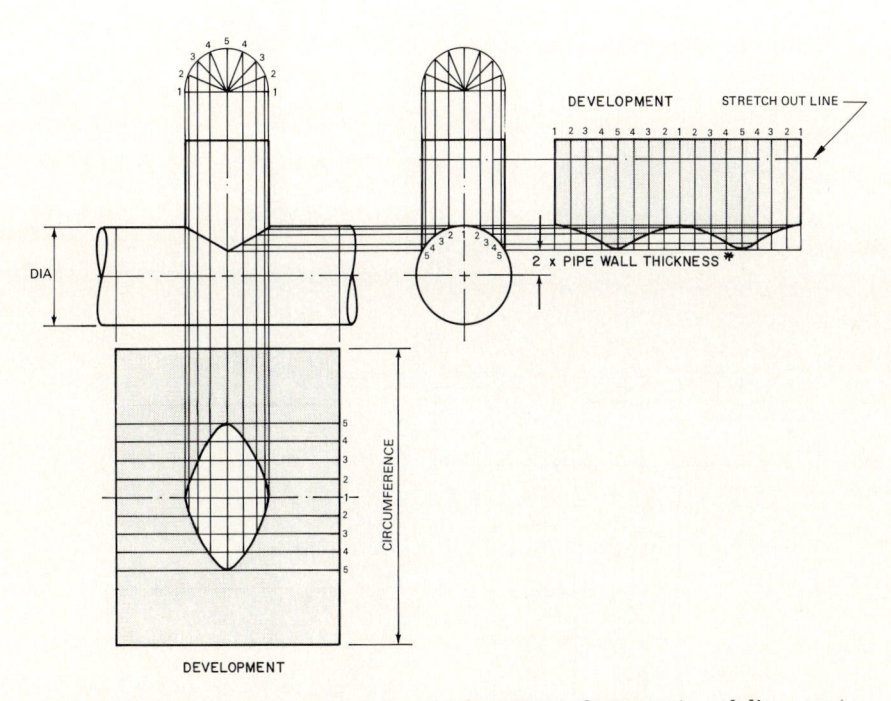

Fig. 11-10 Branch and header development and pattern (equal diameters)

1. Draw the front and side views of the intersecting pipes (excluding the line of intersection).
2. Draw half circles (above each view) corresponding to the branch pipe circumference and divide it into equal parts.
3. Project the points into the views as shown. Where the points intersect the header in the side view label the intersection points as shown and project to the front view.
4. Project the points from the half circle to the front view. Where they intersect their corresponding points extended from the side view it will establish intersection points along the line of intersection. Note that when the pipes are the same diameter the lowest point, 5, will be established by calculating the distance from the header's centerline (2 × pipe wall thickness of the branch pipe). This method is used because the branch will fit inside the hole cut from the header.
5. Draw the stretch-out line for both developments perpendicular to their respective pipes. Calculate the circumference of each pipe and establish the length of the developments. Divide the circumferences into 12 equal parts and establish their lengths by projecting the points from the front view as shown.

DEVELOPMENTS AND PATTERNS

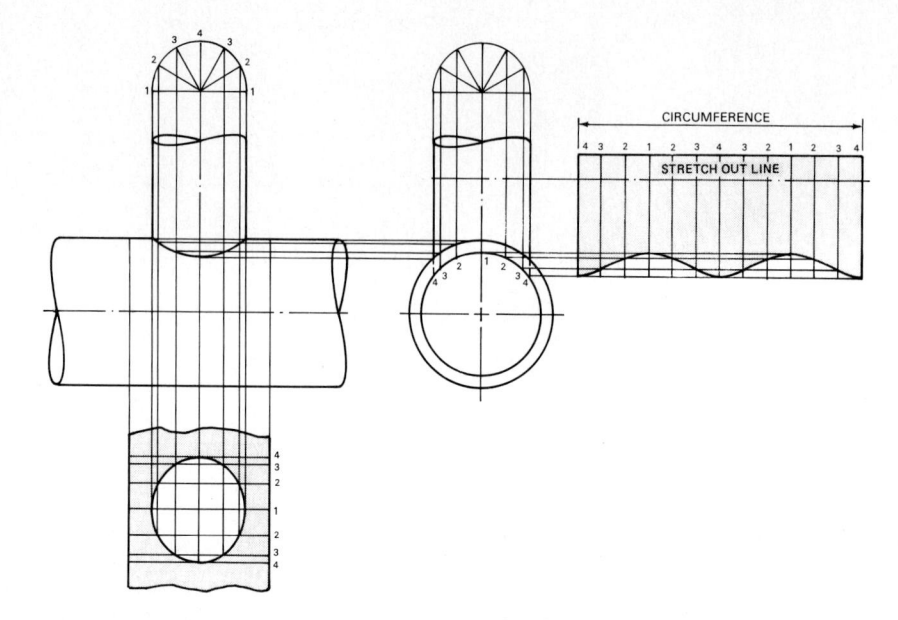

Fig. 11-11 Branch and header development and pattern (unequal diameters)

1. Same as Fig. 11-10. Make sure that the header points are taken from the inside diameter.

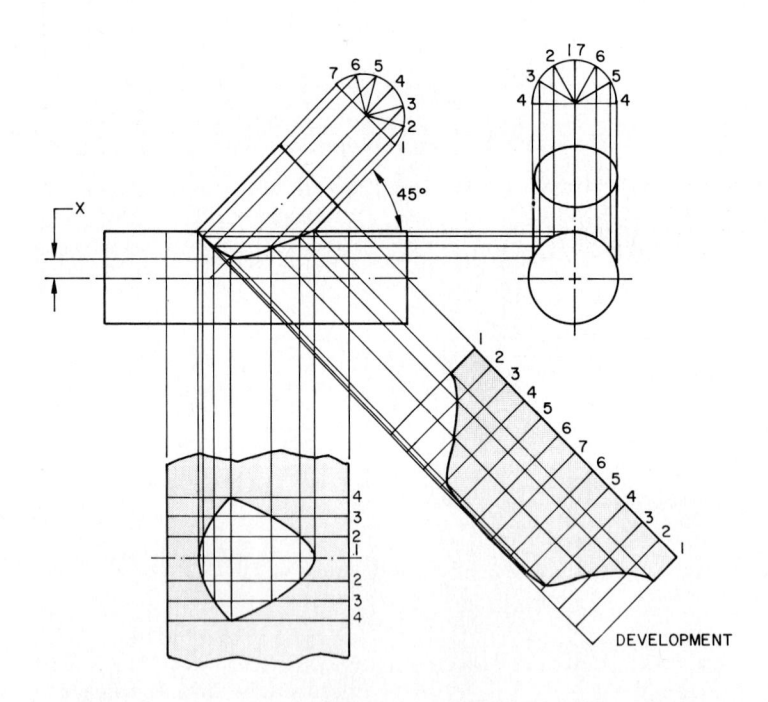

Fig. 11-12 Lateral development and pattern (equal diameters)

1. After drawing the front and side views, construct half circle end sections and divide into equal parts.
2. Project the end sections divisions (points) to the front view to establish the line of intersection. Note that when the pipes are of equal diameters the distance from the header centerline and the lowest point must be calculated, point 4. Dimension X equals 2 × the pipe wall thickness of the branch pipe.
3. Draw the stretch-out lines perpendicular to the pipes and calculate their respective circumferences.
4. Divide the circumference length into 12 equal parts, project the related points to the development, and connect the points with a smooth curve.

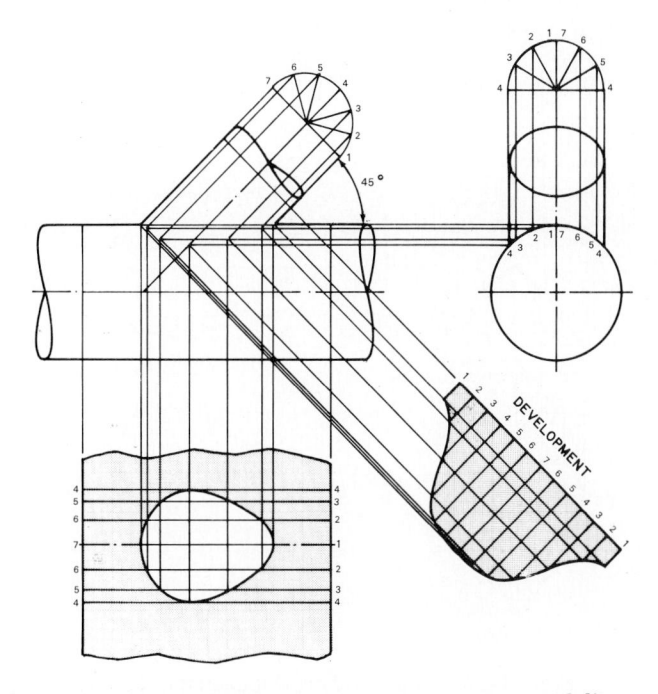

Fig. 11-13 Lateral development and pattern (unequal diameters)

1. Same as Fig. 11-12.

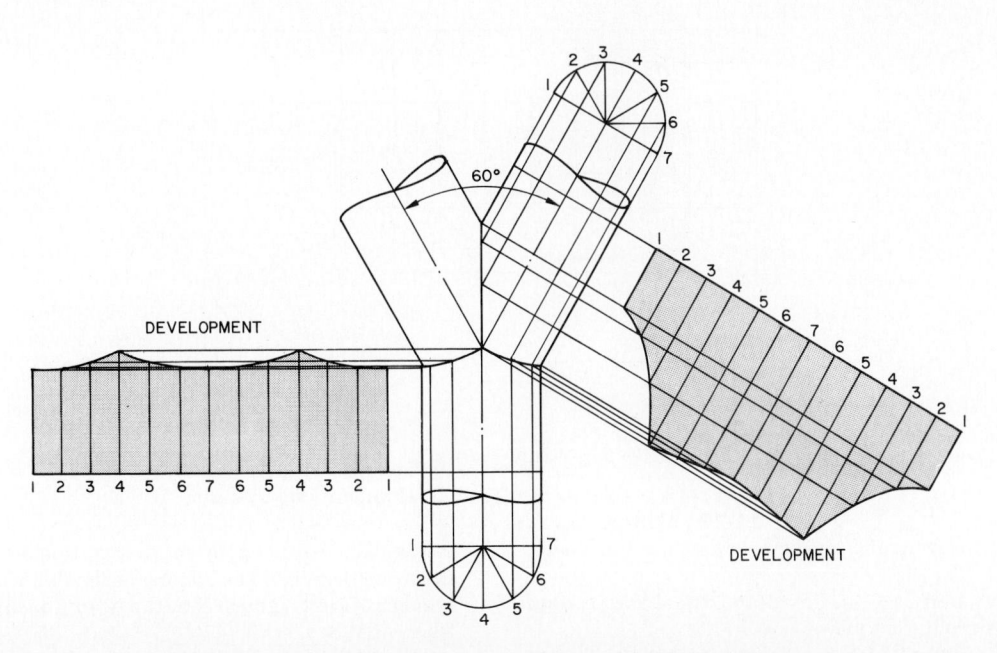

Fig. 11-14 60-degree WYE connection development and pattern

1. The same procedure can be applied here as in the previous examples.

DEVELOPMENTS AND PATTERNS

121

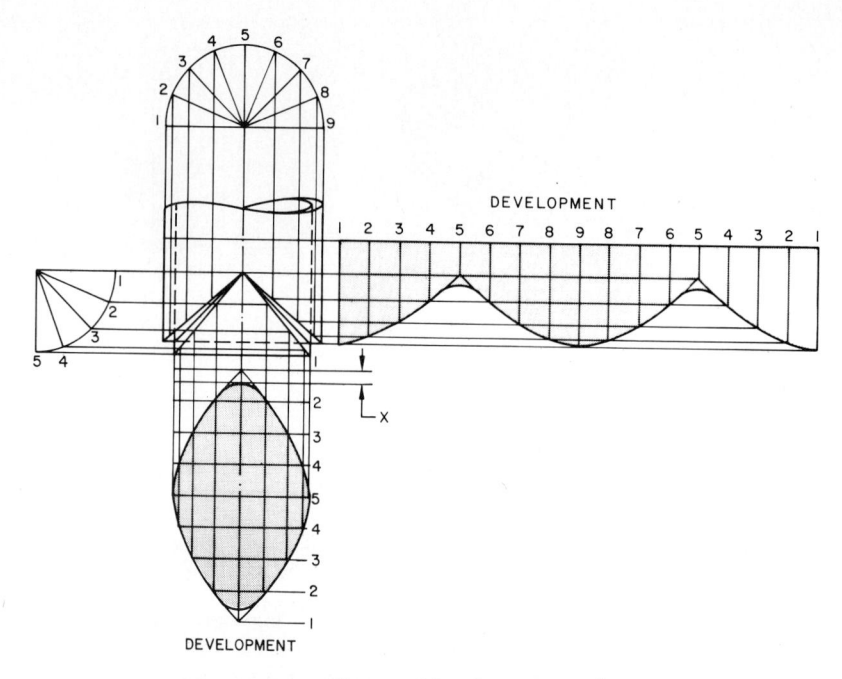

Fig. 11-15 Blunt end development and pattern

1. Draw the front view of the blunt end as shown. Show the ID of the pipe.
2. The blunt end can be established by drawing it in the front view with 45 degree lines extended from the intersection of the two inside diameters.
3. Construct a half and quarter section corresponding to pipe diameter and the blunt end respectively. Divide each equally as shown.
4. Draw the stretch-out lines and divide them into equal parts.
5. Project the points of intersection from the front view and complete the developments.

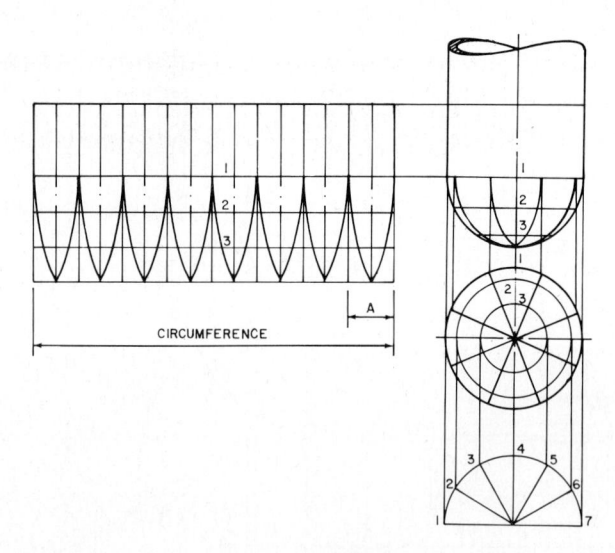

Fig. 11-16 Orange peel head development and pattern

1. Draw the front and top views of the pipe and head as shown.
2. Draw a half circle and divide into equal parts, project each point to the adjacent view.
3. Each point will establish a circular section on the pipe head (sections 1, 2, and 3). Divide the head into equal parts and project to the top view.
4. Extend a stretch-out line from the pipe and divide it into equal parts along its length (the stretch-out line is equal to the circumference). The number of divisions corresponds to the number of parts (gores) that the head was divided into in the front view, 8.
5. Dimension A is the width of one gore along section line 1 in the front view. Use this and the widths of each gore along the other two section lines to lay out the development.

CHAPTER 11

12
WELDING

Fig. 12-1

WELDED PIPE SYSTEMS

Welded piping systems are possible for almost all services. This type of system is a closed container with fittings, valves, and equipment forming a leakproof, maintenance-free, safe piping system. Where internal pressures are high, the welded unit offers a great safety margin. Welded fittings form smooth joints that are easy to insulate. The most common weld is the butt weld, where both ends are machine-beveled to form a groove between the mating parts. This forms an efficient joint when fitted together and fused by welding. Socket-welded fittings are also used, but in less critical situations.

Welding is preferred for systems that involve infrequent dismantling and require strong leakproof connections. A drafter must take care to design a system which will allow adequate room for on-the-job welding construction and also for convenience—especially when welded valves and fittings or other equipment must be taken from the line for servicing and replacement.

Welding provides for cheaper, lighter, stronger, safer joints and connections. Other types of connections that are easier to assemble and dismantle must also be taken into account, however. Flanged connections for valves, instruments, and the like utilize welding to connect the pipeline to the flange. Welded lines require more working space in fabication.

WELDING METHODS

Welding is the procedure by which two pieces of metal are fused together along a line or a surface between them or at a certain point. Welding can be classified by process. *Nonpressure welding* (fusion and brazing) predominates in the piping field. No mechanical pressure is applied. The pieces of metal are welded at the point of contact by heat which is created by an electric arc or by a gas or oxyacetylene flame. *Pressure welding* (forging), or resistance welding, is used to form a joint by the passage of electrical current through the area of the joint as mechanical pressure is applied. This type of welding is usually done by machines only and is not used very often in the construction of piping systems. It is usually more convenient, however, to classify welds into three separate categories: *resistance welding, gas welding,* and *arc welding.*

Resistance Welding

Resistance welding is the process by which heat and pressure are applied at the same time, usually by a machine. Two or more parts can be welded together by passing an electrical current through the work as pressure is applied. This process is not as common as the others in the piping field. The main types of resistance welds are spot, seam, projection, flash, and upset.

Gas Welding

In the form of fusion welding known as gas welding, heat is created by the combustion of a gas and air or pure oxygen. A welding filler is sometimes used as a flux. In *oxyacetylene welding* (sometimes called autogenous welding or gas welding), a flame is produced by the combustion of oxygen and acetylene gases. This type of welding is used less often.

Arc Welding

The third type of welding, which is the most commonly used at present, especially in the piping field, is arc welding. This section examines several arc welding processes: submerged arc welding, shielded arc welding, and gas-metal arc welding.

Submerged Arc Welding In this welding process, coalescence is produced by the heating of an electric arc between the electrodes and the work or between the electrode and the base metal. The work is shielded by a blanket of granular, fusible material (*flux*). In most cases, no pressure is used, and some filler material is created by the melting flux, which also forms a slag shield that coats the molten metal. The slag must be removed at the end of the process.

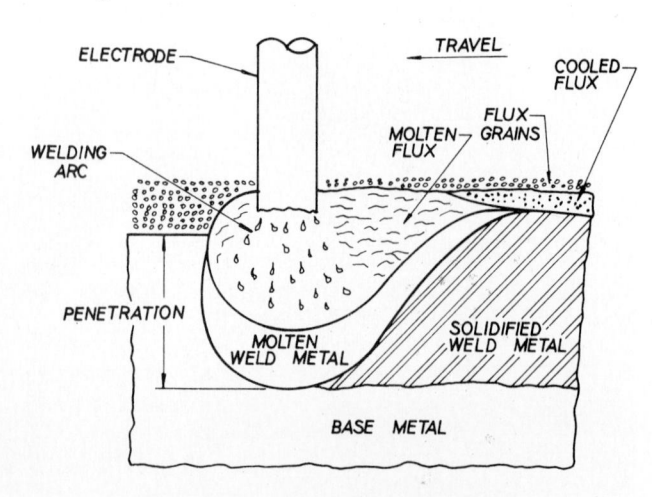

Fig 12-2 Submerged arc welding

The filler material can also be obtained from a supplementary welding rod or from the electrode itself. In this process, loose flux (also called melt or welding composition) is placed over the joint to be welded. The welding zone is completely covered by a blanket of this material. After the arc is established, the flux melts to form a shield which coats the molten metal (Fig. 12-2). A bare wire electrode is used in this process instead of a coated electrode, and the flux is supplied separately. In the manufacture of pipes, especially circumferential pipe joints, this type of welding is often automated.

Shielded Arc Welding In shielded arc welding, the metal arc is produced by the contact of a coated metal electrode and the material to be welded. Fusion takes place by heating the metal in the electrode to a temperature that causes them to melt together. Figure 12-3 shows how the manual shielded arc weld proceeds. Note how slag is formed on top of the base metal or on top of the solidified weld metal. This slag must be removed after the welding process is completed. Figure 12-2, by way of contrast, shows how the electrode in submerged arc welding extends into the work itself and how the flux and base materials are fused together in molten weld metal; note that the penetration is much deeper than in shielded arc welding.

Shielded metal arc welding is almost always accomplished manually by a trained welder rather than an automated machine. It is used quite often in the structural fabrication of piping systems because of its ease of application in the field. It is also used for *tack welding*—holding parts in position prior to welding up a series of deposits in the joint manually. Tack welds are used on almost all pipes to be joined. After the pipes are aligned, they are joined by four or more tack welds before undergoing the complete welding process.

Gas-Metal Arc Welding This process is used for fabricating pipe supports and connecting them to structural members in a piping system. For gas arc welding, heat is created electrically as in the process just described, but the shielding is accomplished by a blanket of gas. Pressure may or may not be used; in the piping field, welding is generally pressureless. A filler metal may or may not be used; in most cases it is added during the process itself. The filler metal may also be added to the welding zone prior to welding. In this process, inert gases are released onto the welding area to form a blanket. In the construction of pipes or piping joints, gas also is applied to the inside of the pipe until it is sealed off after the first pass. This welding procedure is used for exotic metals such as magnesium, aluminum, alloyed steels, and mild steel pipes, especially where pipes are to be used for high-pressure service.

Gas-metal arc welding is also referred to as *gas-tungsten arc welding* (GTAW). Since the electrode in this case is usually of tungsten and not a filler metal, it does not melt. The GTAW process produces root beads of high quality and is seldom used for the entire weld unless there are stringent standards to be met. Characteristics of this process include high-quality welds of nearly all metals, no filler metal across arc stream (consequently no splatter), little postweld cleaning, multiposition welding, and no slag. The base metal is supplemented by a filler metal except where unnecessary (as in the welding of thin-walled pipe).

Figure 12-4 shows gas-metal arc welding, a process in which the electrode wire is inserted

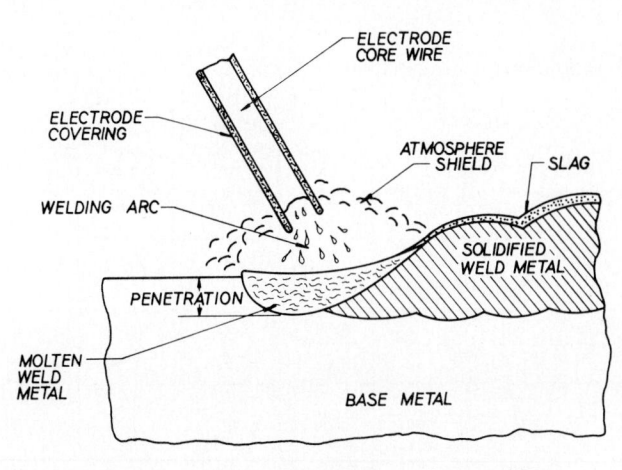

Fig. 12-3 Manual shielded metal arc welding

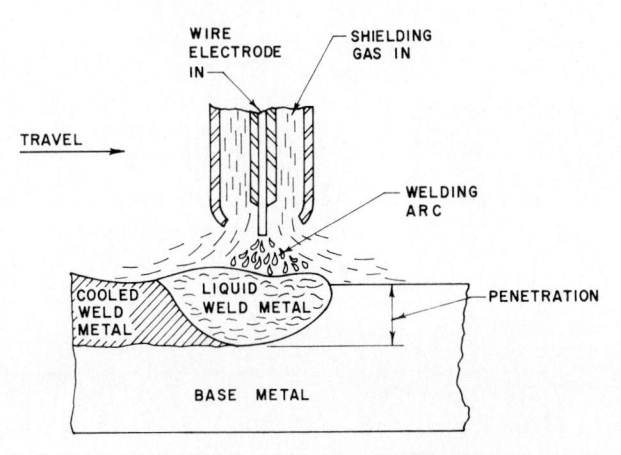

Fig. 12-4 Gas-metal arc welding

through to the arc area. This type of welding does not create a deep penetration. The figure shows how the shielding gas—an inert gas such as helium or argon—is used to shield the work. Filler metal is sometimes added.

WELDING MATERIALS

Weldability is the capacity of a metal to be welded in relation to its suitability to the design and service requirements. Metals that become fused during the welding process undergo changes similar to those which occur during manufacture. Chemical, thermal, physical, and metallurgical changes make it essential for the engineer to understand the nature of the materials to be fused. The metallurgy of welds is the metallurgy of the material.

Since the weldability of different materials varies greatly, so does the process by which the weld is completed. When welding cast iron to steel, for instance, cast iron rods are used as the welding material and the steel must be preheated before an adequate weld can be made. When welding steel castings, there is no easy rule for the process because carbon content can vary drastically between types of steel. Steel welding rods usually produce an adequate weld.

In the piping field, brass is usually brazed instead of welded because of the high temperature produced by the welding process and the low melting point of brass in comparison to other metals. In welding copper, that metal must be used as a filler and care must be taken not to produce oxidation.

Carbon steel welding is usually completed by the use of shielded metal arc welding with a covered electrode. Rod iron has characteristics similar to those of mild steel, and a similar process is used in its welding. Carbon molybdenum steel is becoming less popular because of its unfavorable reactions to high temperatures.

In aluminum and aluminum alloys, most of the commercial welding and brazing processes can be used, although the most common are the inert gas-tungsten arc and inert gas-consumable metal arc processes. Various problems are encountered when using the acetylene process to weld aluminum because of an oxide film that prevents metal flow at welding temperatures. Aluminum is characterized by its low melting point and high thermal conductivity. Great care therefore must be taken when welding this material. In the nickel and nickel alloy group, adequate welding can be accomplished by welding similar materials to themselves.

When dissimilar metals are welded, the weld deposit can be of a different chemical composition than the parts to be welded or the pieces themselves can be of different metallurgical composition. In addition to differentiating between the weld deposit and the joined pieces or between one piece and another, it must also be determined whether the metals themselves have dissimilar major constituents and dissimilar alloying metals. Three different metals can result in a joint where the weld metal becomes a composite of the filler metal and the base metals.

Dissimilar metals are usually welded by fusion welding, resistance welding, soldering, brazing, or pressure gas welding. The major concern is the melting temperature of the metals to be welded. If the melting temperatures are similar, it is usually possible to use gas or arc welding with the addition of a filler metal of the same composition as one of the two pieces to be joined. If the melting points are quite different, then soldering, braze welding, and brazing are used.

Fig. 12-5 Location of elements for standard welding symbols

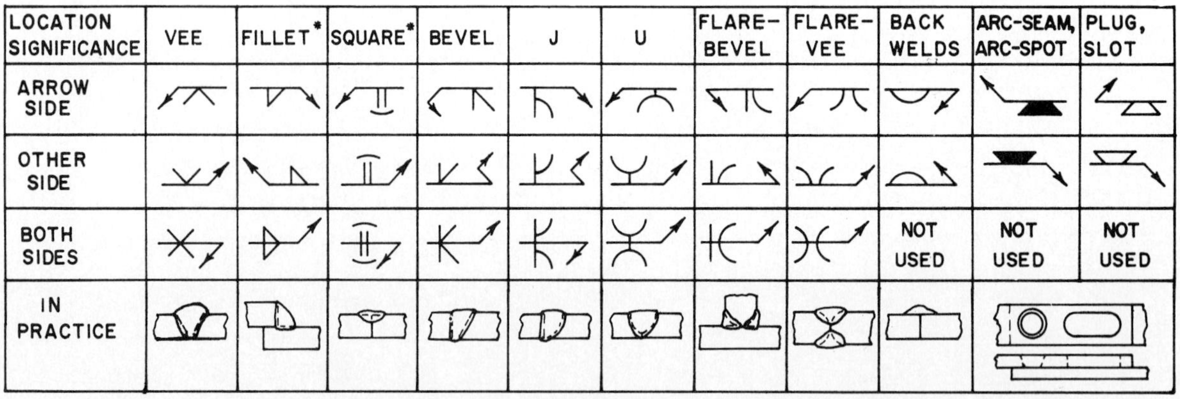

LOCATION SIGNIFICANCE	VEE	FILLET*	SQUARE*	BEVEL	J	U	FLARE-BEVEL	FLARE-VEE	BACK WELDS	ARC-SEAM, ARC-SPOT	PLUG, SLOT
ARROW SIDE											
OTHER SIDE											
BOTH SIDES									NOT USED	NOT USED	NOT USED
IN PRACTICE											

* NOTE: THERE IS NO JOINT PREPARATION ON FILLET & SQUARE WELDS. THE SYMBOLS FOR THESE REPRESENT THE WELDS; OTHERS REPRESENT THE PREPARATION FOR THE WELD.

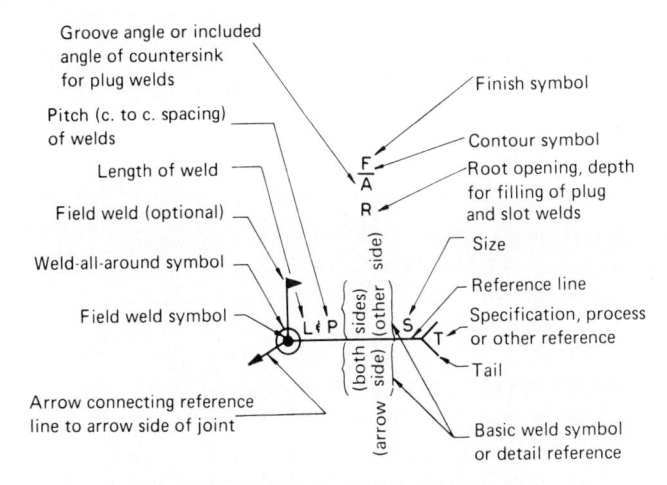

Fig. 12-6 Gas and arc weld symbols

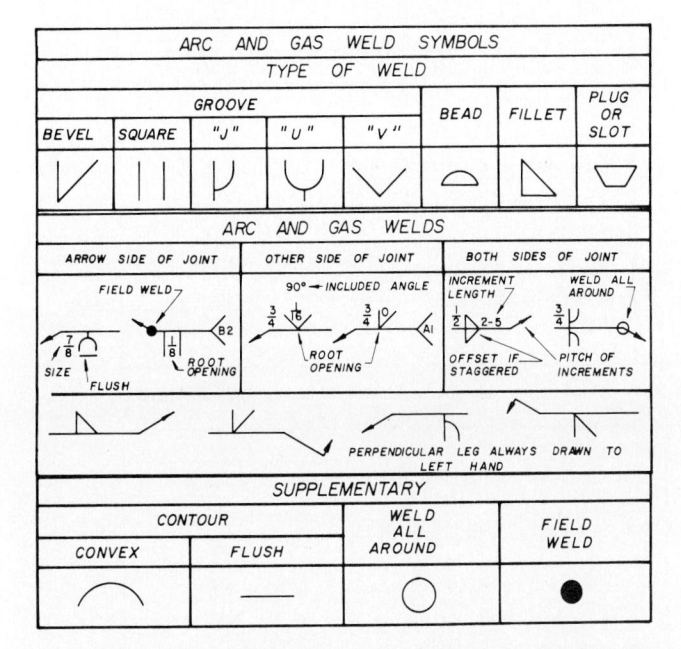

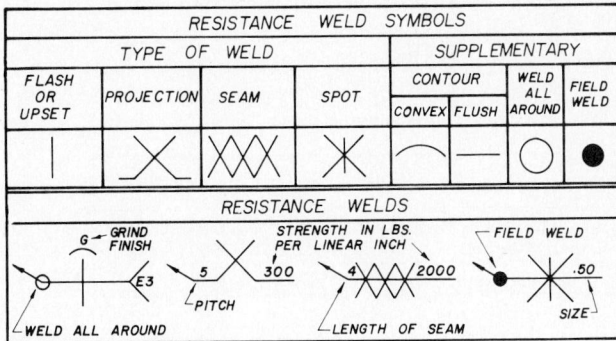

Fig. 12-7 Weld symbols

Note
1. The side of the joint to which the arrow points is the arrow side.
2. Welds on both sides of a joint of the same type are of the same size.
3. Symbols apply between abrupt changes in direction of joint or as dimensioned unless the all-around symbol is used.
4. Welds are continuous and of user's standard proportions.
5. The tail of an arrow may be used for specification reference.
6. Dimensions of weld size, increment length, and spacing are in inches.

TABLE 12-1   ELECTRODE DATA *(COURTESY ITT GRINNELL CORPORATION.)*

COATED ARC WELDING ELECTRODES
Types or Styles

A. W. S. Classification

E 6010 **Direct Current, Reverse Polarity, All Positions.**
All purpose. Moderately smooth finish. Good penetration. This is the electrode used for most carbon steel pipe welding.

E 6011 **Alternating Current, All Positions.**
All purpose. Moderately smooth finish. Good penetration.

E 6012 **Direct Current, Straight Polarity, All Positions.**
High bead. Smooth. Fast. "Cold rod".

E 6013 **Alternating Current, All Positions.**
High bead. Smooth. Fast. "Cold rod".

E 6015 **Direct Current, Reverse Polarity, All Positions.**
"Low hydrogen" electrode.

E 6016 **Direct Current or Alternating Current, All Positions.**
"Low hydrogen" electrode.

E 6018 **Direct Current, All Positions.**
"Low hydrogen" iron powder electrodes

E 6020 **Direct Current, Straight Polarity, Flat Position Only.**
Flat bead. Smooth. Fast. Deep penetration. Can be used with A.C. also. "Hot rod".

E 6024 and **E 6027** **Direct Current, Straight Polarity or Alternating Current, Flat Position Only.**
Flat bead. Smooth. Fast. Deep penetration. "Iron powder electrodes".

NOTE:
This information also applies to E 70, E 80, E 90, and E 100 Series.
The last two numbers (in bold type) designate the types or styles and the first two numbers the minimum specified tensile strength in 1000 psi of the weld deposit as welded.

PHYSICAL PROPERTIES OF E60 AND E70 SERIES ELECTRODES

AWS - ASTM ELECTRODE	TENS. STRENGTH	YIELD STRENGTH	ELONGATION	RED. IN AREA MIN. %
TYPICAL VALUES				
E6010	62,000-70,000	52,000-58,000	22 to 28%	35
E6011	62,000-73,000	52,000-61,000		
E6012	68,000-78,000	55,000-65,000	17 to 22%	25
MINIMUM VALUES				
E7010	70,000	57,000	22	
E7011	70,000	57,000	22	
E7015	70,000	57,000	22	
E7016	70,000	57,000	22	
E7020	70,000	52,000	25	

WELDING

TABLE 12-2 TROUBLESHOOTING ARC WELDING EQUIPMENT (*COURTESY ITT GRINNELL CORPORATION.*)

Trouble	Cause	Remedy
Welder will not start (Starter not operating)	Power circuit dead.	Check voltage.
	Broken power lead.	Repair.
	Wrong supply voltage.	Check nameplate against supply.
	Open power switches.	Close.
	Blown fuses.	Replace.
	Overload relay tripped.	Let set cool. Remove cause of overloading.
	Open circuit to starter button.	Repair.
	Defective operating coil.	Replace.
	Mechanical obstruction in contactor.	Remove.
Welder will not start (Starter operating)	Wrong motor connections.	Check connection diagram.
	Wrong supply voltage.	Check nameplate against supply.
	Rotor stuck.	Try turning by hand.
	Power circuit single-phased.	Replace fuse; repair open line.
	Starter single-phased.	Check contact of starter tips.
	Poor motor connection.	Tighten.
	Open circuit in windings.	Repair.
Starter operates and blows fuse	Fuse too small.	Should be two to three times rated motor current.
	Short circuit in motor connections.	Check starter and motor leads for insulation from ground and from each other.

Trouble	Cause	Remedy
Welder runs but soon stops	Wrong relay heaters.	Renewal part recommendations.
	Welder overloaded.	Considerable overload can be carried only for a short time.
	Duty cycle too high.	Do not operate continually at overload currents.
	Leads too long or too narrow in cross section.	Should be large enough to carry welding current without excessive voltage drop.
	Power circuit single-phased.	Check for one dead fuse or line.
	Ambient temperature too high.	Operate at reduced loads where temperature exceeds 100° F.
	Ventilation blocked.	Check air inlet and exhaust openings.
Welding arc is loud and spatters excessively	Current setting too high.	Check setting and output with ammeter.
	Polarity wrong.	Check polarity, try reversing, or an electrode of opposite polarity.
Welding arc sluggish	Current too low.	Check output, and current recommended for electrode being used.
	Poor connections.	Check all electrode-holder, cable, and ground-cable connections. Strap iron is poor ground return.
	Cable too long or too small.	Check cable voltage drop and change cable.
Touching set gives shock	Frame not grounded.	Ground solidly.
Generator control fails to vary current	Any part of field circuit may be short circuited or open circuited.	Find faulty contact and repair.

TABLE 12-2 (contd.)

Trouble	Cause	Remedy
Welder starts but will not deliver welding current	Wrong direction of rotation.	See INITIAL STARTING.
	Brushes worn or missing.	Check that all brushes bear on commutator with sufficient tension.
	Brush connections loose.	Tighten.
	Open field circuit.	Check connection to rheostat, resistor, and auxiliary brush studs.
	Series field and armature circuit open.	Check with test lamp or bell ringer.
	Wrong driving speed.	Check nameplate against speed of motor or belt drive.
	Dirt, grounding field coils.	Clean and reinsulate.
	Welding terminal shorted.	Electrode holder or cable grounded.
Welder generating but current falls off when welding	Electrode or ground connection loose.	Clean and tighten all connections.
	Poor ground.	Check ground-return circuit
	Brushes worn off.	Replace with recommended grade. Sand to fit. Blow out carbon dust.
	Weak brush spring pressure.	Replace or readjust brush springs.
	Brush not properly fitted.	Sand brushes to fit.
	Brushes in backwards.	Reverse.
	Wrong brushes used.	Renewal part recommendations.
	Brush pigtails damaged.	Replace brushes.
	Rough or dirty commutator.	Turn down or clean commutator.
	Motor connection single-phased.	Check all connections.

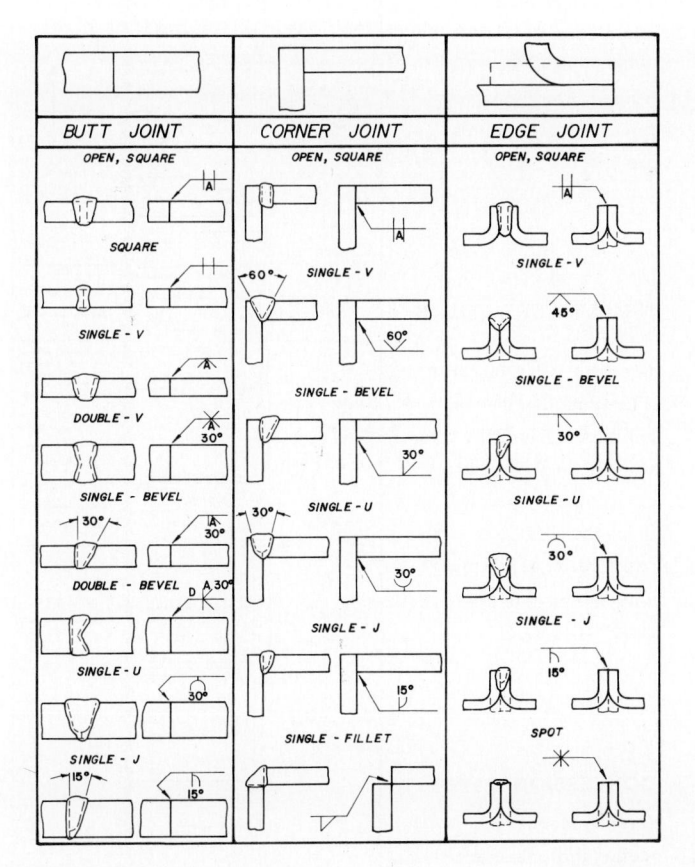

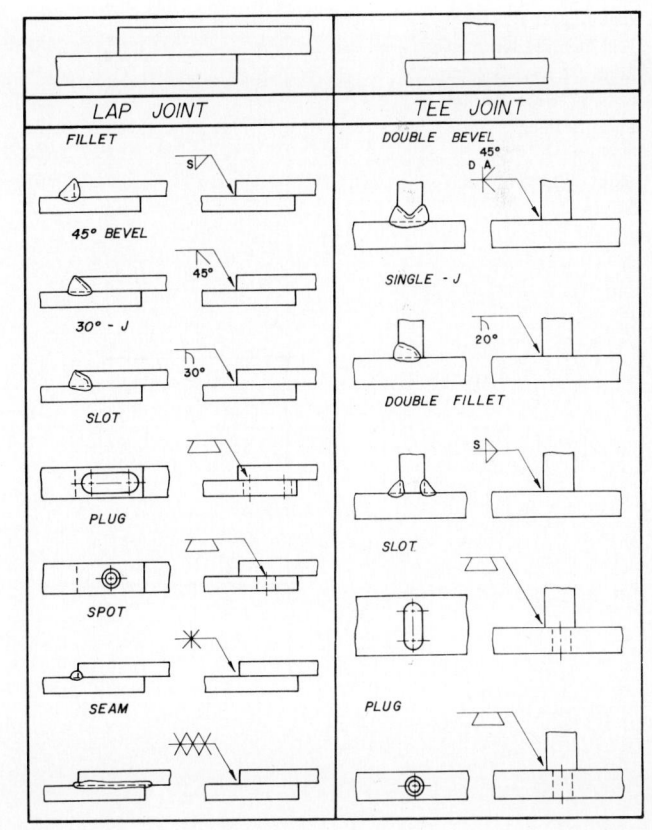

Fig. 12-8 Welding joints

WELDING

TABLE 12-3 WELDING AND BRAZING TEMPERATURES (*COURTESY ITT GRINNELL CORPORATION.*)

Carbon Steel Welding	2700-2790°F
Stainless Steel Welding	2490-2730°F
Cast Iron Welding	1920-2500°F
Copper Welding and Brazing	1980°F
Brazing Copper-Silicon with Phosphor-Bronze	1850-1900°F
Brazing Naval Bronze with Manganese Bronze	1600-1700°F
Silver Solder	1175-1600°F
Low Temperature Brazing	1175-1530°F
Soft Solder	200-730°F
Wrought Iron	2700-2750°F

TABLE 12-4 TEMPERATURE DATA CHART (*COURTESY ITT GRINNELL CORPORATION.*)

COLORS AND APPROXIMATE TEMPERATURE FOR CARBON STEEL

Black Red	990°F
Dark Blood Red	1050
Dark Cherry Red	1175
Medium Cherry Red	1250
Full Cherry Red	1375
Light Cherry, Scaling	1550
Salmon, Free Scaling	1650
Light Salmon	1725
Yellow	1825
Light Yellow	1975
White	2220

MELTING POINTS

This chart contains basic information on working with metals at elevated temperatures. The most commonly used metals are listed.

TEMPERATURE COLOR SCALE

Another use for the chart is in estimating the temperature of metals by color when no heat measuring devices are available. Using the chart is, in most cases, faster, while maintaining a good degree of accuracy.

CONVERSION DATA

A ready means for converting fahrenheit to centigrade is also provided.

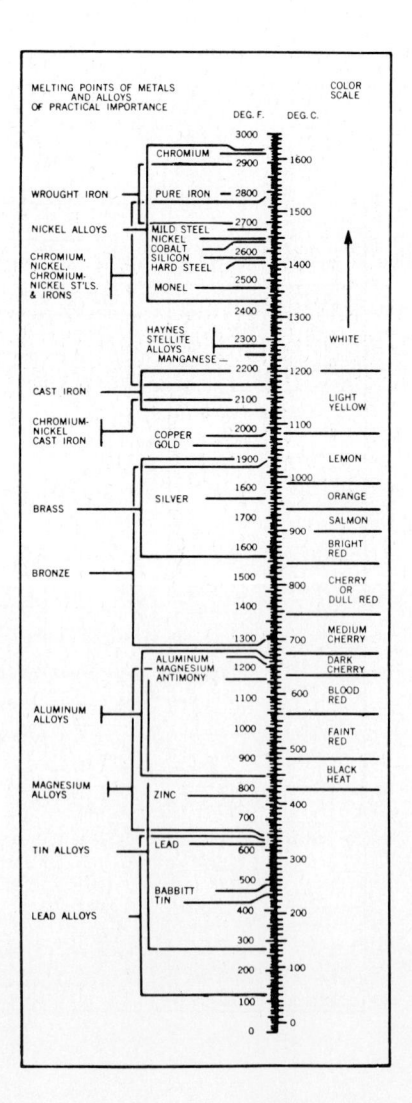

13
PIPE FLOW

Fig. 13-1

FLOW OF FLUIDS*

In the field of hydraulics an important engineering problem is the calculation of the energy loss caused by the resistance of fluids to movement through pipe, valves, fittings, or other enclosed channels. Many empirical equations have been developed to determine the energy loss for specific fluids and flow conditions. Most of these empirical equations are based on the following equations:

$$h = fL \frac{v^x}{d^y}$$

The friction factor (f), the exponential coefficient (x) for the velocity, and the exponential coefficient (y) for the diameter are based on experimental data. These empirical equations give very good results when they are used for fluids and flow conditions within the limits for which they were developed. However, large errors may result if an empirical equation is used for conditions beyond the limits for which it was developed.

The pipe-friction equation was developed by the use of fundamental hydraulic principles and dimensional analysis to determine the energy loss for the steady flow of incompressible fluids through circular cross section pipe. The names Fanning, Darcy, and Weisbach have been associated with variations of the following pipe-friction equation:

$$h = f \frac{L v^2}{D \, 2g}$$

The friction factor was determined experimentally, and it was considered to be constant for a limited range of conditions. The pipe-friction equation gives good results when it is used within the limits for which the friction factor was determined. Investigations have shown that the friction factor is actually a function of the pipe diameter, roughness of the pipe, average flow velocity, fluid density, and absolute viscosity of the fluid. These variables, except for the roughness of the pipe, can be expressed as a dimensionless ratio known as the Reynolds number (R) as shown by the following equation:

$$R = \frac{Dv\rho}{\mu}$$

No simple mathematical equations have been developed to show the relationship between friction factor and Reynolds number for all flow conditions. Curves developed by R. S. Pigott and Emery Kemler to show the relationship between Reynolds number and friction factor are generally used to determine the friction factor.

When the Reynolds number is less than 1200, the flow is considered viscous or laminar, and the friction factor is independent of the roughness of the pipe. Therefore, the friction factor is proportional to the Reynolds number as shown by the following equation:

$$f = \frac{64}{R}$$

When the Reynolds number is between 1200 and 2500, the flow can be either laminar, turbulent, or a combination of both. This region is called the critical zone. Since the friction factor depends on the condition of flow, it is impossible to determine accurately the friction factor in this region. However, it is general practice to consider the flow as being turbulent in the critical zone. When the Reynolds number is less than 2500, it is general practice to consider the flow as being turbulent in the critical zone. When the Reynolds number is greater than 2500, the flow is considered turbulent and the friction factor depends not only on the Reynolds number but also on the size and roughness of the pipe. The friction factor is usually obtained from the curves developed by Pigott and Kemler.

The pipe-friction equation has been modified to include the use of Reynolds number for obtaining the friction factor as shown by the following equation:

$$h = \text{function of } \frac{Dv\rho}{\mu} \frac{L v^2}{D \, 2g}$$

The use of the modified equation is generally called the rational method. Although these equations were developed for steady flow of incompressible fluids, they can be used for steady flow of compressible fluids, providing the change in density of the fluid does not exceed 10 percent of the fluid condition. There are many variations of the pipe-friction equation; two equations frequently used are as follows:

- For viscous flow:
$$\Delta p = .000668 \frac{\mu L v}{d^2}$$

---

*(Courtesy the Lunkenheimer Company, a Division of Conval Corporation.)

- For turbulent flow:
$$\Delta p = .00129\ f\ \frac{L\rho v^2}{d}$$

Δp = pressure loss (In pounds per square inch)

L = Length of pipe (In feet)

v = average velocity of fluid in pipe (In feet per second)

d = inside diameter of pipe (In inches)

f = friction factor

K = flow resistance coefficient

h = head loss (In feet of fluid)

μ = absolute viscosity (In pounds per foot per second)

ρ = density (In pounds per cubic foot)

D = inside diameter of pipe (In feet)

$R = \dfrac{Dv\rho}{\mu}$ (Reynolds number)

g = acceleration of gravity (32.2 ft/sec²)

Investigations have shown that the energy loss for steady flow through valves and fittings can be expressed by the following valve and fittings equation:

$$h = K\ \frac{v^2}{2g}$$

The flow resistance coefficient (K) is related to the velocity of the fluid in nominal size pipe connected to the valve or fitting, and it remains relatively constant for a specific type of valve or fitting.

It is generally accepted that the energy loss through valves and fittings is affected by the same variables that affect the energy loss for flow through straight pipe. This fact makes it possible to express the flow resistance of valves and fittings as being equivalent to a length of straight pipe. By combining the pipe-friction equation with the valve and fitting equation, the flow resistance coefficient can be expressed by the following equation.

$$K = f\ \frac{L}{D}$$

TABLE 13-1 FLOW RESISTANCE FOR VALVES AND FITTINGS (*COURTESY LUNKENHEIMER COMPANY, CINCINNATI, OHIO 45214.*)

	GATE VALVE	ANGLE VALVE	60°-Y VALVE	GLOBE VALVE	45° ELBOW	90° ELBOW	TEE THRU RUN	TEE THRU BRANCH	180° CLOSE RETURN
COEFFICIENT	.2	5	6	10	.4	.9	.5	1.8	2
RATIO	7	180	216	360	14	32	18	64	72

SCHEDULE 40 PIPE										
Final Pipe Size Inches	Inside Diameter inches	\multicolumn{9}{c}{L—EQUIVALENT LENGTH OF SCHEDULE 40 PIPE IN FEET}
½	.622	.4	10	12	20	.8	1.8	1.0	3.6	4.0
¾	.824	.5	13	15	25	1.0	2.3	1.3	4.5	5.0
1	1.049	.6	15	18	30	1.2	2.7	1.5	5.4	6.0
1¼	1.380	.8	20	24	40	1.6	3.6	2.0	7.2	8.0
1½	1.610	1.0	25	30	50	2.0	4.5	2.5	9.0	10
2	2.067	1.2	30	36	60	2.4	5.4	3.0	11	12
2½	2.469	1.5	38	45	75	3.0	6.8	3.8	14	15
3	3.068	2.0	50	60	100	4.0	9.0	5.0	18	20
4	4.026	2.4	60	72	120	4.8	11	6.0	22	24
5	5.047	3.0	75	90	150	6.0	14	7.5	28	30
6	6.065	3.6	90	108	180	7.2	16	9.0	32	36
8	7.981	4.8	120	144	240	9.6	22	12	44	48
10	10.02	6.0	150	180	300	12	27	15	54	60
12	11.94	7.2	180	216	360	14	32	18	64	72
14	13.13	8.0	200	240	400	16	36	20	72	80
16	15.00	9.0	225	270	450	18	40	23	80	90
18	16.88	10	250	300	500	20	45	25	90	100
20	18.81	11	280	336	560	22	50	28	100	110
24	22.63	14	340	408	680	27	60	34	120	140

Note: Flow Resistance Factors are average values based on data for flow of water through valves and fittings.

PIPE FLOW

By using an average friction factor for turbulent flow, this equation can be used to determine the approximate length of pipe that will be equivalent to any valve or fitting for which the flow resistance coefficient has been established. The equivalent length of straight pipe can be used to determine the approximate pressure drop for the required flow conditions by using equations or tables developed for determining the pressure drop for flow through straight pipe.

TABLE 13-2 FLOW CONVERSION CHART *(COURTESY ITT GRINNELL CORPORATION.)*

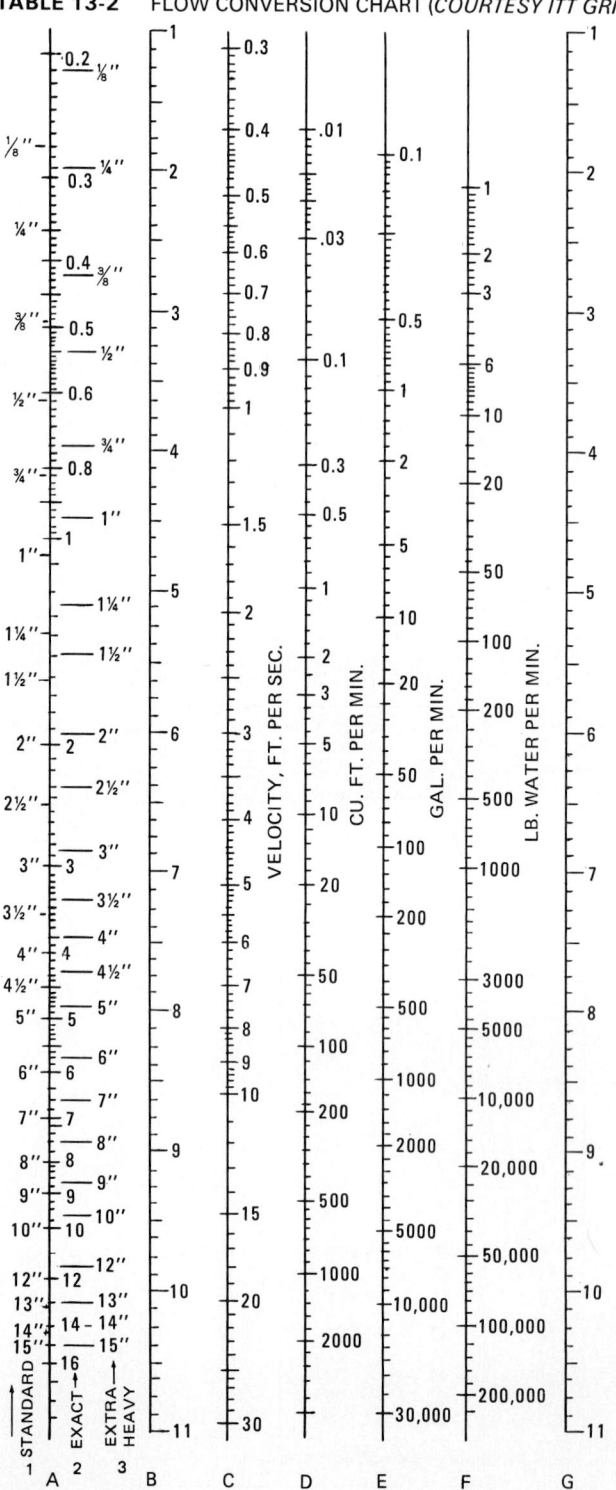

FLOW CONVERSION CHART

The accompanying chart provides fast answers to many problems that may confront the pipe fitter. Procedures for using the chart are as follows:

Note that there are three sets of figures shown in connection with the extreme left-hand column A. The column marked "1 in. standard" gives the internal diameter of standard pipe (somewhat greater than 1 for 1 in. standard pipe). The column marked "2 exact" gives the exact diameter. The column marked "3 extra heavy" gives the internal diameter of extra heavy pipe.

EXAMPLE: How much water is passing through a pipe having an I.D. of exactly 1 in. the velocity of the water being 3 F.P.S.? To apply the chart to the problem locate 1 in. in column A over the word "exact" and run a straight line from the point through the 3 in column C. From the intersection of this line with column B, run a straight line horizontally to column G. The intersection of this line at columns D, E and F gives the following information:

Column D shows the cubic feet/minute flowing through the pipe; column E shows the volume of flow in gallons/minute; column F gives the weight of the water in pounds/minute. (For liquids other than water, multiply the value of column F by the specific gravity of the liquid for accurate weight conversion.) See chart page 31.

If a quantity in columns D, E or F is known then velocity may be determined by reversing the procedure. Draw a horizontal line from the known point to column G. From this intersection draw a line to the exact I.D. of the pipe in column A and extend this line to cross column C. The intersection with column C gives the velocity in feet/second.

The chart can be used as a conversion chart to determine the number of gallons in a certain number of cubic feet of liquid. The horizontal line already drawn to determine answers in columns C and D will provide the answer to the conversion in column E.

A little practice will prove this chart to be a real time-saver.

CHAPTER 13

TABLE 13-3 WATER PRESSURE TO FEET HEAD (*COURTESY ITT GRINNELL CORPORATION.*)

POUNDS PER SQUARE INCH	FEET HEAD	POUNDS PER SQUARE INCH	FEET HEAD
1	2.31	100	230.90
2	4.62	110	253.93
3	6.93	120	277.07
4	9.24	130	300.16
5	11.54	140	323.25
6	13.85	150	346.34
7	16.16	160	369.43
8	18.47	170	392.52
9	20.78	180	415.61
10	23.09	200	461.78
15	34.63	250	577.24
20	46.18	300	692.69
25	57.72	350	808.13
30	69.27	400	922.58
40	92.36	500	1154.48
50	115.45	600	1385.39
60	138.54	700	1616.30
70	161.63	800	1847.20
80	184.72	900	2078.10
90	207.81	1000	2309.00

NOTE: One pound of pressure per square inch of water equals 2.309 feet of water at 62° Fahrenheit. Therefore, to find the feet head of water for any pressure not given in the table above, multiply the pressure pounds per square inch by 2.309.

TABLE 13-4 FEET HEAD OF WATER TO PSI (*COURTESY ITT GRINNELL CORPORATION.*)

FEET HEAD	POUNDS PER SQUARE INCH	FEET HEAD	POUNDS PER SQUARE INCH
1	.43	100	43.31
2	.87	110	47.64
3	1.30	120	51.97
4	1.73	130	56.30
5	2.17	140	60.63
6	2.60	150	64.96
7	3.03	160	69.29
8	3.46	170	73.63
9	3.90	180	77.96
10	4.33	200	86.62
15	6.50	250	108.27
20	8.66	300	129.93
25	10.83	350	151.58
30	12.99	400	173.24
40	17.32	500	216.55
50	21.65	600	259.85
60	25.99	700	303.16
70	30.32	800	346.47
80	34.65	900	389.78
90	38.98	1000	433.00

NOTE: One foot of water at 62° Fahrenheit equals .433 pound pressure per square inch. To find the pressure per square inch for any feet head not given in the table above, multiply the feet head by .433.

TABLE 13-5 PROPERTIES OF LIQUIDS AND GASES (*COURTESY ITT GRINNELL CORPORATION.*)

SPECIFIC GRAVITY OF GASES

Dry Air (1 cu. ft. at 60°F. and 29.92" Hg. weighs .07638 pound) 1.000

Acetylene	C_2H_2	0.91
Ethane	C_2H_6	1.05
Methane	CH_4	0.554
Ammonia	NH_3	0.596
Carbon-dioxide	CO_2	1.53
Carbon-monoxide	CO	0.967
Butane	C_4H_{10}	2.067
Butene	C_4H_8	1.93
Chlorine	Cl_2	2.486
Helium	He	0.138
Hydrogen	H_2	0.0696
Nitrogen	N_2	0.9718
Oxygen	O_2	1.1053

SPECIFIC GRAVITY OF LIQUIDS

LIQUID	TEMP °F	SPECIFIC GRAVITY
Water (1 cu.-ft. weighs 62.41 lb.)	50	1.00
Brine (Sodium Chloride 25%)	32	1.20
Pennsylvania Crude Oil	80	0.85
Fuel Oil No. 1 and 2	85	0.95
Gasoline	80	0.74
Kerosene	85	0.82
Lubricating Oil SAE 10-20-30	115	0.94

TYPICAL BTU VALUES OF FUELS

ASTM RANK
SOLIDS — BTU VALUES PER POUND

Anthracite Class I	11,230
Bituminous Class II Group 1	14,100
Bituminous Class II Group 3	13,080
Sub-Bituminous Class III Group 1	10,810
Sub-Bituminous Class III Group 2	9,670

LIQUIDS — BTU VALUES PER GAL.

Fuel Oil No. 1	138,870
Fuel Oil No. 2	143,390
Fuel Oil No. 4	144,130
Fuel Oil No. 5	142,720
Fuel Oil No. 6	137,275

GASES — BTU VALUES PER CU. FT.

Natural Gas	935 to 1132
Producers Gas	163
Illuminating Gas	534
Mixed (Coke oven and water gas)	545

TABLE 13-6 BOILING POINTS OF WATER AT VARIOUS PRESSURES (*COURTESY ITT GRINNELL CORPORATION.*)

VACUUM, IN INCHES OF MERCURY	BOILING POINT	VACUUM, IN INCHES OF MERCURY	BOILING POINT
29	76.62	7	198.87
28	99.93	6	200.96
27	114.22	5	202.25
26	124.77	4	204.85
25	133.22	3	206.70
24	140.31	2	208.50
23	146.45	1	210.25
22	151.87	Gauge Lbs.	
21	156.75	0	212.0
20	161.19	1	215.6
19	165.24	2	218.5
18	169.00	4	224.4
17	172.51	6	229.8
16	175.80	8	234.8
15	178.91	10	239.4
14	181.82	15	249.8
13	184.61	25	266.8
12	187.21	50	297.7
11	189.75	75	320.1
10	192.19	100	337.9
9	194.50	125	352.9
8	196.73	200	387.9

PIPE FLOW

TABLE 13-7 FLOW OF WATER—EQUALIZATION OF PIPE DISCHARGE RATES FOR PIPE AND COPPER TUBING (*COURTESY ITT GRINNELL CORPORATION.*)

Pipe Size, in.	1/8	1/4	3/8	1/2	3/4	1	1 1/4	1 1/2	2	2 1/2	3	3 1/2	4	5	6	8	10	12
1/8		2.1	4.5	8.1	16	30	60	88	164	255	439	632	867	1525	2414	4795	8468	13292
1/4	2.3		2.1	3.8	7.7	14	28	41	77	120	206	297	407	716	1133	2251	3976	6240
3/8	5.4	2.3		1.8	3.6	6.6	13	19	36	56	97	139	191	335	531	1054	1862	2923
1/2	10	4.4	1.9		2.0	3.7	7.3	11	20	31	54	78	107	188	297	590	1042	1635
3/4	22	9.5	4.1	2.2		1.8	3.6	5.3	10	16	27	38	53	93	147	292	516	809
1	42	18	7.7	4.1	1.9		2.0	2.9	5.5	8.5	15	21	29	51	80	160	282	443
1 1/4	86	37	16	8.4	3.9	2.1		1.5	2.7	4.3	7.4	11	15	26	40	80	142	223
1 1/2	129	55	24	13	5.8	3.1	1.5		1.9	2.9	5.0	7.2	9.9	17	28	55	97	152
2	244	104	45	24	11	5.8	2.8	1.9		1.6	2.7	3.9	5.3	9.3	15	29	52	81
2 1/2	384	164	71	37	17	9.2	4.5	3.0	**1.6**		1.7	2.5	3.4	6.0	9.5	19	33	52
3	668	286	123	65	30	16	7.8	5.2	**2.7**	**1.7**		1.4	2.0	3.5	5.5	11	19	30
3 1/2	968	414	178	94	44	23	11	7.5	**4.0**	**2.5**	**1.4**		1.4	2.4	3.8	7.6	13	21
4	1336	571	246	130	60	32	16	10	**5.5**	**3.5**	**2.0**	**1.4**		1.8	2.8	5.5	9.8	15
5	2371	1014	437	231	107	57	28	18	**9.7**	**6.2**	**3.5**	**2.4**	**1.8**		1.6	3.1	5.6	8.7
6	3717	1589	685	362	168	89	43	29	**15**	**9.7**	**5.6**	**3.8**	**2.8**	**1.6**		2.0	3.5	5.5
8	7490	3203	1380	729	339	179	87	58	**31**	**20**	**11**	**7.7**	**5.6**	**3.2**	**2.0**		1.8	2.8
10	13849	5922	2551	1348	626	331	161	108	**57**	**36**	**21**	**14**	**10**	**5.8**	**3.7**	**1.8**		1.4
12	22079	9442	4067	2148	998	528	256	172	**90**	**58**	**33**	**23**	**17**	**9.3**	**5.9**	**2.9**	**1.4**	

STANDARD WALL / STANDARD WALL / EXTRA STRONG WALL

The tabulated values show the number of pipes of one size required to equal the delivery of a larger pipe under the same head. Thus it requires 29 2" Std. pipes to equal one 8" Std. pipe.

The light face figures pertain to Standard Wall Pipe, and the bold face figures cover the Extra Strong Wall Pipe. (Under the same head the velocity is less in the smaller pipes and the discharge varies as the 5/2 power of the inside diameter.)

FLOW OF WATER—EQUALIZATION OF COPPER TUBING DISCHARGE RATES*

Tube Size, in.	3/8	1/2	3/4	1	1 1/4	1 1/2	2	2 1/2	3	3 1/2	4
3/8		1.8	4.5	8.8	15	23	46	79	124	180	250
1/2	2.0		2.5	4.8	8.2	13	25	43	68	99	137
3/4	4.7	2.4		2.0	3.3	5.1	10	17	27	40	55
1	9.6	4.9	2.1		1.7	2.6	5.2	9.0	14.1	20	28
1 1/4	17	8.6	3.6	1.8		1.5	3.1	5.3	8.3	12	17
1 1/2	25	13	5.4	2.6	1.5		2.0	3.4	5.4	7.8	11
2	52	27	11	5.4	3.1	2.1		1.7	2.7	3.9	5.4
2 1/2	90	46	19	9.4	5.4	3.6	**1.7**		1.6	2.3	3.2
3	140	71	30	15	8.3	5.6	**2.7**	**1.6**		**1.5**	2.0
3 1/2	205	104	44	21	12	8.2	**3.9**	**2.3**	**1.5**		1.4
4	284	145	61	30	17	11	**5.5**	**3.2**	**2.0**	**1.4**	

TYPE K / TYPE L

*The tabulated values show the number of tubes of one size required to equal the delivery of a larger tube under the same head. Thus it requires 28 1" Type L tubes to equal one 4" Type L tube.

The light face figures pertain to Type L tube and the bold face figures cover type K tube. (Under the same head the velocity is less in the smaller tubes and the discharge varies at the 5/2 power of the inside diameter.)

14
INSULATION

FIG. 14-1   (*Courtesy Manville Products Corporation.*)

TABLE 14-1 THERMAL INSULATION (COURTESY MANVILLE PRODUCTS CORPORATION.)

Insulation material	Insulation form	Insulation use	Insulation covering	Insulation Operation Temperature	Pipe diameter	Widths of Blocks	Length	Thickness
Asbestos fiber with diatomaceous silica	Rigid sheets	Breechings, ovens, housings, driers	Not required	Up to 900°F	—	3'-4'	8'-10'	$\frac{1}{2}''-2''$
Asbestos paper, laminated, corrugated	Formed pipe covering Blocks Sheets	Steam lines and underground lines Cold-water lines Boilers, ovens	Canvas jacket	Up to 300°F	3''-12''	6''-36''	36''	$\frac{1}{2}''-2''$
Calcined diatomaceous silicate-bonded with asbestos fibers	Molded sections Blocks	Boilers, furnaces, hot-blast stoves, producer gas mains	Not required. May have cement coating	Up to 1900°F	—	3''-12''	18'' and 36''	1''-4''
85% Magnesia	Molded sections Blocks	Piping and heating equipment	Canvas jacket, Asbestos cement, asphalt-saturated asbestos roofing felt, metal jacket, asbestos cloth	Up to 600°F	—	3''-12''	18'' and 36''	1''-5''
Cork (natural)	Molded pipe covering Sheets Granulated	Cold pipelines and tanks, cold-storage rooms	Plaster mastic coating Portland cement plaster	−25°F to +60°F	—	18''	36''	1''-6''
Rock cork	Mineral wool with water-resistant asphaltic binder	Hot and cold piping and equipment	Asphalt-saturated and -coated asbestos jacket	−300°F to +350°F	$\frac{1}{2}''-14''$	18''	18'' and 36''	1''-4''
Expanded perlite	Loose powder	Insulating fill	—	−400°F to +350°F	—	—	—	—
Foamed plastic	Flexible molded tubing and sheets Rigid molded sections, blocks, boards	Heating and cooling lines, refrigerant lines, air ducts Freezers, coolers, buildings	Facing may be part of insulating material Surfaced with cement, plaster, asphalt emulsion	−100°F to +220°F −100°F to +180°F	$\frac{3}{8}''-4''$ $\frac{1}{2}''-30''$	36'' 12''-24''	60'' 48'' 18'' and 36'' 48''-144''	$\frac{1}{8}''-1''$ $\frac{1}{5}''-4''$ 1''-8''

138

Glass fiber	Rigid boards	Cold-storage rooms	Sheathed in cold-storage grade asphalt	−50°F to +100°F	—	12"	36"	1"–4"
	Rigid blocks	Tanks and heated equipment	Not required. Can be covered	Up to 450°F	—	6" and 12"	36"	$\frac{1}{2}$"–2"
	Molded half-cylinders and segments	Hot and cold pipes	Vapor barrier	Up to 450°F	To 33"	—	36"	1"–5$\frac{1}{2}$"
	Semirigid board with hard aluminum facing	Forming rectangular ducts	Hard aluminum, shaped to form duct lined with glass fibers	Up to 250°F	—	48"	120"	1" and 1$\frac{1}{2}$"
	One-piece tubing	Lightweight covering for pipes	Jackets of canvas, kraft foil, laminate with self-extinguishing adhesive and fiber glass reinforcement, or roofing felt.	−30°F to +400°F	$\frac{1}{2}$"–12"	—	36"	$\frac{1}{2}$"–2"
	Prefabricated round tubes	Air ducts	Aluminum or vinyl plastic casing	Up to 250°F	4"–36"	—	72"	1" and 1$\frac{1}{2}$"
	Semirigid blanket of glass fiber and thermosetting resin	Duct covering Duct lining Heating and cooling equipment	Plain or coated; with vapor barrier and flame-retardant facings.	Up to 350°F	—	24"–48"	200"	$\frac{1}{2}$"–3"
	Wraparound blanket	Pipe covering	Roofing felt Asbestos-finished felt Corrugated aluminum Sheet metal	−100°F to +600°F	—	36"	100'	$\frac{1}{2}$"–1$\frac{1}{2}$"
	Blanket faced with wire mesh or wire lath	Heating and cooling equipment	Cement or mastic	−100°F to +1000°F	—	24"	96"	1"–4$\frac{1}{2}$"
	Rolls, batts, bulk, shredded	Ovens Prefabricated equipment Expansion joints and irregular spaces	Metal lath or sheet metal	Up to 1000°F	—	6"–48"	54"–120"	$\frac{3}{4}$"–3$\frac{1}{4}$"

TABLE 14-1 (CONT'D.)

Insulation material	Insulation form	Insulation use	Insulation covering	Insulation Operation Temperature	Pipe diameter	Widths of Blocks	Length	Thickness
Gilsonite	Granular fill	Underground hotlines	—	Up to 520°F	—	—	—	To 6"
Hydrous calcium silicate-bonded with asbestos fibers	Molded pipe coverings Blocks	Piping and equipment Underground piping	Not required indoors Canvas facing Outside can be pre-jacketed with aluminum, roofing felt, or coat of weatherproof asphalt mastic	Up to 1200°F	To 33"	6"–18"	24" and 36" 36"	1"–3"
Refractory fiber-felt	Alumina and silica fiber blanket	Gas turbines Lining for oil-burner combustion chamber High-temperature packing	Prefabricated with metal foils or meshes	Up to 2000°F	—	42"	4'–26'	$\frac{1}{4}$"–2"
Silica aggregate bonded	Rigid pipe covering	Hot and cold lines Low-pressure steam pipes	Canvas, kraft bonded to aluminum	Up to 300°F	$\frac{3}{8}$"–12"	—	36"	$\frac{3}{4}$"–2"
Siliceous fibers felted	Rigid board Wraparound blanket	Air ducts Housings and plenum chambers	Vapor barriers of kraft paper and asphalt, aluminum foil scrimkraft Plain asbestos paper	Up to 450°F	—	15" and 30"	48"–120"	1"–3"
Spun mineral fibers	Felted blankets	Air ducts, breechings, ovens, driers	Wire mesh, ribbed metal lath, saturated asbestos felt	Up to 1000°F	—	24"	48" and 96"	1"–6"
	Felted fibers with heat-resistant binder formed in blocks	Furnaces, boilers, heating vessels, tanks	For outside use, covered with cement finish over wire mesh	Up to 1900°F	—	6"–24"	18" and 36"	1"–5"

TABLE 14-2 PROPERTIES OF INSULATING MATERIALS *(COURTESY MANVILLE PRODUCTS CORPORATION.)*

Material	Fire Rating	k Factor at Mean Temperature	Density, Pounds per Cubic Foot	Water Absorption Percentage of Volume	Compressive Strength, Psi
Asbestos fiber with diatomaceous silica	IC	0.55 at 100 F: 1.80 at 600 F	23-65	3.5	13000-16000
Asbestos paper, laminated, corrugated	IC	0.47 at 100 F: 0.80 at 300 F	—	—	—
Calcined diatomaceous silica	IC	0.67 at 300 F: 0.82 at 1000 F	12-24	0.7	85
85% Magnesia	IC	0.35 at 100 F: 0.46 at 400 F	12	—	70
Cork	F	0.29 to 0.33 at 75 F	7-14	—	4
Rock Cork	IC	0.27 at −40 F: 0.33 at 80 F	15	3	4
Expanded perlite	SE	0.26 at 75 F: 0.30 at 140 F	3	—	—
Foamed plastic	SB-SE	0.21 at −20 F: 0.30 at 125 F	1-7	0-4	10-25
Gilsonite	...	0.5 to 0.6 at 180 F	40	—	—
Glass fiber	IC	0.17 at −50 F: 0.73 at 400 F	0.5-3.5	—	—
Hydrous calcium silicate	IC	0.33 at 100 F: 0.65 at 800 F	11	—	117-165
Refractory-fiber felt	IC	0.31 at 300 F: 1.16 at 1000 F	3-24	—	—
Silica aggregate	IC	0.34 at 50 F: 0.41 at 200 F	9	—	—
Siliceous fibers, felted	IC	0.23 at 50 F: 0.28 at 100 F	2.5-8	—	—
Spun mineral fibers, felted	IC	0.24 at 100 F: 0.65 at 700 F	8	—	—

SE—Self-extinguishing
F—Flammable
IC—Incombustible
SB—Slow-burning

INSULATION

TABLE 14-3 RECOMMENDED THICKNESS OF HYDROUS CALCIUM SILICATE PIPE INSULATION (*COURTESY MANVILLE PRODUCTS CORPORATION.*)

Nominal Pipe Size, Inches	\multicolumn{11}{c}{Temperature of Pipe, Degrees Fahrenheit}
	100 to 199	200 to 299	300 to 399	400 to 499	500 to 599	600 to 699*	700 to 799*	800 to 899*	900 to 999*	1000 to 1099*	1100 to 1200*
\multicolumn{12}{c}{Nominal Thickness of Insulation, Inches}
\multicolumn{12}{c}{Process}
1½ & less	1	1	1	1	1	1	1½	1½	1½	2	2
2	1	1	1	1	1	1½	1½	1½	2	2	2
2½	1	1	1	1	1	1½	1½	1½	2	2	2
3	1	1	1	1	1	1½	1½	2	2	2	2½
3½	1	1	1	1	1	1½	1½	2	2	2	2½
4	1	1	1	1	1	1½	1½	2	2	2	2½
4½	1	1	1	1	1½	1½	1½	2	2	2½	2½
5	1	1	1	1	1½	1½	1½	2	2	2½	2½
6	1	1	1	1	1½	1½	2	2	2	2½	2½
7	1½	1½	1½	1½	1½	1½	2	2	2	2½	2½
8	1½	1½	1½	1½	1½	1½	2	2	2½	2½	2½
9	1½	1½	1½	1½	1½	1½	2	2	2½	2½	2½
10	1½	1½	1½	1½	1½	1½	2	2	2½	2½	2½
11	1½	1½	1½	1½	1½	1½	2	2	2½	2½	3
12	1½	1½	1½	1½	1½	1½	2	2	2½	2½	3
14 & up	1½	1½	1½	1½	1½	2	2	2	2½	2½	3
\multicolumn{12}{c}{Utility—Steam Generation}
1½ & less	1	1	1½	2**	2**	2½	2½	2½	3	3	3
2	1	1	1½	2**	2**	2½	2½	3	3	3	3½
2½	1	1	1½	2**	2**	2½	2½	3	3	3½	3½
3	1	1	1½	2**	2**	2½	3	3	3	3½	3½
3½	1	1	1½	2**	2½**	2½	3	3	3½	3½	3½
4	1	1	1½	2**	2½**	2½	3	3	3½	3½	4
4½	1	1	1½	2	2½*	2½	3	3	3½	3½	4
5	1	1½	1½	2	2½	2½	3	3½	3½	4	4
6	1	1½	2	2	2½	3	3	3½	3½	4	4
7	1½	1½	2	2	2½	3	3	3½	3½	4	4
8	1½	1½	2	2	2½	3	3½	3½	4	4	4½
9	1½	1½	2	2½	2½	3	3½	3½	4	4	4½
10	1½	1½	2	2½	2½	3	3½	3½	4	4	4½
11	1½	1½	2	2½	2½	3	3½	3½	4	4½	4½
12	1½	1½	2	2½	3**	3	3½	4	4	4½	4½
14 & up	1½	1½	2	2½	3**	3	3½	4	4	4½	4½
\multicolumn{12}{c}{Commercial}
1½ & less	1	1½	2	2½**	3**						
2	1	1½	2	2½**	3**						
2½	1	2	2½	3**	3**						
3	1	2	2½	3**	3½**						
3½	1	2	2½	3**	3½**						
4	1	2	2½	3**	3½**						
4½	1	2	2½	3**	3½**						
5	1	2	2½	3**	3½**						
6	1	2	3	3½**	4**						
7	1½	2	3	3½**	4**						
8	1½	2	3	3½**	4**						
9	1½	2½	3	3½**	4**						
10	1½	2½	3	3½**	4**						
11	1½	2½	3	3½**	4**						
12	1½	2½	3	3½**	4**						
14 & up	1½	2½	3½	4**	4½						

*Double-layer construction recommended.
**Available in single or double layer.

TABLE 14-4 RECOMMENDED THICKNESS OF 85 PERCENT MAGNESIA PIPE INSULATION (*COURTESY MANVILLE PRODUCTS CORPORATION.*)

Nominal Pipe size, Inches	\multicolumn{15}{c}{Temperature of Pipe, Degrees Farenheit}
	100 to 199	200 to 299	300 to 399	400 to 499	500 to 500	100 to 199	200 to 299	300 to 399	400 to 499	500 to 600	100 to 199	200 to 299	300 to 399	400 to 499	500 to 600
	\multicolumn{15}{c}{Nominal Thickness of Insulation, Inches}
	\multicolumn{5}{c}{Utility Steam Generation}	\multicolumn{5}{c}{Process}	\multicolumn{5}{c}{Commercial}
1½ & less	1	1	1½	2	2	1	1	1	1	1	1	1½	2	2½	3
2	1	1	1½	2	2	1	1	1	1	1	1	1½	2	2½	3
2½	1	1	1½	2	2	1	1	1	1	1	1	2	2½	3	3
3	1	1	1½	2	2	1	1	1	1	1	1	2	2½	3	3½
3½	1	1	1½	2	2½	1	1	1	1	1	1	2	2½	3	3½
4	1	1	1½	2	2½	1	1	1	1	1	1	2	2½	3	3½
4½	1	1	1½	2	2½	1	1	1	1	1½	1	2	2½	3	3½
5	1	1½	1½	2	2½	1	1	1	1	1½	1	2	2½	3	3½
6	1	1½	2	2	2½	1	1	1	1	1½	1	2	3	3½	4
7	1½	1½	2	2	2½	1½	1½	1½	1½	1½	1½	2	3	3½	4
8	1½	1½	2	2	2½	1½	1½	1½	1½	1½	1½	2	3	3½	4
9	1½	1½	2	2½	2½	1½	1½	1½	1½	1½	1½	2½	3	3½	4
10	1½	1½	2	2½	2½	1½	1½	1½	1½	1½	1½	2½	3	3½	4
11	1½	1½	2	2½	2½	1½	1½	1½	1½	1½	1½	2½	3	3½	4
12	1½	1½	2	2½	3	1½	1½	1½	1½	1½	1½	2½	3	3½	4
14 & up	1½	1½	2	2½	3	1½	1½	1½	1½	1½	1½	2½	3½	4	4½

TABLE 14-5 RECOMMENDED THICKNESS OF BLOCK INSULATION (*COURTESY MANVILLE PRODUCTS CORPORATION.*)

Material and Use	\multicolumn{11}{c}{Temperature of Surface, Degrees Fahrenheit}
	100 to 199	200 to 299	300 to 399	400 to 499	500 to 599	600 to 699	700 to 799	800 to 899	900 to 999	1000 to 1099	1100 to 1200
	\multicolumn{11}{c}{Nominal Thickness of Insulation, Inches}
Hydrous Calcium Silicate											
Utility Steam Generation	1½	1½	2	2½	3	3	3½	4	4	4½	4½
Process	1½	1½	1½	1½	1½	2	2	2	2½	2½	3
Commercial	1½	2½	3½	4	4½	...	...	...	...	...	...
85% Magnesia											
Utility Steam Generation	1½	1½	2	2½	3	...	...	...	...	...	...
Process	1½	1½	1½	1½	1½	...	...	...	...	...	...
Commercial	1½	2½	3½	4	4½	...	...	...	...	...	...

INSULATION

15
PIPE SUPPORTS

Fig. 15-1

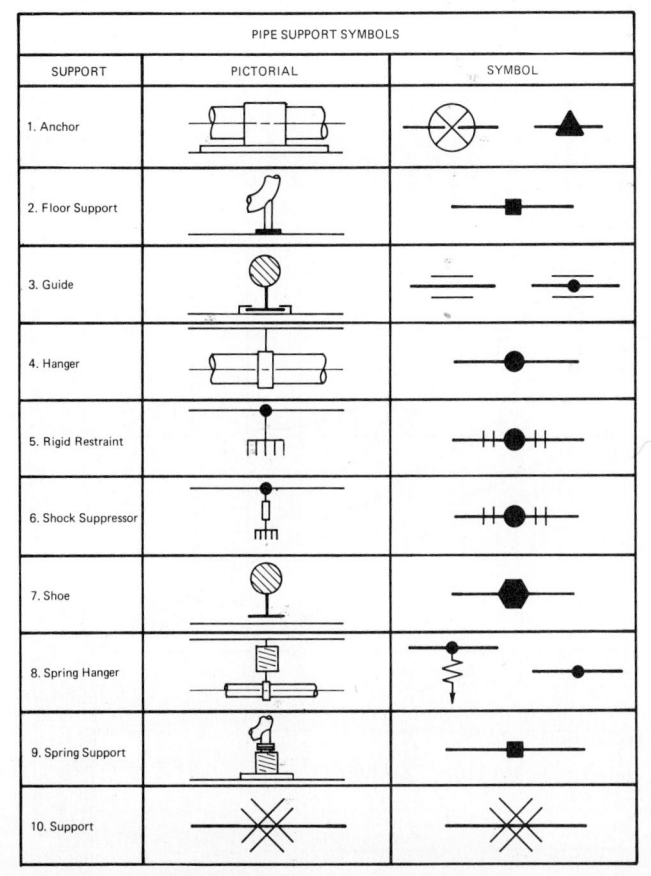

SUPPORT	PICTORIAL	SYMBOL
1. Anchor		
2. Floor Support		
3. Guide		
4. Hanger		
5. Rigid Restraint		
6. Shock Suppressor		
7. Shoe		
8. Spring Hanger		
9. Spring Support		
10. Support		

Fig. 15-2 Pipe support symbols

SPACING OF PIPE SUPPORTS*

When a horizontal pipeline is supported at intermediate points, sagging of the pipe occurs between these supports, the amount of sag being dependent upon the weight of the pipe, fluid, insulation, and valves or fittings which may be included in the line. If the pipeline is installed with no downward pitch, pockets will be formed in each span in which case condensation may collect if the line is transporting steam. In order to eliminate these pockets, the line must be pitched downward so that the outlet of each span is lower than the maximum sag.

Crane has conducted tests to determine the deflection of horizontal standard pipe lines filled with water, in pipe sizes 3/4 in. to 4 in. inclusive, the results of which have indicated that for pipes larger than 2 in. and with supports having center-to-center dimensions greater than 10 ft, the resultant deflection is less than that determined by the

---
(Courtesy Crane Company.)

PIPE SUPPORTS

use of the formula for a uniformly loaded pipe fixed at both ends. For pipe sizes 2 in. and smaller, the test deflection was in excess of that determined by the formula for pipe having fixed ends and approached, for the shorter spans, the deflection as determined by the use of the formula for pipelines having unrestrained ends.

Figure 15-3 gives the deflection of horizontal standard pipelines filled with water, for varying spans, based upon the results obtained from tests for sizes 2 in. and smaller and upon the formula for fixed ends for the larger sizes of pipe. The deflection values given Figure 15-3 are twice those obtained from test or calculation, to compensate for any variables, including weight of insulation, etc.

The formula given below indicates the vertical distance that the span must be pitched so that the outlet is lower than the maximum sag of the pipe.

$$h = \frac{144\, S2\, y}{36\, S^2 - y^2}$$

where

h = difference in elevation of span ends (in inches)

S = length of one span (in feet)

y = deflection of one span (in inches)

By eliminating the inconsequential term y^2 from the denominator, the formula reduces to:

$$h = 4y$$

The pitch of pipe spans, called the average gradient, is a ratio between the drop in elevation and the length of the span. This is expressed as so many inches in a certain number of feet.

$$\text{Average gradient} = \frac{4y}{S}$$

The dotted lines as shown on the chart on the opposite page are plotted from the above formula and indicate average gradients of 1″ in 10′, 1″ in 15′, 1″ in 20′, 1″ in 30′, and 1″ in 40′.

EXAMPLE: What is the maximum distance between supports for a 4-in. standard pipeline assuming a pitch or average gradient of 1″ in 30 feet?

Using Table 15-1, find the point where the diagonal dotted line for an average gradient of 1″ in 30 ft intersects the diagonal solid line for 4-in. pipe. From this point, proceed downward to the bottom line, where the maximum span is noted to be approximately 22 ft.

145

Fig. 15-4 Thermal expansion of pipe (*Courtesy Crane Company.*)

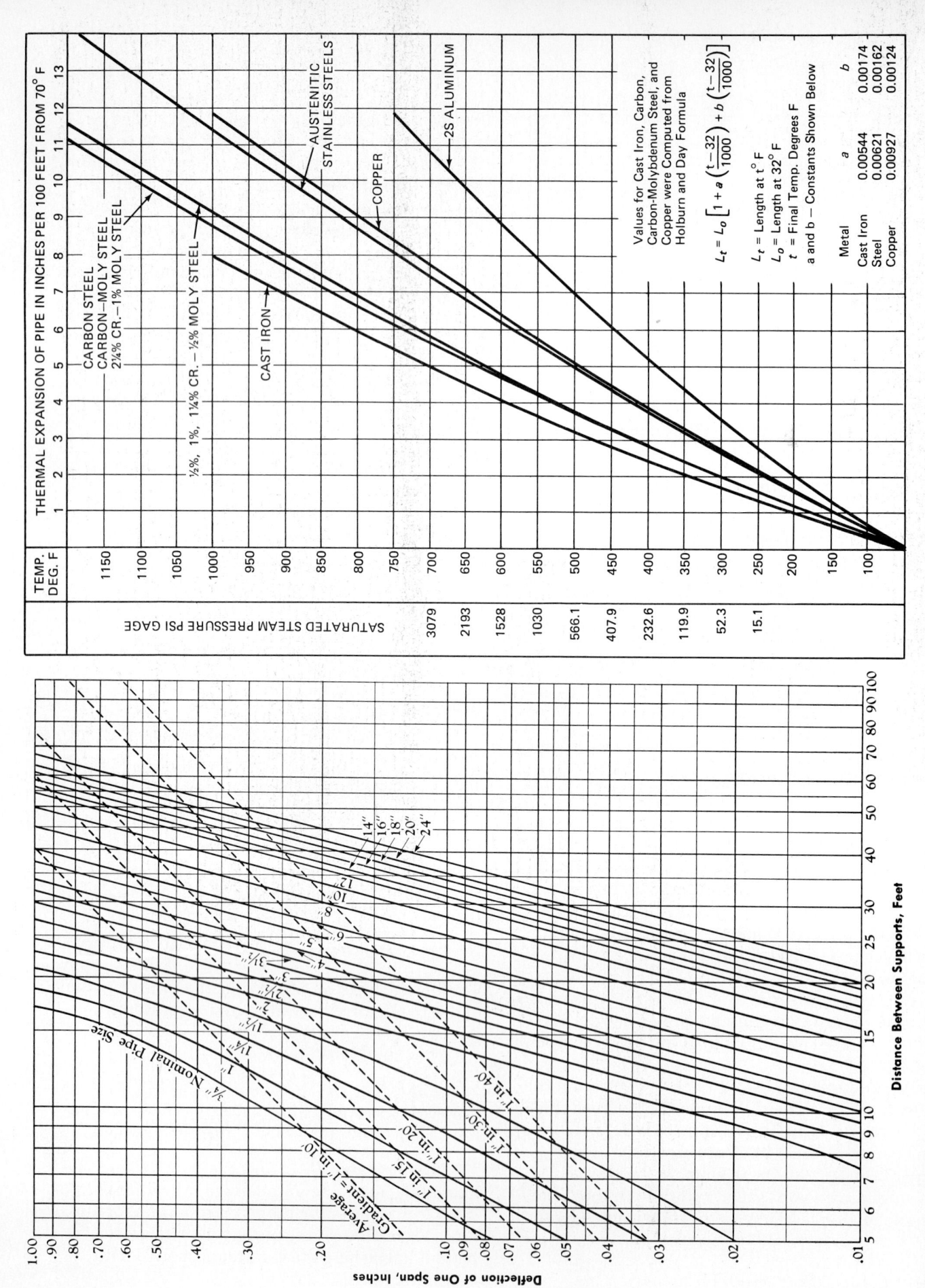

Fig. 15-3 Spacing of pipe supports, deflection of horizontal pipelines (based on standard pipe filled with water) (*Courtesy Crane Company.*)

The Code for Pressure Piping, ANSI B31.1, makes the following statements relative to installations within the scope of the code:

> *121.1.4 Hanger Spacing:* Supports for piping with the longitudinal axis in approximately a horizontal position shall be spaced to prevent excessive sag, bending, and shear stresses in the piping, with special consideration given where components, such as flanges and valves, impose concentrated loads.
>
> Where calculations are not made, suggested maximum spacing of supports for standard and heavier pipe are given in Table 15-1.
>
> Vertical supports shall be spaced to prevent the pipe from being overstressed from the combination of all loading effects.

Figure 15-5 shows the calculations necessary for determining rod swing angle acceptability. The swing angle of a hanger rod is the rotation angle of the rod due to pipe movement from cold to hot position. In most cases, the specifications allow for a maximum swing angle of 4 degrees,

CALCULATIONS FOR DETERMINING ROD SWING ANGLE ACCEPTABILITY

A = Vertical pin distance
B = Horizontal movement
C = Rod length
angle B = Rod swing angle

$$\tan \text{angle B} = \frac{B}{A}$$

Maximum allowable rod swing angle without offset is 4 degrees.

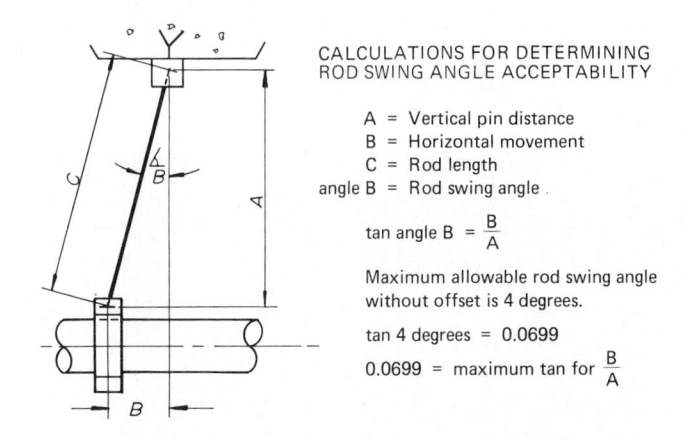

tan 4 degrees = 0.0699

0.0699 = maximum tan for $\frac{B}{A}$

Fig. 15-5 Rod swing angle acceptability.

depending on the service. This specification should be checked for each job. In cases where the swing angle is 2 degrees or more, or the total movement is in excess of 2 in., the hanger itself should be offset two-thirds of the thermal movement in cold position. Figure 15-5 shows dimension *A*, the vertical pin distance; dimension *B*, the horizontal linear movement; dimension *C*, the rod length; and angle *B*, the rod swing angle. The calculations are set up so that the tangent of angle *B* is equal to *B/A*. The maximum allowable swing angle without an offset is 4 degrees, so we need a tangent of 4 degrees, which equals 0.0699. This is the maximum tangent for *B/A*. When calculating the conditions for swing angle acceptability, dimensions for *A*, *B*, and *C* may be plugged into this calculation, thereby determining the angle at *B*. At no time should the maximum tangent for *B/A* equal more than the tangent of 4 degrees or 0.0699.

The offset should be 66 percent (two-thirds) of the thermal movement in cold position. In other words, if the total movement exceeds 2 degrees (or 2 in.) the hanger should be moved so that the total rod swing angle in the hot position becomes 2 degrees or less with a total movement of less than 2 in. This is done by moving the rod attachment in the direction of thermal movement (66 percent, or two-thirds of the total thermal movement), which is the *B* dimension.

Figure 15-6 gives the calculations for determining the length of a structural brace. This is a simple trigonometric calculation, and for each problem dimension *B*, *A1*, and *A2* should be available. When determining the calculations for the structural brace, the *A1* and *B* dimensions will be known and *A2* will be derived by calculating one-half the height of the structural member for

TABLE 15-1 SUGGESTED MAXIMUM SPACING BETWEEN PIPE SUPPORTS FOR HORIZONTAL STRAIGHT RUNS OF STANDARD AND HEAVIER PIPE (AT MAXIMUM OPERATING TEMPERATURE OF 750°F) *(COURTESY CRANE COMPANY.)*

Nominal Pipe Size Inches	Water Service	Steam, Gas, or Air Service
1	7 Feet	9 Feet
2	10 Feet	13 Feet
3	12 Feet	15 Feet
4	14 Feet	17 Feet
6	17 Feet	21 Feet
8	19 Feet	24 Feet
12	23 Feet	30 Feet
16	27 Feet	35 Feet
20	30 Feet	39 Feet
24	32 Feet	42 Feet

Table Notes

1. The values in the table do not apply where span calculations are made or where there are concentrated loads between supports such as flanges, valves, specialties, etc.

2. The spacing is based on a maximum combined bending and shear stress of 1500 psi and insulated pipe filled with water, or the equivalent weight of steel pipe for steam, gas, or air service, and the pitch of the line is such that a sag of 0.1″ between supports is permissible.

PIPE SUPPORTS

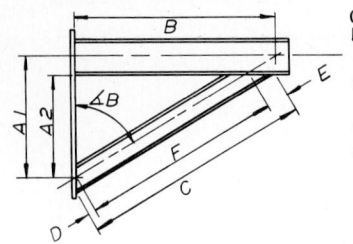

CALCULATIONS FOR DETERMINING LENGTH OF STRUCTURAL BRACE

tan angle B = $\frac{B}{A1}$

then find cos angle B
A2 = A1 − 1/2 thickness of upper brace

$F = \frac{A2}{\cos B}$

E = 2 × tan B
D = 2/tan B
C = D + F + E = brace length

Fig. 15-6 Determining lengths of structural braces

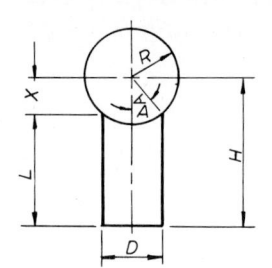

CALCULATIONS FOR DETERMINING STANCHION LENGTH

D = diameter of stanchion
H = distance from pipe centerline to base of stanchion
R = radius of pipe
X = $R^2 - (D/2)^2$
L = H − X = length of stanchion

Fig. 15-7 Stanchion length determination

the *B* dimension that covers the horizontal structural components.

Figure 15-7 provides the calculations for determining stanchion length. In this case, the radius of the pipe will be known, as will the height dimension or dimension from the centerline of the pipe to the bottom of the stanchion. The important calculation will be solving for *X*, which will equal

$$R^2 - \frac{(\text{diameter of stanchion})^2}{2}$$

After *X* is determined, the length of the stanchion can be found by taking distance *H* and subtracting the calculated distance *X*.

TABLE 15-2 THERMAL EXPANSION OF PIPE MATERIALS (*COURTESY CRANE COMPANY.*)

INCHES PER FOOT − FROM 70°

Saturation Gage Pressure Vacuum in. of Mercury	Saturation Gage Pressure Press. psi	Temp. °F.	Carbon Carbon-Moly Low Chrome-Moly (Thru 3% Cr)	Intermediate Alloy Steels (5 thru 9 CrMO)	Austenitic Stainless Steels	Straight Chromium Stainless Steels 12Cr, 17Cr, 27Cr
29.18	−	70	0.00	0.00	0.00	0.00
27.99	−	100	.0022	.0021	.0033	.0019
22.35	−	150	.0060	.0057	.0089	.0052
6.46	−	200	.0099	.0093	.0146	.0085
	15.1	250	.0140	.0131	.0203	.0120
	52.3	300	.0182	.0170	.0261	.0156
	119.9	350	.0225	.0210	.0321	.0192
	232.6	400	.0270	.0250	.0380	.0230
	407.9	450	.0316	.0292	.0440	.0269
	666.1	500	.0362	.0335	.0501	.0308
	1030	550	.0411	.0379	.0562	.0349
	1528	600	.0460	.0424	.0624	.0390
	2193	650	.0511	.0469	.0687	.0431
	3079	700	.0563	.0514	.0750	.0473
		750	.0616	.0562	.0815	.0516
		800	.0670	.0610	.0880	.0560
		850	.0725	.0658	.0946	.0604
		900	.0781	.0707	.1012	.0649
		950	.0835	.0756	.1080	.0694
		1000	.0889	.0806	.1148	.0740
		1050	.0946	.0855	.1216	.0785
		1100	.1004	.0905	.1284	.0831
		1150	.1057	.0952	.1352	.0875
		1200	.1110	.1000	.1420	.0920
		1250	.1166	.1053	.1488	.0965
		1300	.1222	.1106	.1556	.1011
		1350	.1278	.1155	.1624	.1056
		1400	.1334	.1205	.1692	.1101

TABLE 15-3 SPACING OF PIPE SUPPORTS, STRESSES CALCULATED FOR STANDARD WEIGHT PIPE

(a) Stress Due to Sag—Pipe Filled with Water

Pipe Size (in.)	10	12	14	16	18	20	24	30	36	42	48	54	60	Pounds of Water per Lineal Foot
½	2387	3438	4680	6113	7736	9551								0.13164
1	1531	2205	3002	3921	4962	6126	8822	13785						0.37345
1½	1342	1932	2630	3436	4349	5369	7731	12081						0.88260
2	903	1301	1771	2313	2927	3614	5205	8133	11711	15941				1.4541
3	620	892	1215	1587	2008	2480	3571	5580	8035	10936	14284			3.0032
4	503	725	986	1288	1631	2013	2900	4531	6525	8881	11600	14681		5.5172
5	424	610	831	1085	1373	1696	2442	3816	5495	7480	9769	12365	15265	8.6666
6	368	530	721	942	1192	1472	2120	3313	4771	6495	8483	10736	13255	12.530
8	319	459	625	816	1033	1276	1837	2871	4135	5628	7351	9304	11487	22.206
10	283	407	554	724	917	1132	1630	2547	3668	4993	6522	8254	10191	35.454
12	238	343	467	610	772	954	1374	2146	3091	4207	5496	6955	8587	49.760
14 OD	213	308	419	547	701	855	1232	1925	2772	3774	4929	6239	7702	60.000
16 OD	197	284	387	506	640	790	1138	1778	2561	3486	4554	5763	7115	79.187
18 OD	181	261	355	464	588	726	1045	1633	2353	3201	4182	5292	6534	100.48
20 OD	173	250	340	444	562	694	1000	1563	2251	3064	4003	5066	6254	125.30

(b) Stress Due to Sag—Pipe Empty

Pipe Size (in.)	10	12	14	16	18	20	24	30	36	42	48	54	60
½	2063	2971	4044	5282	6685	8254	11886						
1	1249	1799	2449	3199	4049	4998	7198	11247					
1½	1010	1454	1979	2586	3273	4040	5818	9091	13092				
2	644	977	1262	1649	2087	2577	3711	5798	8350	11365	14844		
3	434	625	851	1111	1406	1736	2500	3907	5627	7659	10003	12661	15631
4	331	477	650	849	1075	1327	1911	2986	4390	5853	7644	9675	11945
5	265	381	519	678	858	1060	1526	2385	3435	4675	6106	7728	9541
6	220	317	432	656	667	823	1186	1853	2669	3893	5085	6436	7945
8	167	240	327	427	541	668	962	1503	2165	2998	3915	4956	6118
10	131	189	258	337	426	526	758	1185	1706	2323	3034	3840	4741
12	110	159	217	283	359	443	638	997	1436	1955	2554	3232	3990
14 OD	101	145	198	259	332	405	583	911	1312	1786	2333	2953	3646
16 OD	88	127	172	225	285	352	508	793	1143	1555	2032	2571	3175
18 OD	78	112	153	200	253	313	450	704	1014	1380	1802	2281	2817
20 OD	70	100	137	179	226	230	403	630	907	1235	1613	2041	2520

PIPE SUPPORTS

16
RIGGING AND HOISTING

Fig. 16-1 (*Courtesy Chicago Bridge and Iron Company.*)

WIRE ROPE

Wire rope has largely displaced manila rope in hauling and hoisting heavy loads. As with manila rope, the care of wire rope has a direct bearing on its safe use. Some of the reasons responsible for the use of wire rope in place of manila are

1. Greater strength for equal diameter and weight
2. Equal strength either wet or dry
3. Constant length regardless of weather conditions
4. Greater uniformity in strength throughout
5. Greater number of types for various uses
6. Lower cost per unit of strength
7. Greater durability, with equal care in use

Strength of wire ropes vary, depending on the material from which the individual strands are made and the method used in forming the cable, ranging from 30 to 100 tons per square inch.

Primarily there are three classes of wire rope: iron, cast steel, and plow steel.

Iron wire is soft and of low tensile strength, around 30 to 40 tons per square inch. Commonly used for drum-type elevator cables and to some extent for derrick guys; being replaced by low-carbon steel wire in these uses.

Cast steel may have a tensile strength up to 90 tons per square inch and because of its greater strength is generally used for hoisting purposes. To check quickly whether a piece of wire is iron or cast steel, bend it. Iron will bend easily and take a long time to regain its original shape; cast steel will be harder to bend and will snap back to its original shape very quickly.

Plow steel wire rope is made from high-grade, open hearth furnace steel and has an average tensile strength of 110 tons per square inch. This is the best and safest wire rope for cranes, derricks, dredges, and slings or straps for heavy loads.

Lubrication—Wire Rope

All wire rope, whether used indoors or out, should, in the course of regular work, be considered as a group of moving wires constantly rubbing against one another, with friction resulting. This friction causes incessant wear on the moving parts of the wire rope or cable and will shorten its life very rapidly unless lubricants are used to overcome the friction.

Cable or wire rope should be treated at regular intervals with a lubricant to prevent rusting and to overcome the friction. Lubricating intervals will depend on the types and amount of work encountered. Under average conditions, if worked steadily on equipment, wire rope or cable will require lubrication once every three weeks. Where heavy abrasive dusts exist, more frequent lubrication is in order. Rusty ropes may break without warning.

Sheaves

The life of wire rope or cable is directly affected by the condition and size of the sheaves over which it is used. Sheaves should be at least 16 times the diameter of the rope or cable that is used over them. In passing over a sheave, the inside portion of the cable, which is against the sheave, is shortened and compression is developed in that section of the cable. The outside portion (away from the sheave) is lengthened or stretched, causing tension in that section. These compressive and tensional stresses combine to create bending stresses which increase rapidly as the diameter of the sheaves decrease. As these bending stresses cause much undue wear and directly shorten the safe working life of the rope or cable, the ratio mentioned between sheaves and rope should be maintained.

New wire rope may be badly injured and will not work properly in sheaves that have become worn or in which the grooves have become irregular in shape. When sheaves are worn or damaged, it is more economical to renew the sheaves rather than to allow excessive wear on the cable.

One cause of very severe wear in wire rope or cables is reverse bending, which will shorten the life of the rope by approximately one-half. Reverse bending refers to the bending of a cable or rope over sheaves, first in one direction, then in another.

Another cause of severe rope wear is twisting of the fall rope. When the fall rope is twisted and a hoist is made, the wear produced is equal to more than that resulting from weeks of normal use. The person in charge of lifting operations should guard against twisting of the fall rope and should not allow a lift to be made if the fall rope is twisted.

Handling Cable or Wire Rope

Cable or wire rope cannot and must not be coiled or uncoiled like manila rope. Cable or wire rope must be taken off the reel in a straight line, avoiding kinking. The reel may be mounted on a heavy pipe or roller to facilitate unwinding. If space is limited, the cable as it comes off the reel may be layed out in a figure eight, after which it can be reeved into the line for which it is intended.

TABLE 16-1 SAFE LOAD (IN POUNDS) ON IMPROVED PLOW STEEL WIRE ROPE (6 STRANDS, 19 OR 37 WIRES PER STRAND, HEMP CORE) (*COURTESY ITT GRINNELL CORPORATION.*)

Diam. Inches	Circum. Inches	Single Vertical Wirerope	Two Part Sling 60°	Two Part Sling 45°	Two Part Sling 30°	Weight Per Foot, Lbs.	Breaking Strength Tons, (2000 lbs)
1/4	3/4	1,100	1,900	1,550	1,100	.10	2.74
3/8	1 1/8	2,500	4,230	3,460	2,450	.23	6.10
1/2	1 1/2	4,300	7,450	6,080	4,300	.40	10.70
5/8	2	6,600	11,600	9,430	6,670	.63	16.70
3/4	2 1/4	9,400	16,500	13,450	9,520	.90	23.8
7/8	2 3/4	12,800	22,300	18,200	12,800	1.23	32.2
1	3	16,000	29,000	23,690	16,790	1.60	41.8
1 1/8	3 1/2	21,000	36,450	29,780	21,040	2.03	52.6
1 1/4	4	26,000	44,700	36,570	25,870	2.50	64.6
1 3/8	4 1/4	31,000	53,800	43,900	31,050	3.03	77.7
1 1/2	4 3/4	37,000	63,700	52,000	36,800	3.60	92.0
1 5/8	5	43,000	74,400	60,700	42,900	4.23	107.0
1 3/4	5 1/2	49,600	86,000	70,260	49,700	4.90	124.0
2	6 1/4	64,000	110,700	90,400	64,000	6.40	160.0
2 1/8	6 5/8	63,000	125,200	102,200	72,200	7.22	181.0
2 1/4	7 1/8	81,000	140,300	114,600	79,000	8.10	202.0
2 1/2	7 7/8	98,000	170,000	139,100	98,400	10.00	246.0
2 3/4	8 5/8	117,600	203,500	166,700	117,700	12.1	294.0

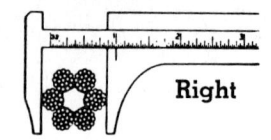

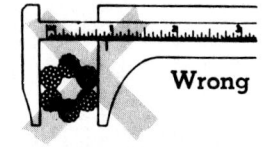

Right

Wrong

Wire rope is usually manufactured slightly larger than the nominal diameter. The diameter of a new rope may exceed the nominal diameter by the amounts shown in the United States Federal Specification for Wire Rope.

TABLE 16-2 SAFE WORKING LOAD OF WIRE ROPE IN TONS OF 2000 LBS (SAFETY FACTOR 5-OSHA) (*COURTESY ITT GRINNELL CORPORATION.*)

Rope Diameter (inches)	6 × 19 Class Wire Rope Extra Improved Plow Fiber Core	6 × 19 Class Wire Rope Extra Improved Plow IWRC	6 × 19 Class Wire Rope Improved Plow Fiber Core	6 × 19 Class Wire Rope Improved Plow IWRC	6 × 37 Class Wire Rope Extra Improved Plow Fiber Core	6 × 37 Class Wire Rope Extra Improved Plow IWRC	6 × 37 Class Wire Rope Improved Plow Fiber Core	6 × 37 Class Wire Rope Improved Plow IWRC	8 × 19 Spin-Resistant Rope Extra Improved Plow IWRC	8 × 19 Spin-Resistant Rope Improved Plow IWRC
1/4	.60	.68	.54	.58	.57	.64	.51	.55		
5/16	.93	1.05	.85	.91	.88	.99	.80	.86		
3/8	1.34	1.51	1.22	1.31	1.27	1.42	1.15	1.24		
7/16	1.81	2.04	1.65	1.77	1.72	1.93	1.56	1.68		
1/2	2.36	2.66	2.14	2.30	2.24	2.52	2.04	2.20	2.34	2.04
9/16	2.98	3.36	2.70	2.90	2.82	3.18	2.58	2.78	2.94	2.56
5/8	3.66	4.12	3.34	3.58	3.48	3.92	3.16	3.40	3.62	3.14
3/4	5.24	5.88	4.76	5.12	4.96	5.58	4.52	4.86	5.18	4.52
7/8	7.08	7.96	6.44	6.92	6.74	7.56	6.12	6.58	7.00	6.10
1	9.20	10.34	8.36	8.98	8.74	9.82	7.96	8.56	9.10	7.92
1 1/8	11.58	13.00	10.52	11.30	11.00	12.38	10.02	10.78	11.46	9.96
1 1/4	14.20	15.98	12.92	13.88	13.54	15.22	12.30	13.22	14.10	12.26
1 3/8	17.08	19.20	15.54	16.70	16.30	18.34	14.82	15.94	16.98	14.76
1 1/2	20.20	22.80	18.40	19.78	19.32	21.60	17.58	18.90	20.00	17.46
1 5/8	23.60	26.40	21.40	23.00	22.60	25.40	20.60	22.20		
1 3/4	27.20	30.60	24.80	26.60	26.00	29.20	23.80	25.60		
1 7/8	31.00	34.80	28.20	30.40	29.80	33.60	27.20	29.20		
2	35.20	39.60	32.00	34.40	33.80	38.00	30.80	33.00		

When it is necessary to make a short bend, as in attaching wire rope or when it is to be looped, thimbles should always be used. U BOLTS OF ALL CLAMPS MUST BE ON THE DEAD END OF THE ROPE. In clamping a strap or an eye, the loose or "dead" end is clamped against the main part of the rope, with the clamps spaced apart a distance equal to 6 x diameter of the rope. Clamp fastenings seldom develop more than 4/5 of rope strength at best.

The point of greatest fatigue and/or wear in a rope usually develops at or near the end where it is clamped around the boom or where attached to the becket on the block. Clamps should be inspected at least once weekly and tightened if they show signs of loosening. All clamped or spliced fastenings, especially those on cranes or derricks, should be shifted and changed at least once every six months.

Number of Crosby or Safety Clips and Distance Between Clips Needed for Safety

Diameter of Rope, Inches	Number of Clips	Distance Between Clips, Inches
1/4 – 3/8	3	2 1/4
7/16 – 5/8	3	3 3/4
3/4 – 1 1/8	4	6 3/4
1 1/4 – 1 1/2	5	9
1 5/8 – 1 3/4	6	10 1/2
2 and over	7	6 times diam. of cable

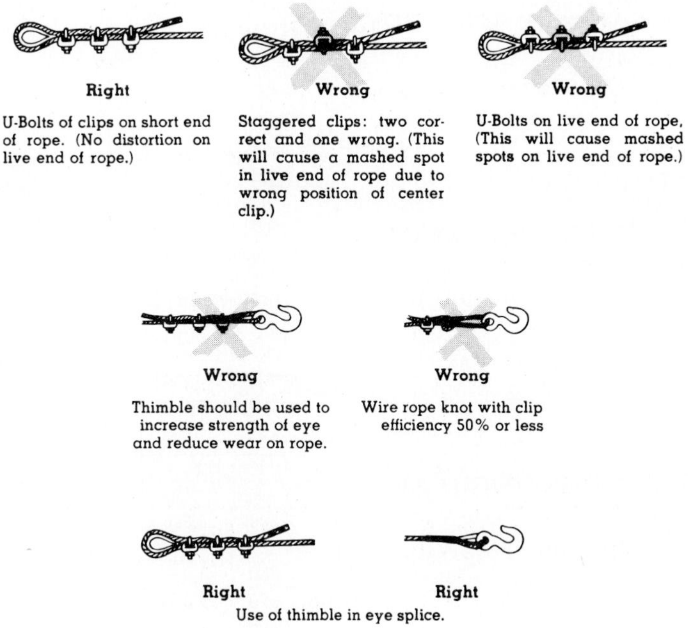

Right
U-Bolts of clips on short end of rope. (No distortion on live end of rope.)

Wrong
Staggered clips: two correct and one wrong. (This will cause a mashed spot in live end of rope due to wrong position of center clip.)

Wrong
U-Bolts on live end of rope. (This will cause mashed spots on live end of rope.)

Wrong
Thimble should be used to increase strength of eye and reduce wear on rope.

Wrong
Wire rope knot with clip efficiency 50% or less

Right **Right**
Use of thimble in eye splice.

Fig. 16-2 Clamp fastenings (*Courtesy ITT Grinnell Corporation.*)

SLIPPERY RING

Used to secure a rope temporarily to a pie or ring; easy to untie.

1

DOUBLE HITCH

Used to secure a rope to a post or ring.

2

KILLICK HITCH

Used to secure a rope to a post or a pipe.

3

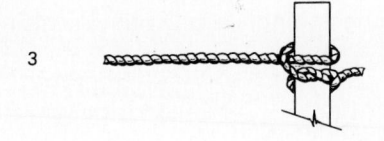

TIMBER HITCH

Used to secure a rope to a post or pipe. This knot will loosen easily when the load is removed. It will not slip when loaded.

4

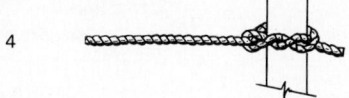

CLOVE HITCH

Used to secure a rope to a post or a pipe.

5

SLIPPERY HITCH

Good for quick release by pulling the end of the rope.

6

Fig. 16-3 Knots for rigging and hoisting (Illustration By John J. Higgins.)

RIGGING AND HOISTING 153

MARLIN SPIKE HITCH

Used to secure a rope to a stake.

7

OVERHAND KNOT

Used to prevent unraveling part of certain types of knots. It reduces strength of rope.

8

FIGURE EIGHT KNOT

Prevents unraveling of rope.

9

SQUARE KNOT

Used to join two pieces of rope that are of equal diameter. This knot will slip when wet.

10

SINGLE CARRICK BEND

Used to join two ropes of any diameter.

11

SHEET BEND

Used to join two ropes of any diameter.

12

DOUBLE SHEET BEND

Used to join ropes of different diameters.

13

SURGEON'S KNOT

Used for tying bundles which can be compressed. The extra twist will hold while forming the second loop.

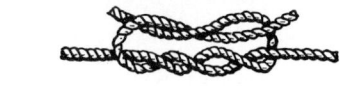

14

SLIP KNOT

This knot is unsafe for heavy loads.

15

BOWLINE

This knot has a nonslipping eye and is easily untied.

16

RUNNING BOWLINE

This knot has a strong running loop.

17

EYE SPLICE

Each strand should be tucked at least four times.

18

FLEMISH EYE KNOT

This knot is safer than a double overhand knot.

19

SHEEP SHANK

Used for shortening a rope without cutting it.

20

Fig. 16-3 (contd.)

TABLE 16-3 SAFE WORKING LOAD (SWL) OF NEW FIBER ROPES (ROPE USED SIX MONTHS WILL HAVE A SWL OF 50 PERCENT) *(COURTESY ITT GRINNELL CORPORATION.)*

Nominal Rope Diameter (inches)	Manila	Nylon	Polypropylene	Polyester	Polyethylene
3/16	100	200	150	200	150
1/4	120	300	250	300	250
5/16	200	500	400	500	350
3/8	270	700	500	700	500
1/2	530	1,250	830	1,200	800
5/8	880	2,000	1,300	1,900	1,050
3/4	1,080	2,800	1,700	2,400	1,500
7/8	1,540	3,800	2,200	3,400	2,100
1	1,800	4,800	2,900	4,200	2,500
1 1/8	2,400	6,300	3,750	5,600	3,300
1 1/4	2,700	7,200	4,200	6,300	3,700
1 1/2	3,700	10,200	6,000	8,900	5,300
1 5/8	4,500	12,400	7,300	10,800	6,500
1 3/4	5,300	15,000	8,700	12,900	7,900
2	6,200	17,900	10,400	15,200	9,500

Three-strand ropes; safety factor = 5.

TABLE 16-4 SAFE WORKING LOADS OF NEW BRAIDED SYNTHETIC FIBER ROPE-POUNDS (ROPE USED SIX MONTHS WILL HAVE A SWL OF 50%)

Nominal Rope Diameter (inches)	Nylon Cover Nylon Core	Nylon Cover Polypropylene Core	Polyester Cover Polypropylene Core
1/4	420	—	380
5/16	640	—	540
3/8	880	680	740
7/16	1,200	1,000	1,060
1/2	1,500	1,480	1,380
9/16	2,100	1,720	—
5/8	2,400	2,100	2,400
3/4	3,500	3,200	2,860
7/8	4,800	4,150	3,800
1	5,700	4,800	5,600
1 1/8	8,000	7,000	—
1 1/4	8,800	8,000	—
1 1/2	12,800	12,400	—
1 5/8	16,000	14,000	—
1 3/4	19,400	18,000	—
2	23,600	20,000	—

Safety factor = 5.

RIGGING AND HOISTING

1. **Hoist Load**
   With forefinger of one hand extended and pointed in an upward direction, making a circular motion.

2. **Lower Load**
   With forefinger of one hand extended and pointed downward, making a circular motion.

3. **Stop**
   Forearm and hand extended in a horizontal position, moving back and forth.

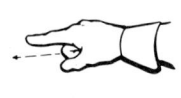

4. **Swing or House**
   One arm extended with forefinger indicating direction of swing.

5. **Boom Up**
   Fingers of hand closed with thumb extended, pointed in an upward position.

6. **Boom Down**
   Fingers of hand closed with thumb extended, pointed in a down position.

7. **Travel Forward**
   Use both hands, making a circular motion as a wheel rolling forwards.

8. **Travel Backward**
   Use both hands, making a circular motion as if wheel is rolling backwards.

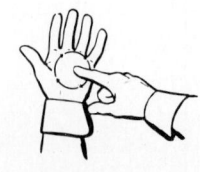

9. **Make Movement Slowly**
   Extend one hand, open, with palm facing operator. With extended forefinger of other hand draw circle on palm. Give signal for next movement.

10. **Emergency Stop**
    Extend both forearms and hands horizontal and move them back and forth rapidly.

11. **Boom Up And Lower Load**
    Make boom-up signal with one hand and hook-lowering signal with other hand.

12. **Boom Up And Raise Load**
    Make boom-up signal with one hand and hook-raising signal with other hand.

13. **Use Burden Block**
    Tap fist on head and make signal for raising or lowering hook.

14. **Use Whip Line**
    Tap elbow with one hand and then make signal for raising or lowering hook.

15. **Move Load Very Slightly**
    Make motion as if pulling a hair out of your head, follow with a signal which you want.

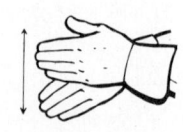

16. **Cut The Load Off**
    Rub the flats of hands together in an up and down movement.

17. **Dog Off Load And Boom**
    Clasp fingers of one hand with fingers of others, palms facing.

Fig. 16-4 Standard hoisting signals (*Courtesy ITT Grinnell Corporation.*)

17
FORMULA

Fig. 17-1

CIRCLES	$\text{Area} = A = \pi r^2$ $A = 3.141\, r^2$ $A = .7854\, d^2$ $\text{Radius} = r = d/2$ $\text{Diameter} = d = 2 \times r$ $\text{Circumference} = C = \pi \times d$ $C = 3.141\, d$
SECTOR OF CIRCLE	$\text{Area} = A = \dfrac{3.141 \times r \times r \times \alpha}{360}$ $\text{Arc (length)} = L = .01745 \times r \times \alpha$ $\text{Angle} = \alpha = \dfrac{L}{.01745 \times r}$ $\text{Radius} = r = \dfrac{L}{.01745 \times \alpha}$
SEGMENT OF CIRCLE	$\text{Area} = A = 1/2\,[r \times L - c\,(r - h)]$ $\text{Arc (length)} = L = .01745\, r\alpha$ $\text{Angle} = \alpha = \dfrac{57.296\, L}{r}$ $\text{Height} = h = r - 1/2\sqrt{4\, r^2 - c^2}$
CIRCULAR RING	$\text{Ring Area} = A = .7854\, (\text{OD}^2 - \text{ID}^2)$ $\text{Ring Sector Area} = a = .00873\, \alpha(\text{OD}^2 - \text{ID}^2)$ OD = outside diameter ID = inside diameter α = ring sector angle OR = outside radius IR = inside radius

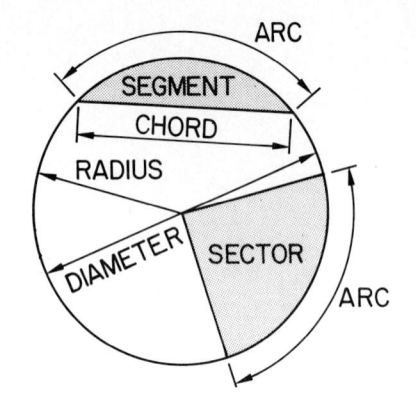

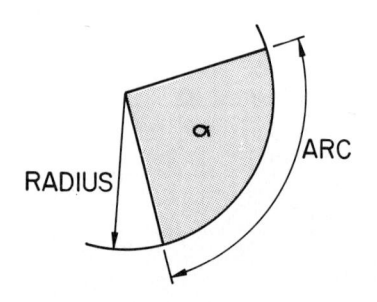

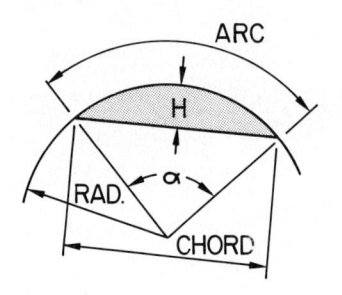

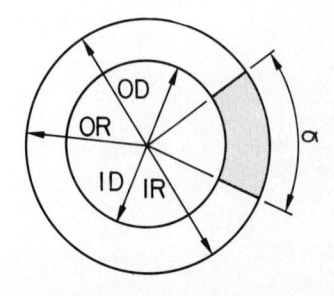

Fig. 17-2 Formulas for circles

EQUILATERAL TRIANGLE	Area = $A = a^2 \dfrac{\sqrt{3}}{4} = .433\, a^2$ $A = .577\, h^2$ $A = \dfrac{a^2}{2}$ or $\dfrac{a \times h}{2}$ Perimeter = $P = 3a$ Height = $h = \dfrac{a}{2}\sqrt{3} = .866\, a$
RIGHT TRIANGLE	Area = $A = \dfrac{a \times b}{2}$ Perimeter = $P = a + b + c$ Height = $a = \sqrt{b^2 - c^2}$ Base = $b = \sqrt{a^2 - c^2}$ Hypotenuse = $c = \sqrt{a^2 + b^2}$
ACUTE ANGLE TRIANGLE	Area = $A = \dfrac{h \times b}{2}$ $A = \sqrt{S(S-a)(S-b)(S-c)}$ S is equal to $1/2\,(a + b + c)$ Perimeter = $P = a + b + c$ Height = $h = \dfrac{2}{b}\sqrt{S(S-a)(S-b)(S-c)}$
OBTUSE ANGLE TRIANGLE	Area = $A = \dfrac{b \times h}{2}$ $A = \sqrt{S(S-a)(S-b)(S-c)}$ Perimeter = $P = a + b + c$ Height = $h = \dfrac{2}{b}\sqrt{S(S-a)(S-b)(S-c)}$

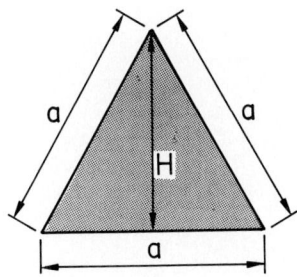

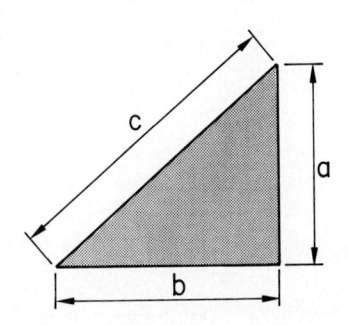

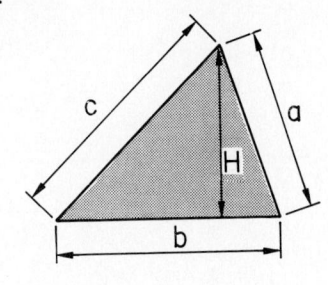

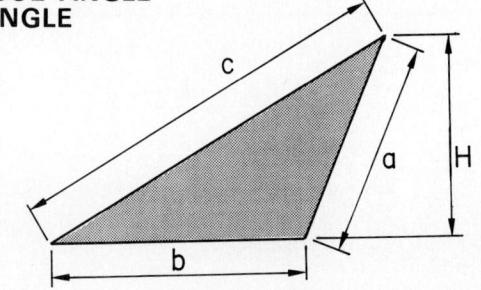

Fig. 17-3 Formulas for triangles

FORMULA

SQUARES	Area = $A = s^2$
	$A = .5\, d^2$
	Side = $s = .707\, d$
	Diagonal = $d = 1.414\, s$
	Perimeter = $P = 4\, s$
RECTANGLES	Area = $A = ab$
	Side $a = a = \sqrt{d^2 - b^2}$
	Side $b = b = \sqrt{d^2 - a^2}$
	Diagonal = $d = \sqrt{a^2 + b^2}$
	Perimeter = $P = 2\,(a + b)$
PARALLELOGRAM	Area = $A = ab$
	Height = $a = \dfrac{A}{b}$
	Base = $b = \dfrac{A}{a}$
TRAPEZOID	Area = $A = h \times \dfrac{a + b}{2}$
TRAPEZIUM	Area = $A = \dfrac{(H + h) \times a + cH + dh}{2}$
	Area = A = Divide the figure into two triangles. Compute the area of each. Add the areas together.

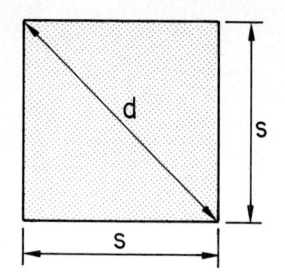

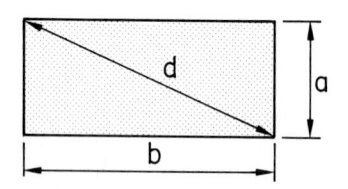

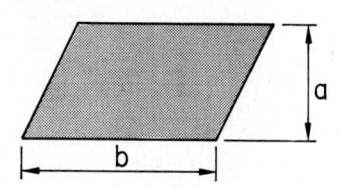

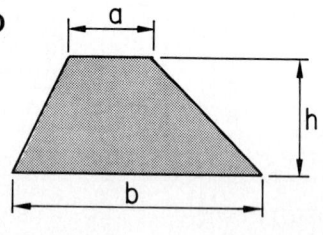

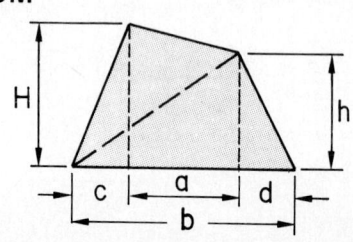

Fig. 17-4 Formulas for polygons (four-sided)

HEXAGON

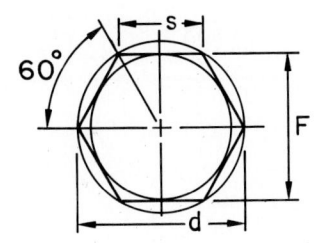

Area = $A = .866\, f^2$ Flats = $f = 1.732\, s$

$A = .650\, d^2$ $f = .866\, d$

$A = 2.598\, s^2$ Diagonal = $d = 2 \times s$

Side = $s = .577\, f$ $d = 1.155\, f$

$s = .5\, d$ Perimeter = $P = 6\, s$

OCTAGON

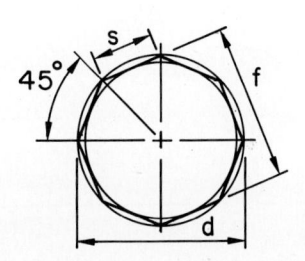

Area = $A = .828\, f^2$ Flats = $f = 2.414\, s$

$A = .707\, d^2$ $f = .924\, d$

$A = 4.828\, s^2$ Diagonal = $d = 2.613\, s$

Side = $s = .414\, f$ $d = 1.083\, f$

$s = .383\, d$ Perimeter = $P = 8\, s$

REGULAR POLYGON (MULTISIDED)

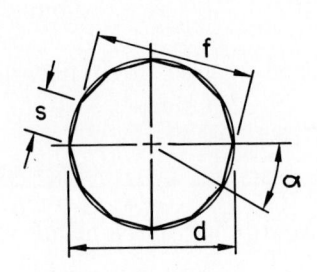

Area = $A = \dfrac{n \times s \times 1/2\, f}{2}$

Angle = $\alpha = \dfrac{360°}{n}$

Side = $s = 2\sqrt{1/2\, d^2 - 1/2\, f^2}$

Perimeter = P = Sum of sides

n = Number of sides

Fig. 17-5   Formulas for regular polygons (six-, eight-, and multisided)

FORMULA

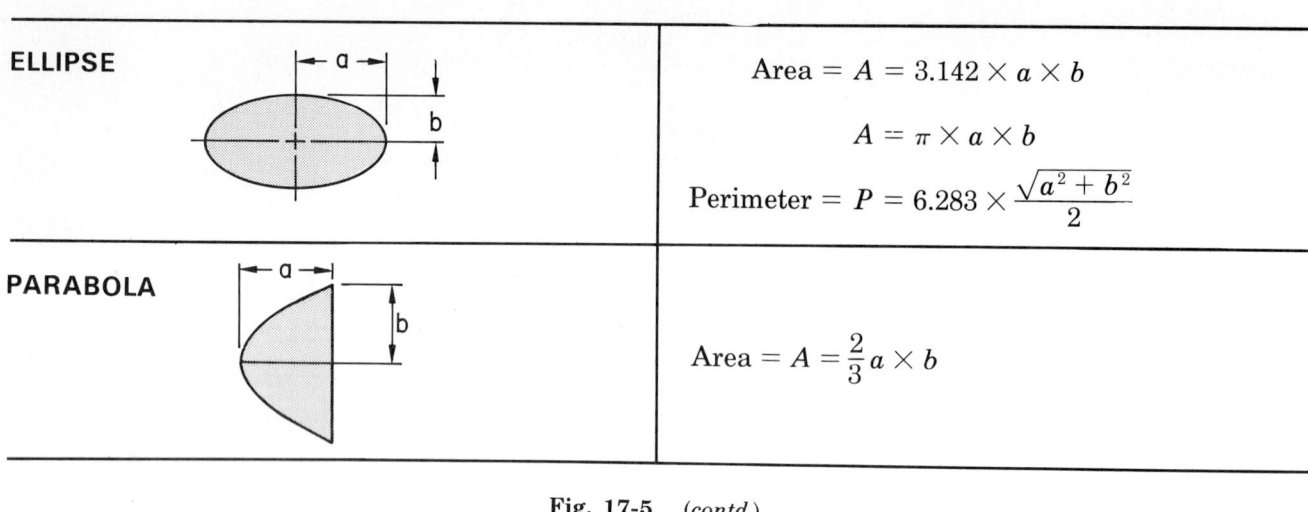

ELLIPSE		Area = $A = 3.142 \times a \times b$ $A = \pi \times a \times b$ Perimeter = $P = 6.283 \times \dfrac{\sqrt{a^2 + b^2}}{2}$
PARABOLA		Area = $A = \dfrac{2}{3} a \times b$

Fig. 17-5 (contd.)

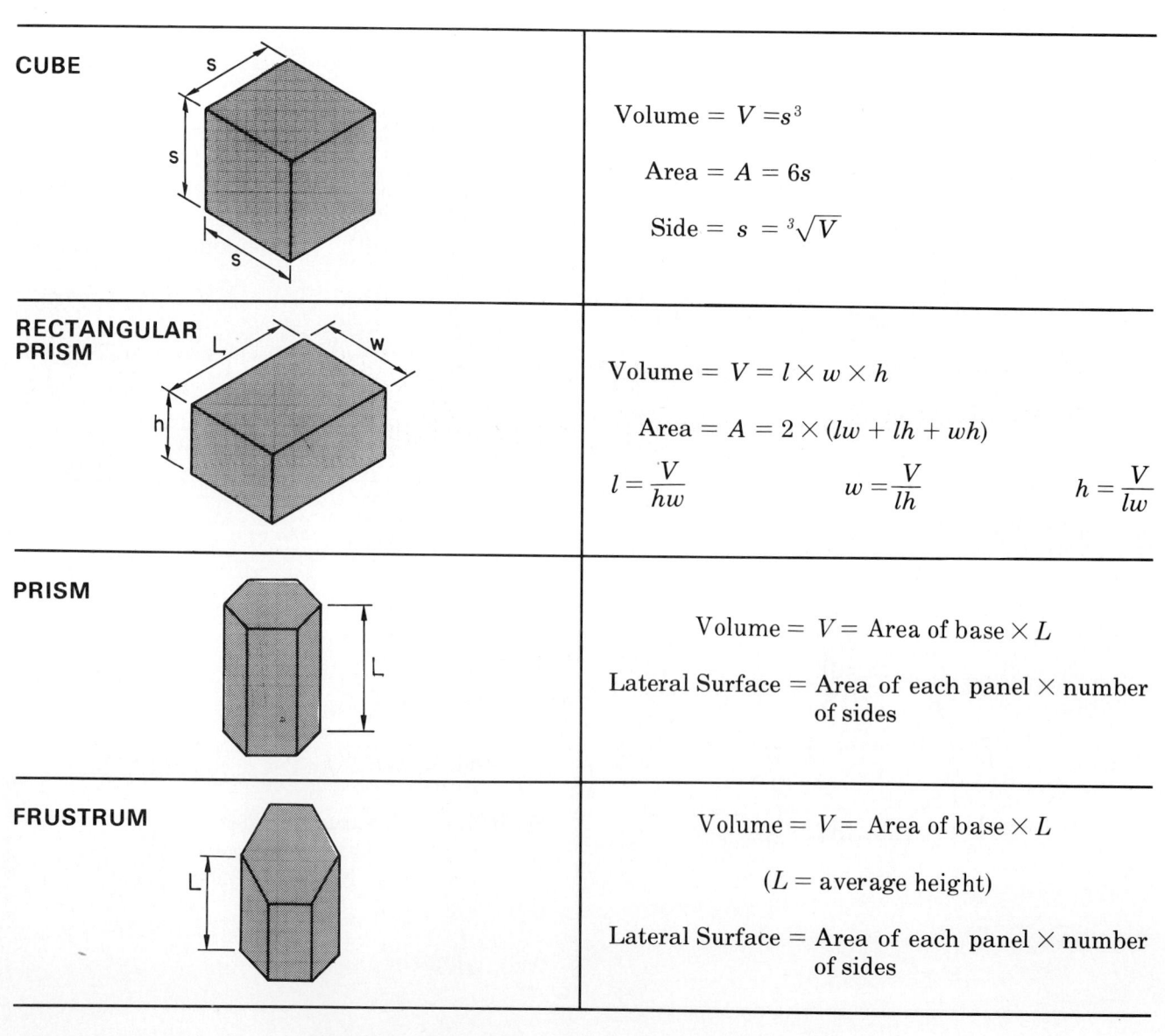

CUBE		Volume = $V = s^3$ Area = $A = 6s$ Side = $s = \sqrt[3]{V}$
RECTANGULAR PRISM		Volume = $V = l \times w \times h$ Area = $A = 2 \times (lw + lh + wh)$ $l = \dfrac{V}{hw} \qquad w = \dfrac{V}{lh} \qquad h = \dfrac{V}{lw}$
PRISM		Volume = $V =$ Area of base $\times L$ Lateral Surface = Area of each panel $\times$ number of sides
FRUSTRUM		Volume = $V =$ Area of base $\times L$ ($L =$ average height) Lateral Surface = Area of each panel $\times$ number of sides

Fig. 17-6 Formulas for cubes, prisms, cones, and pyramids.

CONES

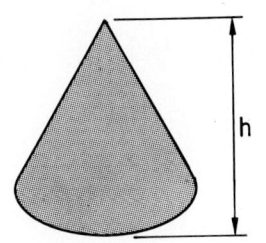

Volume = V = 1/3 Area of base × h

Lateral Surface = 1/2 perimeter of base × slant height

= 1/2 × π × r × sh

(r = radius of base, sh = slant height)

Slant Height = sh = $\sqrt{r^2 + h^2}$

FRUSTRUM OF A CONE

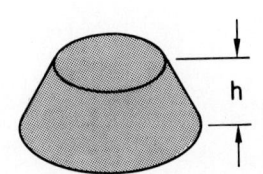

Volume = $V = .262 \times h\,(D^2 + d^2 + Dd)$

(D = diameter of large base; d = diameter of small base)

Lateral Surface = Average perimeter of base × slant height

Slant Height = $sh = \sqrt{(R-r)^2 + h^2}$

PYRAMIDS

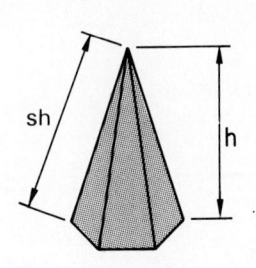

Volume = $V = 1/3$ area of base × h

Lateral Surface = 1/2 perimeter of base × slant height

Slant Height = $sh = \sqrt{r^2 + h^2}$

FRUSTRUM OF A PYRAMID

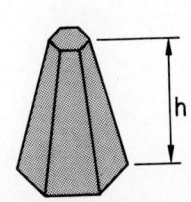

Volume = $V = 1/3h\,(A1 + A2 + \sqrt{A1 \times A2})$

($A1$ = Area of large base; $A2$ = Area of small base)

Lateral Surface = Average perimeter of bases × slant height

Fig. 17-6   (contd.)

FORMULA

CYLINDERS

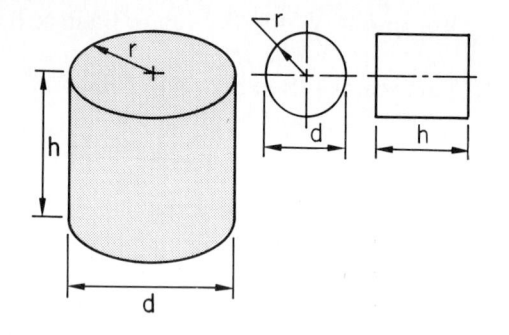

$$\text{Volume} = V = \pi \times r^2 \times h$$

$$V = .7854 \times d^2 \times h$$

$$\text{Lateral Surface} = 6.2832 \times r \times h$$

$$= \pi \times d \times h$$

FRUSTRUM OF A CYLINDER

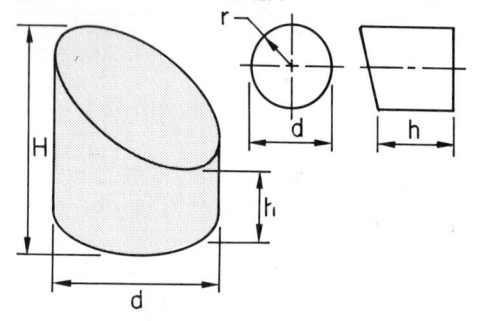

$$\text{Volume} = V = 1.5708 \times r^2 \times (H + h)$$

$$V = .3927 \times d^2 \times (H + h)$$

$$\text{Lateral Surface} = \pi \times r \times (H + h)$$

$$= 1.5708 \times d \times (H + h)$$

ELLIPTICAL TANKS

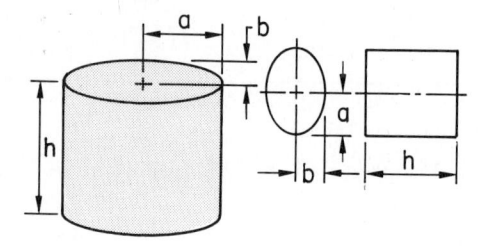

$$\text{Volume} = V = \pi \times a \times b \times h$$

$$\text{Lateral Surface} = \pi \times \sqrt{2 \times (a + b)} \times h$$

SPHERE

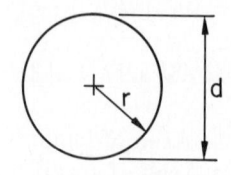

$$\text{Volume} = V = 4.188 \times r^3$$

$$V = \frac{4 \times \pi \times r^3}{3}$$

$$V = \frac{\pi \times d^3}{6}$$

$$\text{Surface} = 4 \times \pi \times r^2$$

$$= 12.566 \times r^2$$

$$= \pi \times d^2$$

Fig. 17-7 Formulas for cylinders, spheres, torus, and ellipsoids.

SPHERICAL SECTOR

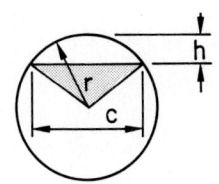

Volume $= V = \dfrac{2 \times \pi \times r^2 \times h}{3}$

$V = 2.094 \times r^2 \times h$

Area $= A = \pi \times r \times (2h + 1/2\, c)$

(A = area of spherical and conical surfaces)

$c = 2 \times \sqrt{h \times (2r - h)}$

SPHERICAL SEGMENT

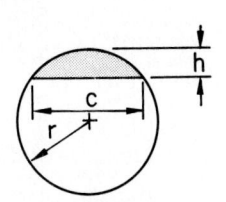

Volume $= V = \pi \times h^2 \times \left(r - \dfrac{h}{3}\right)$

Area $= A = 2 \times \pi \times r \times h$

$A = 6.283 \times r \times h$

(A = area of spherical surface)

$c = 2 \times \sqrt{h \times (2r - h)}$

SPHERICAL ZONE

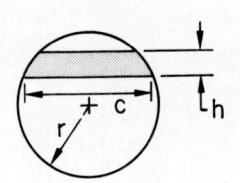

Volume $= V = .532 \times h \times \left(\dfrac{3C^2}{4} + \dfrac{3c^2}{4} + h^2\right)$

Area $= A = 2 \times \pi \times r \times h$

Area $= A = 6.283 \times r \times h$

TORUS

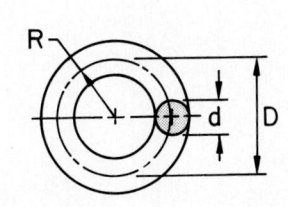

Volume $= V = 2 \times \pi^2 \times R \times 1/2\, d^2$

$V = \dfrac{\pi^2}{4} \times D \times d^2$

$V = 2.467 \times D \times d^2$

Area $= A = \pi^2 \times D \times d$

$A = 9.869 \times D \times d$

ELLIPSOID

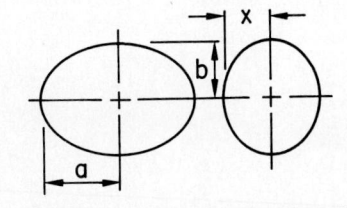

Volume $= V = \dfrac{4 \times \pi}{3} \times a \times b \times x$

$V = 4.188 \times a \times b \times x$

Fig. 17-7 (*Contd.*)

FORMULA

TABLE 17-1 CIRCUMFERENCE, AREA, VOLUME OF CIRCLES AND CYLINDERS
(COURTESY CHICAGO BRIDGE AND IRON COMPANY.)

Diam. in Feet	Circumference Feet	Circumference Meters	Area of Circle Sq. Feet	Area of Circle Sq. Meters	Volume U.S. Gals.	Volume Imperial Gals.	Volume U.S. Bbls. (42 Gals.)	Diam. in Feet
1	3.14	0.9576	0.785	.0730	5.9	4.9	0.140	1
2	6.28	1.9151	3.142	.2919	23.5	19.6	0.560	2
3	9.42	2.8727	7.069	.6567	52.9	44.0	1.259	3
4	12.57	3.8302	12.566	1.1675	94.0	78.3	2.238	4
5	15.71	4.7878	19.635	1.8241	146.9	122.3	3.497	5
6	18.85	5.7454	28.274	2.6268	211.5	176.1	5.04	6
7	21.99	6.7029	38.485	3.5753	287.9	239.7	6.85	7
8	25.13	7.6605	50.266	4.6698	376.0	313.1	8.95	8
9	28.27	8.6180	63.617	5.9102	475.9	396.3	11.33	9
10	31.42	9.5756	78.540	7.2966	587.5	489.2	13.99	10
11	34.56	10.5331	95.033	8.8289	710.9	591.9	16.93	11
12	37.70	11.4907	113.097	10.5071	846.0	704.5	20.14	12
13	40.84	12.4482	132.732	12.3312	992.9	826.8	23.64	13
14	43.98	13.4058	153.938	14.3013	1,151.5	958.9	27.42	14
15	47.12	14.3634	176.715	16.4173	1,321.9	1,100.7	31.47	15
16	50.27	15.3209	201.062	18.6793	1,504.0	1,252.4	35.81	16
17	53.41	16.2785	226.980	21.0871	1,697.9	1,413.8	40.43	17
18	56.55	17.2360	254.469	23.6409	1,903.6	1,585.1	45.32	18
19	59.69	18.1936	283.529	26.3407	2,120.9	1,766.1	50.50	19
20	62.83	19.1511	314.159	29.1864	2,350.1	1,956.9	55.95	20
21	65.97	20.1087	346.361	32.1780	2,591.0	2,157.4	61.69	21
22	69.12	21.0663	380.133	35.3155	2,843.6	2,367.8	67.70	22
23	72.26	22.0238	415.476	38.5989	3,108.0	2,587.9	74.00	23
24	75.40	22.9814	452.389	42.0283	3,384.1	2,817.9	80.57	24
25	78.54	23.9389	490.874	45.6037	3,672.0	3,057.6	87.43	25
26	81.68	24.8965	530.929	49.3249	3,971.6	3,307.1	94.56	26
27	84.82	25.8541	572.555	53.1921	4,283.0	3,566.4	101.98	27
28	87.97	26.8116	615.752	57.2052	4,606.1	3,835.4	109.67	28
29	91.11	27.7692	660.520	61.3643	4,941.0	4,114.3	117.64	29
30	94.25	28.7267	706.858	65.6693	5,287.7	4,402.9	125.90	30
31	97.39	29.6843	754.768	70.1202	5,646.1	4,701.4	134.43	31
32	100.53	30.6418	804.248	74.7171	6,016.2	5,009.6	143.24	32
33	103.67	31.5994	855.299	79.4598	6,398.1	5,327.5	152.34	33
34	106.81	32.5570	907.920	84.3486	6,791.7	5,655.3	161.71	34
35	109.96	33.5145	962.113	89.3832	7,197.1	5,992.9	171.36	35
36	113.10	34.4721	1,017.88	94.5638	7,614.2	6,340.2	181.29	36
37	116.24	35.4296	1,075.21	99.8903	8,043.1	6,697.4	191.50	37
38	119.38	36.3872	1,134.11	105.3627	8,483.8	7,064.3	201.99	38
39	122.52	37.3447	1,194.59	110.9811	8,936.2	7,441.0	212.77	39
40	125.66	38.3023	1,256.64	116.7454	9,400.3	7,827.4	223.82	40
41	128.81	39.2599	1,320.25	122.6556	9,876.2	8,223.7	235.15	41
42	131.95	40.2174	1,385.44	128.7118	10,363.8	8,629.7	246.76	42
43	135.09	41.1750	1,452.20	134.9139	10,863.2	9,045.6	258.65	43
44	138.23	42.1325	1,520.53	141.2619	11,374.4	9,471.2	270.82	44
45	141.37	43.0901	1,590.43	147.7559	11,897.2	9,906.6	283.27	45
46	144.51	44.0476	1,661.90	154.3958	12,431.9	10,351.8	296.00	46
47	147.65	45.0052	1,734.94	161.1816	12,978.3	10,806.8	309.01	47
48	150.80	45.9628	1,809.56	168.1134	13,536.4	11,271.5	322.30	48
49	153.94	46.9203	1,885.74	175.1911	14,106.3	11,746.0	335.86	49
50	157.08	47.8779	1,963.50	182.4147	14,688.0	12,230.4	349.71	50
51	160.22	48.8354	2,042.82	189.7842	15,281.4	12,724.5	363.84	51
52	163.36	49.7930	2,123.72	197.2997	15,886.5	13,228.4	378.25	52
53	166.50	50.7505	2,206.18	204.9611	16,503.4	13,742.0	392.94	53
54	169.65	51.7081	2,290.22	212.7685	17,132.0	14,265.5	407.91	54
55	172.79	52.6657	2,375.83	220.7218	17,772.4	14,798.7	423.15	55
56	175.93	53.6232	2,463.01	228.8210	18,424.6	15,341.8	438.68	56
57	179.07	54.5808	2,551.76	237.0661	19,088.5	15,894.6	454.49	57
58	182.21	55.5383	2,642.08	245.4572	19,764.1	16,457.2	470.57	58
59	185.35	56.4959	2,733.97	253.9942	20,451.5	17,029.6	486.94	59
60	188.50	57.4534	2,827.43	262.6772	21,150.7	17,611.7	503.59	60
61	191.64	58.4110	2,922.47	271.5060	21,861.6	18,203.7	520.51	61
62	194.78	59.3686	3,019.07	280.4808	22,584.2	18,805.4	537.72	62
63	197.92	60.3261	3,117.25	289.6016	23,318.6	19,416.9	555.21	63
64	201.06	61.2837	3,216.99	298.8682	24,064.8	20,038.2	572.97	64
65	204.20	62.2412	3,318.31	308.2808	24,822.7	20,669.3	591.02	65
66	207.35	63.1988	3,421.19	317.8394	25,592.3	21,310.2	609.34	66
67	210.49	64.1563	3,525.65	327.5438	26,373.7	21,960.9	627.95	67
68	213.63	65.1139	3,631.68	337.3942	27,166.9	22,621.3	646.83	68
69	216.77	66.0715	3,739.28	347.3905	27,971.8	23,291.5	665.99	69

TABLE 17-1 (CONTD.)

Diam. in Feet	Circumference Feet	Circumference Meters	Area of Circle Sq. Feet	Area of Circle Sq. Meters	Volume of Cylinder Per Foot of Height U.S. Gals.	Volume of Cylinder Per Foot of Height Imperial Gals.	Volume of Cylinder Per Foot of Height U.S. Bbls. (42 Gals.)	Diam. in Feet
70	219.91	67.0290	3,848.45	357.5328	28,788.4	23,971.5	685.44	70
71	223.05	67.9866	3,959.19	367.8210	29,616.8	24,661.3	705.16	71
72	226.19	68.9441	4,071.50	378.2551	30,457.0	25,360.9	725.17	72
73	229.34	69.9017	4,185.39	388.8352	31,308.9	26,070.3	745.45	73
74	232.48	70.8593	4,300.84	399.5611	32,172.5	26,789.4	766.01	74
75	235.62	71.8168	4,417.86	410.4331	33,047.9	27,518.3	786.86	75
76	238.76	72.7744	4,536.46	421.4509	33,935.1	28,257.0	807.98	76
77	241.90	73.7319	4,656.63	432.6147	34,834.0	29,005.5	829.38	77
78	245.04	74.6895	4,778.36	443.9244	35,744.6	29,763.8	851.06	78
79	248.19	75.6470	4,901.67	455.3800	36,667.0	30,531.9	873.02	79
80	251.33	76.6046	5,026.55	466.9816	37,601.2	31,309.7	895.27	80
81	254.47	77.5622	5,153.00	478.7291	38,547.1	32,097.4	917.79	81
82	257.61	78.5197	5,281.02	490.6226	39,504.7	32,894.8	940.59	82
83	260.75	79.4773	5,410.61	502.6619	40,474.1	33,702.0	963.67	83
84	263.89	80.4348	5,541.77	514.8472	41,455.3	34,519.0	987.03	84
85	267.04	81.3924	5,674.50	527.1785	42,448.2	35,345.8	1,010.67	85
86	270.18	82.3499	5,808.80	539.6556	43,452.9	36,182.3	1,034.59	86
87	273.32	83.3075	5,944.68	552.2787	44,469.3	37,028.7	1,058.79	87
88	276.46	84.2651	6,082.12	565.0478	45,497.4	37,884.8	1,083.27	88
89	279.60	85.2226	6,221.14	577.9627	46,537.3	38,750.7	1,108.03	89
90	282.74	86.1802	6,361.73	591.0236	47,589.0	39,626.4	1,133.07	90
91	285.88	87.1377	6,503.88	604.2304	48,652.4	40,511.9	1,158.39	91
92	289.03	88.0953	6,647.61	617.5832	49,727.6	41,407.1	1,183.99	92
93	292.17	89.0528	6,792.91	631.0819	50,814.5	42,312.2	1,209.87	93
94	295.31	90.0104	6,939.78	644.7265	51,913.1	43,227.0	1,236.03	94
95	298.45	90.9680	7,088.22	658.5170	53,023.5	44,151.6	1,262.47	95
96	301.59	91.9255	7,238.23	672.4535	54,145.7	45,086.0	1,289.18	96
97	304.73	92.8831	7,389.81	686.5359	55,279.6	46,030.2	1,316.18	97
98	307.88	93.8406	7,542.96	700.7643	56,425.3	46,984.2	1,343.46	98
99	311.02	94.7982	7,697.69	715.1386	57,582.7	47,947.9	1,371.02	99
100	314.16	95.7557	7,853.98	729.6588	58,751.9	48,921.5	1,398.85	100
101	317.30	96.7133	8,011.85	744.3249	59,932.8	49,904.8	1,426.97	101
102	320.44	97.6709	8,171.28	759.1370	61,125.4	50,897.9	1,455.37	102
103	323.58	98.6284	8,332.29	774.0950	62,329.8	51,900.8	1,484.04	103
104	326.73	99.5860	8,494.87	789.1989	63,546.0	52,913.5	1,513.00	104
105	329.87	100.5435	8,659.01	804.4488	64,773.9	53,935.9	1,542.24	105
106	333.01	101.5011	8,824.73	819.8446	66,013.6	54,968.2	1,571.75	106
107	336.15	102.4586	8,992.02	835.3863	67,265.0	56,010.2	1,601.55	107
108	339.29	103.4162	9,160.88	851.0740	68,528.2	57,062.0	1,631.62	108
109	342.43	104.3738	9,331.32	866.9076	69,803.1	58,123.6	1,661.98	109
110	345.58	105.3313	9,503.32	882.8871	71,089.7	59,195.0	1,692.61	110
111	348.72	106.2889	9,676.89	899.0126	72,388.2	60,276.2	1,723.53	111
112	351.86	107.2464	9,852.03	915.2840	73,698.3	61,367.1	1,754.72	112
113	355.00	108.2040	10,028.75	931.7013	75,020.2	62,467.8	1,786.20	113
114	358.14	109.1615	10,207.03	948.2645	76,353.9	63,578.4	1,817.95	114
115	361.28	110.1191	10,386.89	964.9737	77,699.3	64,698.7	1,849.98	115
116	364.42	111.0767	10,568.32	981.8288	79,056.5	65,828.7	1,882.30	116
117	367.57	112.0342	10,751.31	998.8299	80,425.4	66,968.6	1,914.89	117
118	370.71	112.9918	10,935.88	1,015.9769	81,806.1	68,118.3	1,947.76	118
119	373.85	113.9493	11,122.02	1,033.2698	83,198.5	69,277.7	1,980.92	119
120	376.99	114.9069	11,309.73	1,050.7086	84,602.7	70,446.9	2,014.35	120
121	380.13	115.8645	11,499.01	1,068.2934	86,018.6	71,625.9	2,048.06	121
122	383.27	116.8220	11,689.86	1,086.0241	87,446.3	72,814.7	2,082.05	122
123	386.42	117.7796	11,882.29	1,103.9008	88,885.7	74,013.3	2,116.33	123
124	389.56	118.7371	12,076.28	1,121.9233	90,336.8	75,221.7	2,150.88	124
125	392.70	119.6947	12,271.84	1,140.0918	91,799.8	76,439.8	2,185.71	125
126	395.84	120.6522	12,468.98	1,158.4063	93,274.4	77,667.7	2,220.82	126
127	398.98	121.6098	12,667.68	1,176.8666	94,760.9	78,905.5	2,256.21	127
128	402.12	122.5674	12,867.96	1,195.4729	96,259.0	80,153.0	2,291.88	128
129	405.27	123.5249	13,069.81	1,214.2252	97,769.0	81,410.2	2,327.83	129
130	408.41	124.4825	13,273.23	1,233.1233	99,290.6	82,677.3	2,364.06	130
131	411.55	125.4400	13,478.22	1,252.1674	100,824.0	83,954.2	2,400.57	131
132	414.69	126.3976	13,684.77	1,271.3574	102,369.2	85,240.8	2,437.36	132
133	417.83	127.3551	13,892.91	1,290.6934	103,926.1	86,537.2	2,474.43	133
134	420.97	128.3127	14,102.61	1,310.1753	105,494.8	87,843.4	2,511.78	134
135	424.12	129.2703	14,313.88	1,329.8031	107,075.2	89,159.4	2,549.41	135
136	427.26	130.2278	14,526.72	1,349.5769	108,667.4	90,485.2	2,587.32	136
137	430.40	131.1854	14,741.14	1,369.4965	110,271.3	91,820.7	2,625.51	137
138	433.54	132.1429	14,957.12	1,389.5622	111,887.0	93,166.1	2,663.98	138
139	436.68	133.1005	15,174.67	1,409.7737	113,514.4	94,521.2	2,702.73	139

FORMULA

TABLE 17-1 (CONTD.)

Diam. in Feet	Circumference Feet	Circumference Meters	Area of Circle Sq. Feet	Area of Circle Sq. Meters	Volume of Cylinder Per Foot of Height U.S. Gals.	Volume of Cylinder Per Foot of Height Imperial Gals.	Volume of Cylinder Per Foot of Height U.S. Bbls. (42 Gals.)	Diam. in Feet
140	439.82	134.0580	15,393.80	1,430.1312	115,153.6	95,886.1	2,741.75	140
141	442.96	135.0156	15,614.50	1,450.6346	116,804.6	97,260.8	2,781.06	141
142	446.11	135.9732	15,836.77	1,471.2839	118,467.2	98,645.3	2,820.65	142
143	449.25	136.9307	16,060.60	1,492.0792	120,141.7	100,039.5	2,860.52	143
144	452.39	137.8883	16,286.01	1,513.0204	121,827.8	101,443.6	2,900.66	144
145	455.53	138.8458	16,512.99	1,534.1076	123,525.8	102,857.4	2,941.09	145
146	458.67	139.8034	16,741.54	1,555.3406	125,235.4	104,281.0	2,981.80	146
147	461.81	140.7609	16,971.67	1,576.7196	126,956.9	105,714.4	3,022.78	147
148	464.96	141.7185	17,203.36	1,598.2446	128,690.1	107,157.6	3,064.05	148
149	468.10	142.6761	17,436.62	1,619.9154	130,435.0	108,610.6	3,105.60	149
150	471.24	143.6336	17,671.46	1,641.7322	132,191.7	110,073.3	3,147.42	150
151	474.38	144.5912	17,907.86	1,663.6950	133,960.1	111,545.9	3,189.53	151
152	477.52	145.5487	18,145.84	1,685.8036	135,740.3	113,028.2	3,231.91	152
153	480.66	146.5063	18,385.38	1,708.0582	137,532.2	114,520.3	3,274.58	153
154	483.81	147.4638	18,626.50	1,730.4587	139,335.9	116,022.2	3,317.52	154
155	486.95	148.4214	18,869.19	1,753.0052	141,151.3	117,533.9	3,360.75	155
156	490.09	149.3790	19,113.45	1,775.6976	142,978.5	119,055.3	3,404.25	156
157	493.23	150.3365	19,359.28	1,798.5359	144,817.4	120,586.6	3,448.04	157
158	496.37	151.2941	19,606.68	1,821.5202	146,668.1	122,127.6	3,492.10	158
159	499.51	152.2516	19,855.65	1,844.6503	148,530.6	123,678.4	3,536.44	159
160	502.65	153.2092	20,106.19	1,867.9264	150,404.7	125,239.0	3,581.07	160
161	505.80	154.1667	20,358.30	1,891.3485	152,290.7	126,809.4	3,625.97	161
162	508.94	155.1243	20,611.99	1,914.9165	154,188.4	128,389.5	3,671.15	162
163	512.08	156.0819	20,867.24	1,938.6304	156,097.8	129,979.5	3,716.62	163
164	515.22	157.0394	21,124.06	1,962.4902	158,019.0	131,579.2	3,762.36	164
165	518.36	157.9970	21,382.46	1,986.4960	159,951.9	133,188.7	3,808.38	165
166	521.50	158.9545	21,642.43	2,010.6477	161,896.6	134,808.0	3,854.68	166
167	524.65	159.9121	21,903.96	2,034.9453	163,853.0	136,437.1	3,901.26	167
168	527.79	160.8696	22,167.07	2,059.3889	165,821.2	138,076.0	3,948.13	168
169	530.93	161.8272	22,431.75	2,083.9784	167,801.2	139,724.6	3,995.27	169
170	534.07	162.7848	22,698.00	2,108.7138	169,792.8	141,383.1	4,042.69	170
171	537.21	163.7423	22,965.82	2,133.5952	171,796.3	143,051.3	4,090.39	171
172	540.35	164.6999	23,235.21	2,158.6225	173,811.5	144,729.3	4,138.37	172
173	543.50	165.6574	23,506.18	2,183.7957	175,838.4	146,417.1	4,186.63	173
174	546.64	166.6150	23,778.71	2,209.1149	177,877.1	148,114.7	4,235.17	174
175	549.78	167.5726	24,052.81	2,234.5800	179,927.5	149,822.0	4,283.99	175
176	552.92	168.5301	24,328.49	2,260.1910	181,989.7	151,539.2	4,333.09	176
177	556.06	169.4877	24,605.73	2,285.9480	184,063.7	153,266.1	4,382.47	177
178	559.20	170.4452	24,884.55	2,311.8508	186,149.4	155,002.8	4,432.13	178
179	562.35	171.4028	25,164.94	2,337.8997	188,246.8	156,749.3	4,482.07	179
180	565.49	172.3603	25,446.90	2,364.0944	190,356.0	158,505.6	4,532.29	180
181	568.63	173.3179	25,730.42	2,390.4351	192,476.9	160,271.7	4,582.79	181
182	571.77	174.2755	26,015.52	2,416.9217	194,609.6	162,047.5	4,633.56	182
183	574.91	175.2330	26,302.19	2,443.5543	196,754.1	163,833.1	4,684.62	183
184	578.05	176.1906	26,590.43	2,470.3327	198,910.3	165,628.6	4,735.96	184
185	581.19	177.1481	26,880.25	2,497.2571	201,078.2	167,433.8	4,787.58	185
186	584.34	178.1057	27,171.63	2,524.3275	203,257.9	169,248.8	4,839.48	186
187	587.48	179.0632	27,464.58	2,551.5437	205,449.3	171,073.5	4,891.65	187
188	590.62	180.0208	27,759.11	2,578.9060	207,652.5	172,908.1	4,944.11	188
189	593.76	180.9784	28,055.20	2,606.4141	209,867.5	174,752.4	4,996.85	189
190	596.90	181.9359	28,352.87	2,634.0682	212,094.2	176,606.5	5,049.86	190
191	600.04	182.8935	28,652.10	2,661.8682	214,332.6	178,470.5	5,103.16	191
192	603.19	183.8510	28,952.91	2,689.8141	216,582.8	180,344.1	5,156.74	192
193	606.33	184.8086	29,255.29	2,717.9059	218,844.8	182,227.6	5,210.59	193
194	609.47	185.7661	29,559.24	2,746.1437	221,118.5	184,120.9	5,264.73	194
195	612.61	186.7237	29,864.76	2,774.5275	223,403.9	186,023.9	5,319.14	195
196	615.75	187.6813	30,171.85	2,803.0571	225,701.1	187,936.8	5,373.84	196
197	618.89	188.6388	30,480.51	2,831.7327	228,010.1	189,859.4	5,428.81	197
198	622.04	189.5964	30,790.74	2,860.5542	230,330.8	191,791.8	5,484.07	198
199	625.18	190.5539	31,102.55	2,889.5217	232,663.2	193,734.0	5,539.60	199
200	628.32	191.5115	31,415.93	2,918.6351	235,007.4	195,685.9	5,595.42	200

Notes:
1. If diameters are assumed as meters, values in columns "Circumference Feet" and "Area of Circle Square Feet" will represent circumference in meters and area of circle in square meters respectively.
2. If diameters are assumed as meters, values in column "Area of Circle Square Feet" will represent volume of cylinder in cubic meters per vertical meter of height.

Formula to determine capacity per foot of vertical height of cylinder.
D = Diameter in Feet.
$0.1398854 D^2$ = Barrels of 42 U.S. Gallons per vertical foot.
$5.875185 D^2$ = U.S. Gallons per vertical foot.
$4.892148 D^2$ = Imperial Gallons per vertical foot.
$0.022240 D^2$ = Cubic Meters per vertical foot.
$0.785398 D^2$ = (D = Diameter in Meters) = Cubic meters per vertical meter.

CHAPTER 17

TABLE 17-2 VOLUME AND SURFACE AREA OF SPHERES (*COURTESY CHICAGO BRIDGE AND IRON COMPANY.*)

Diam. in Ft.	Surface of Sphere in Sq. Ft.	Volume of Sphere Cu. Ft.	Volume of Sphere U.S. Gals.	Volume of Sphere U.S. Bbls.	Diam. in Ft.	Surface of Sphere in Sq. Ft.	Volume of Sphere Cu. Ft.	Volume of Sphere U.S. Gals.	Volume of Sphere U.S. Bbls.
1	3.14	0.52	3.92	.09	61	11,690	118,847	889,037	21,168
2	12.57	4.19	31.33	.75	62	12,076	124,788	933,481	22,226
3	28.27	14.14	105.75	2.52	63	12,469	130,924	979,382	23,319
4	50.27	33.51	250.67	5.97	64	12,868	137,258	1,026,764	24,447
5	78.54	65.45	489.60	11.66	65	13,273	143,793	1,075,649	25,611
6	113.10	113.10	846.03	20.14	66	13,685	150,533	1,126,062	26,811
7	153.94	179.59	1,343.46	31.99	67	14,103	157,479	1,178,026	28,048
8	201.06	268.08	2,005.40	47.75	68	14,527	164,636	1,231,565	29,323
9	254.47	381.70	2,855.34	67.98	69	14,957	172,007	1,286,701	30,636
10	314.16	523.60	3,916.79	93.26	70	15,394	179,594	1,343,460	31,987
11	380	697	5,213	124	71	15,837	187,402	1,401,863	33,378
12	452	905	6,768	161	72	16,286	195,432	1,461,935	34,808
13	531	1,150	8,605	205	73	16,742	203,689	1,523,699	36,279
14	616	1,437	10,748	256	74	17,203	212,175	1,587,178	37,790
15	707	1,767	13,219	315	75	17,671	220,893	1,652,397	39,343
16	804	2,145	16,043	382	76	18,146	229,847	1,719,378	40,938
17	908	2,572	19,243	458	77	18,627	239,040	1,788,145	42,575
18	1,018	3,054	22,843	544	78	19,113	248,475	1,858,721	44,255
19	1,134	3,591	26,865	640	79	19,607	258,155	1,931,131	45,979
20	1,257	4,189	31,334	746	80	20,106	268,083	2,005,398	47,748
21	1,385	4,849	36,273	864	81	20,612	278,262	2,081,544	49,561
22	1,521	5,575	41,706	993	82	21,124	288,696	2,159,594	51,419
23	1,662	6,371	47,656	1,135	83	21,642	299,387	2,239,571	53,323
24	1,810	7,238	54,146	1,289	84	22,167	310,339	2,321,498	55,274
25	1,963	8,181	61,200	1,457	85	22,698	321,555	2,405,400	57,271
26	2,124	9,203	68,842	1,639	86	23,235	333,038	2,491,299	59,317
27	2,290	10,306	77,094	1,836	87	23,779	344,792	2,579,219	61,410
28	2,463	11,494	85,981	2,047	88	24,328	356,818	2,669,184	63,552
29	2,642	12,770	95,527	2,274	89	24,885	369,121	2,761,217	65,743
30	2,827	14,137	105,753	2,518	90	25,447	381,704	2,855,341	67,984
31	3,019	15,599	116,685	2,778	91	26,016	394,569	2,951,581	70,276
32	3,217	17,157	128,345	3,056	92	26,590	407,720	3,049,959	72,618
33	3,421	18,817	140,758	3,351	93	27,172	421,161	3,150,499	75,012
34	3,632	20,580	153,946	3,665	94	27,759	434,893	3,253,225	77,458
35	3,848	22,449	167,932	3,998	95	28,353	448,921	3,358,160	79,956
36	4,072	24,429	182,742	4,351	96	28,953	463,247	3,465,327	82,508
37	4,301	26,522	198,397	4,724	97	29,559	477,875	3,574,750	85,113
38	4,536	28,731	214,922	5,117	98	30,172	492,807	3,686,453	87,773
39	4,778	31,059	232,340	5,532	99	30,791	508,048	3,800,459	90,487
40	5,027	33,510	250,675	5,968	100	31,416	523,599	3,916,792	93,257
41	5,281	36,087	269,949	6,427	101	32,047	539,465	4,035,475	96,083
42	5,542	38,792	290,187	6,909	102	32,685	555,647	4,156,531	98,965
43	5,809	41,630	311,412	7,415	103	33,329	572,151	4,279,984	101,904
44	6,082	44,602	333,648	7,944	104	33,979	588,978	4,405,858	104,901
45	6,362	47,713	356,918	8,498	105	34,636	606,131	4,534,176	107,957
46	6,648	50,965	381,245	9,077	106	35,299	623,615	4,664,962	111,071
47	6,940	54,362	406,653	9,682	107	35,968	641,431	4,798,239	114,244
48	7,238	57,906	433,166	10,313	108	36,644	659,584	4,934,030	117,477
49	7,543	61,601	460,807	10,972	109	37,325	678,076	5,072,359	120,771
50	7,854	65,450	489,599	11,657	110	38,013	696,910	5,213,250	124,125
51	8,171	69,456	519,566	12,371	111	38,708	716,090	5,356,726	127,541
52	8,495	73,622	550,732	13,113	112	39,408	735,619	5,502,811	131,019
53	8,825	77,952	583,120	13,884	113	40,115	755,499	5,651,527	134,560
54	9,161	82,448	616,754	14,685	114	40,828	775,735	5,802,900	138,164
55	9,503	87,114	651,656	15,516	115	41,548	796,329	5,956,951	141,832
56	9,852	91,952	687,851	16,377	116	42,273	817,284	6,113,705	145,564
57	10,207	96,967	725,362	17,271	117	43,005	838,603	6,273,185	149,362
58	10,568	102,160	764,213	18,196	118	43,744	860,290	6,435,415	153,224
59	10,936	107,536	804,427	19,153	119	44,488	882,348	6,600,417	157,153
60	11,310	113,097	846,027	20,144	120	45,239	904,779	6,768,217	161,148

Note: If diameters are assumed as meters, values in columns "Surface of Sphere in Square Feet" and "Volume of Sphere—Cubic Feet" will represent Surface of Sphere in Square Meters and Volume of Sphere in Cubic Meters respectively.
Surface area of sphere = 3.141593 D^2 Square Feet.

Volume of sphere $\begin{cases} = 0.523599 \ D^3 \ \text{Cubic Feet.} \\ = 0.093257 \ D^3 \ \text{Barrels of 42 U.S. Gallons.} \end{cases}$

Number of barrels of 42 U.S. Gallons at any inch in a true sphere = $(3d - 2h) \ h^2 \times .0000539681$ where d is diameter of sphere and h is depth of liquid both in inches.

FORMULA

18

TRIANGLES

Fig. 18-1

NONTRIG FORMULAS FOR RIGHT TRIANGLES

Pythagorean Theorem

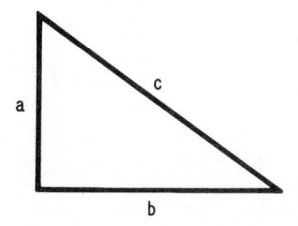

$$c^2 = a^2 + b^2$$

$c = \sqrt{a^2 + b^2}$ Area $= 1/2\, ab$
$a = \sqrt{c^2 - b^2}$ or
$b = \sqrt{c^2 - a^2}$ Area = Hero's Formula

Hero's Formula

Area of any triangle $= \sqrt{S(s-a)(s-b)(s-c)}$

where $S = \dfrac{a+b+c}{2}$

Fig. 18-2 Formulas for right triangles (Pythagorean theorem and Hero's formula)

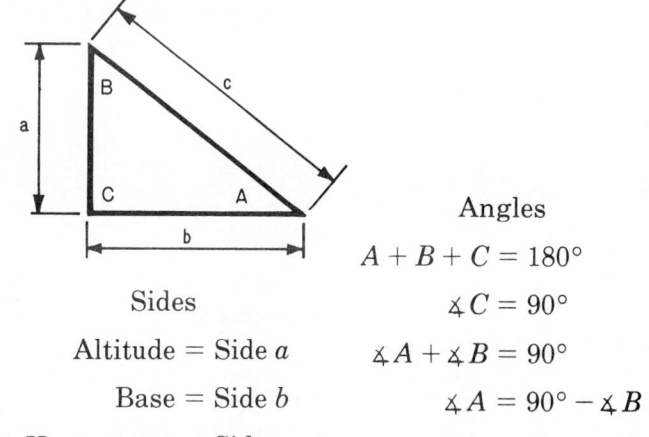

Sides
Altitude = Side a
Base = Side b
Hypotenuse = Side c

Angles
$A + B + C = 180°$
$\angle C = 90°$
$\angle A + \angle B = 90°$
$\angle A = 90° - \angle B$
$\angle B = 90° - \angle A$

$\angle B$ is always opposite side b
$\angle A$ is always opposite side a
$\angle C$ is always opposite side c

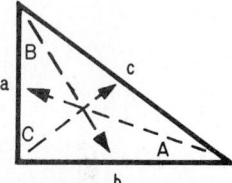

$\angle A$ is always adjacent to side b
$\angle B$ is always adjacent to side a

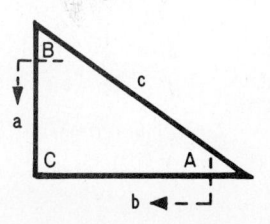

Fig. 18-3 Right triangle setup for trigonometry

TRIGONOMETRY FUNCTIONS

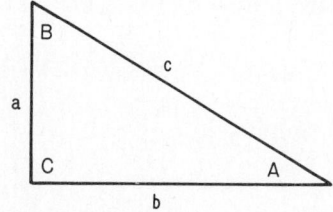

Name	Abbreviation	Formula	Ratio	For $\angle A$	For $\angle B$
Sine	sin	$\dfrac{\text{Opposite Side}}{\text{Hypotenuse}}$	$\dfrac{\text{Opp}}{\text{Hyp}}$	$\sin \angle A = \dfrac{a}{c}$	$\sin \angle B = \dfrac{b}{c}$
Cosine	cos	$\dfrac{\text{Adjacent Side}}{\text{Hypotenuse}}$	$\dfrac{\text{Adj}}{\text{Hyp}}$	$\cos \angle A = \dfrac{b}{c}$	$\cos \angle B = \dfrac{a}{c}$
Tangent	tan	$\dfrac{\text{Opposite Side}}{\text{Adjacent Side}}$	$\dfrac{\text{Opp}}{\text{Adj}}$	$\tan \angle A = \dfrac{a}{b}$	$\tan \angle B = \dfrac{b}{a}$
Cotangent	cot	$\dfrac{\text{Adjacent Side}}{\text{Opposite Side}}$	$\dfrac{\text{Adj}}{\text{Opp}}$	$\cot \angle A = \dfrac{b}{a}$	$\cot \angle B = \dfrac{a}{b}$
Secant	sec	$\dfrac{\text{Hypotenuse}}{\text{Adjacent Side}}$	$\dfrac{\text{Hyp}}{\text{Adj}}$	$\sec \angle A = \dfrac{c}{b}$	$\sec \angle B = \dfrac{c}{a}$
Cosecant	csc	$\dfrac{\text{Hypotenuse}}{\text{Opposite Side}}$	$\dfrac{\text{Hyp}}{\text{Opp}}$	$\csc \angle A = \dfrac{c}{a}$	$\csc \angle B = \dfrac{c}{b}$

Fig. 18-4 Trigonometry functions

TRIANGLES

Solving for Unknowns with Right Angle Trigonometry

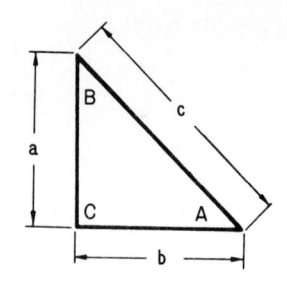

Information known	Unknown angles and sides		Angles
	Sides		
Sides c and a	$b = \sqrt{c^2 - a^2}$	$\sin A = \dfrac{a}{c}$	$B = 90° - A$
Sides c and b	$a = \sqrt{c^2 - b^2}$	$\sin B = \dfrac{b}{c}$	$A = 90° - B$
Sides a and b	$c = \sqrt{a^2 + b^2}$	$\tan A = \dfrac{a}{b}$	$B = 90° - A$
Side c Angle A	$a = c \times \sin A$	$b = c \times \cos A$	$B = 90° - A$
Side c Angle B	$a = c \times \cos B$	$b = c \times \sin B$	$A = 90° - B$
Side a Angle A	$c = \dfrac{a}{\sin A}$	$b = \dfrac{a}{\tan A}$	$B = 90° - A$
Side a Angle B	$c = \dfrac{a}{\cos B}$	$b = a \times \tan B$	$A = 90° - B$
Side b Angle A	$c = \dfrac{b}{\cos A}$	$a = b \times \tan A$	$B = 90° - A$
Side b Angle B	$c = \dfrac{b}{\sin B}$	$a = \dfrac{b}{\tan B}$	$A = 90° - B$

Fig. 18-5 Solving for unknowns with right angle trigonometry

Formulas for Non-Right Triangles

CASE 1

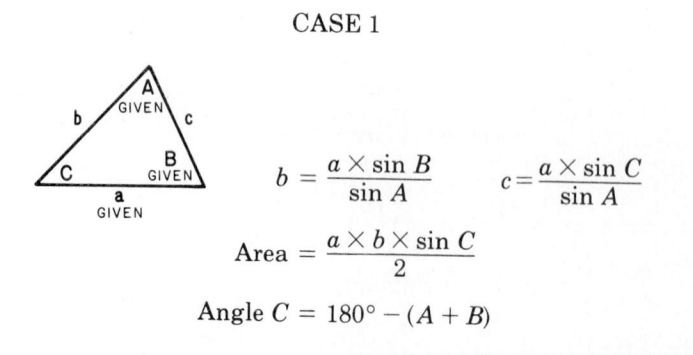

$b = \dfrac{a \times \sin B}{\sin A}$

$c = \dfrac{a \times \sin C}{\sin A}$

Area $= \dfrac{a \times b \times \sin C}{2}$

Angle $C = 180° - (A + B)$

ONE SIDE AND TWO ANGLES GIVEN
Label known side a, the angle opposite A, and the other given angle B, then use formulas provided.

CASE 2

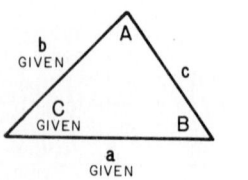

$\tan A = \dfrac{a \times \sin C}{b - (a \times \cos C)}$

c = same as Case 1 or

$c = \sqrt{a^2 + b^2 - 2ab \times \cos C}$

Area = same as Case 1

Angle $B = 180° - (A + C)$

TWO SIDES AND THE ANGLE BETWEEN GIVEN
Label the given sides a and b and the angle between them C, then use the formulas provided.

Fig. 18-6 Formulas for non-right angle triangles (oblique-angled)

CASE 3

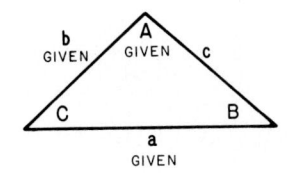

$$\sin B = \frac{b \times \sin A}{a}$$

c = same as Case 1

Area = same as Case 1

Angle $C = 180° - (A + B)$

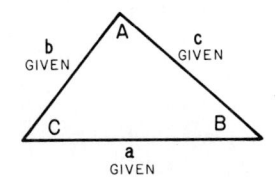

CASE 4

$$\cos A = \frac{b^2 + c^2 - a^2}{2bc}$$

$\sin B$ = same as Case 3

Area = same as Case 1

Angle $C = 180° - (A + B)$

TWO SIDES GIVEN AND THE ANGLE OPPOSITE ONE OF THE GIVEN SIDES

Label given angle A, opposite of angle A label the side a, and the other given side b, then use the formulas provided.

THREE SIDES GIVEN

Label triangle with sides a, b, and c opposite angles A, B, and C, respectively, then apply the formulas provided.

Fig. 18-6 (contd.)

TABLE 18-1 Natural trigonometric functions

Min.	Dec. of Degree	Min.	Dec. of Degree	Min.	Dec. of Degree	Min.	Dec. of Degree	Min.	Dec. of Degree	Min.	Dec. of Degree
¼	0.00416	12¼	0.20416	24¼	0.40416	36¼	0.60416	48¼	0.80416		
½	0.00833	12½	0.20833	24½	0.40833	36½	0.60833	48½	0.80833		
¾	0.01250	12¾	0.21250	24¾	0.41250	36¾	0.61250	48¾	0.81250		
1	0.01666	13	0.21666	25	0.41666	37	0.61666	49	0.81666		
1¼	0.02083	13¼	0.22083	25¼	0.42083	37¼	0.62083	49¼	0.82083		
1½	0.02500	13½	0.22500	25½	0.42500	37½	0.62500	49½	0.82500		
1¾	0.02916	13¾	0.22916	25¾	0.42916	37¾	0.62916	49¾	0.82916		
2	0.03333	14	0.23333	26	0.43333	38	0.63333	50	0.83333		
2¼	0.03750	14¼	0.23750	26¼	0.43750	38¼	0.63750	50¼	0.83750		
2½	0.04166	14½	0.24166	26½	0.44166	38½	0.64166	50½	0.84166		
2¾	0.04583	14¾	0.24583	26¾	0.44583	38¾	0.64583	50¾	0.84583		
3	0.05000	15	0.25000	27	0.45000	39	0.65000	51	0.85000		
3¼	0.05416	15¼	0.25416	27¼	0.45416	39¼	0.65416	51¼	0.85416		
3½	0.05833	15½	0.25833	27½	0.45833	39½	0.65833	51½	0.85833		
3¾	0.06250	15¾	0.26250	27¾	0.46250	39¾	0.66250	51¾	0.86250		
4	0.06666	16	0.26666	28	0.46666	40	0.66666	52	0.86666		
4¼	0.07083	16¼	0.27083	28¼	0.47083	40¼	0.67083	52¼	0.87083		
4½	0.07500	16½	0.27500	28½	0.47500	40½	0.67500	52½	0.87500		
4¾	0.07916	16¾	0.27916	28¾	0.47916	40¾	0.67916	52¾	0.87916		
5	0.08333	17	0.28333	29	0.48333	41	0.68333	53	0.88333		
5¼	0.08750	17¼	0.28750	29¼	0.48750	41¼	0.68750	53¼	0.88750		
5½	0.09166	17½	0.29166	29½	0.49166	41½	0.69166	53½	0.89166		
5¾	0.09583	17¾	0.29583	29¾	0.49583	41¾	0.69583	53¾	0.89583		
6	0.10000	18	0.30000	30	0.50000	42	0.70000	54	0.90000		
6¼	0.10416	18¼	0.30416	30¼	0.50416	42¼	0.70416	54¼	0.90416		
6½	0.10833	18½	0.30833	30½	0.50833	42½	0.70833	54½	0.90833		
6¾	0.11250	18¾	0.31250	30¾	0.51250	42¾	0.71250	54¾	0.91250		
7	0.11666	19	0.31666	31	0.51666	43	0.71666	55	0.91666		
7¼	0.12083	19¼	0.32083	31¼	0.52083	43¼	0.72083	55¼	0.92083		
7½	0.12500	19½	0.32500	31½	0.52500	43½	0.72500	55½	0.92500		
7¾	0.12916	19¾	0.32916	31¾	0.52916	43¾	0.72916	55¾	0.92916		
8	0.13333	20	0.33333	32	0.53333	44	0.73333	56	0.93333		
8¼	0.13750	20¼	0.33750	32¼	0.53750	44¼	0.73750	56¼	0.93750		
8½	0.14166	20½	0.34166	32½	0.54166	44½	0.74166	56½	0.94166		
8¾	0.14583	20¾	0.34583	32¾	0.54583	44¾	0.74583	56¾	0.94583		
9	0.15000	21	0.35000	33	0.55000	45	0.75000	57	0.95000		
9¼	0.15416	21¼	0.35416	33¼	0.55416	45¼	0.75416	57¼	0.95416		
9½	0.15833	21½	0.35833	33½	0.55833	45½	0.75833	57½	0.95833		
9¾	0.16250	21¾	0.36250	33¾	0.56250	45¾	0.76250	57¾	0.96250		
10	0.16666	22	0.36666	34	0.56666	46	0.76666	58	0.96666		
10¼	0.17083	22¼	0.37083	34¼	0.57083	46¼	0.77083	58¼	0.97083		
10½	0.17500	22½	0.37500	34½	0.57500	46½	0.77500	58½	0.97500		
10¾	0.17916	22¾	0.37916	34¾	0.57916	46¾	0.77916	58¾	0.97916		
11	0.18333	23	0.38333	35	0.58333	47	0.78333	59	0.98333		
11¼	0.18750	23¼	0.38750	35¼	0.58750	47¼	0.78750	59¼	0.98750		
11½	0.19166	23½	0.39166	35½	0.59166	47½	0.79166	59½	0.99166		
11¾	0.19583	23¾	0.39583	35¾	0.59583	47¾	0.79583	59¾	0.99583		
12	0.20000	24	0.40000	36	0.60000	48	0.80000	60	1.00000		

TRIANGLES

TABLE 18-2 Conversion of minutes into decimals of a degree (Courtesy ITT Grinnell Corporation.)

Degrees	Sin	Cos	Tan	Cot	Sec	Csc	
0°00'	.0000	1.0000	.0000	—	1.000	—	90°00'
10	029	000	029	343.8	000	343.8	50
20	058	000	058	171.9	000	171.9	40
30	.0087	1.0000	.0087	114.6	1.000	114.6	30
40	116	.9999	116	85.94	000	85.95	20
50	145	999	145	68.75	000	68.76	10
1°00'	.0175	.9998	.0175	57.29	1.000	57.30	89°00'
10	204	998	204	49.10	000	49.11	50
20	233	997	233	42.96	000	42.98	40
30	.0262	.9997	.0262	38.19	1.000	38.20	30
40	291	996	291	34.37	000	34.38	20
50	320	995	320	31.24	001	31.26	10
2°00'	.0349	.9994	.0349	28.64	1.001	28.65	88°00'
10	378	993	378	26.43	001	26.45	50
20	407	992	407	24.54	001	24.56	40
30	.0436	.9990	.0437	22.90	1.001	22.93	30
40	465	989	466	21.47	001	21.49	20
50	494	988	495	20.21	001	20.23	10
3°00'	.0523	.9986	.0524	19.08	1.001	19.11	87°00'
10	552	985	553	18.07	002	18.10	50
20	581	983	582	17.17	002	17.20	40
30	.0610	.9981	.0612	16.35	1.002	16.38	30
40	640	980	641	15.60	002	15.64	20
50	669	978	670	14.92	002	14.96	10
4°00'	.0698	.9976	.0699	14.30	1.002	14.34	86°00'
10	727	974	729	13.73	003	13.76	50
20	756	971	758	13.20	003	13.23	40
30	.0785	.9969	.0787	12.71	1.003	12.75	30
40	814	967	816	12.25	003	12.29	20
50	843	964	846	11.83	004	11.87	10
5°00'	.0872	.9962	.0875	11.43	1.004	11.47	85°00'
10	901	959	904	11.06	004	11.10	50
20	929	957	934	10.71	004	10.76	40
30	.0958	.9954	.0963	10.39	1.005	10.43	30
40	987	951	992	10.08	005	10.13	20
50	.1016	948	.1022	9.788	005	9.839	10
6°00'	.1045	.9945	.1051	9.514	1.006	9.567	84°00'
10	074	942	080	9.255	006	9.309	50
20	103	939	110	9.010	006	9.065	40
30	.1132	.9936	.1139	8.777	1.006	8.834	30
40	161	932	169	8.556	007	8.614	20
50	190	929	198	8.345	007	8.405	10
7°00'	.1219	.9925	.1228	8.144	1.008	8.206	83°00'
10	248	922	257	7.953	008	8.016	50
20	276	918	287	7.770	008	7.834	40
30	.1305	.9914	.1317	7.596	1.009	7.661	30
40	334	911	346	7.429	009	7.496	20
50	363	907	376	7.269	009	7.337	10
8°00'	.1392	.9903	.1405	7.115	1.010	7.185	82°00'
10	421	899	435	6.968	010	7.040	50
20	449	894	465	6.827	011	6.900	40
30	.1478	.9890	.1495	6.691	1.011	6.765	30
40	507	886	524	6.561	012	6.636	20
50	536	881	554	6.435	012	6.512	10
9°00'	.1564	.9877	.1584	6.314	1.012	6.392	81°00'
	Cos	Sin	Cot	Tan	Csc	Sec	Degrees

Degrees	Sin	Cos	Tan	Cot	Sec	Csc	
9°00'	.1564	.9877	.1584	6.314	1.012	6.392	81°00'
10	593	872	614	197	013	277	50
20	622	868	644	084	013	166	40
30	.1650	.9863	.1673	5.976	1.014	6.059	30
40	679	858	703	871	014	5.955	20
50	708	853	733	769	015	855	10
10°00'	.1736	.9848	.1763	5.671	1.015	5.759	80°00'
10	765	843	793	576	016	665	50
20	794	838	823	485	016	575	40
30	.1822	.9833	.1853	5.396	1.017	5.487	30
40	851	827	883	309	018	403	20
50	880	822	914	226	018	320	10
11°00'	.1908	.9816	.1944	5.145	1.019	5.241	79°00'
10	937	811	974	066	019	164	50
20	965	805	2004	4.989	020	089	40
30	.1994	.9799	.2035	4.915	1.020	5.016	30
40	.2022	793	065	843	021	4.945	20
50	051	787	095	773	022	876	10
12°00'	.2079	.9781	.2126	4.705	1.022	4.810	78°00'
10	108	775	156	638	023	745	50
20	136	769	186	574	024	682	40
30	.2164	.9763	.2217	4.511	1.024	4.620	30
40	193	757	247	449	025	560	20
50	221	750	278	390	026	502	10
13°00'	.2250	.9744	.2309	4.331	1.026	4.445	77°00'
10	278	737	339	275	027	390	50
20	306	730	370	219	028	336	40
30	.2334	.9724	.2401	4.165	1.028	4.284	30
40	363	717	432	113	029	232	20
50	391	710	462	061	030	182	10
14°00'	.2419	.9703	.2493	4.011	1.031	4.134	76°00'
10	447	696	524	3.962	031	086	50
20	476	689	555	914	032	039	40
30	.2504	.9681	.2586	3.867	1.033	3.994	30
40	532	674	617	821	034	950	20
50	560	667	648	776	034	906	10
15°00'	.2588	.9659	.2679	3.732	1.035	3.864	75°00'
10	616	652	711	689	036	822	50
20	644	644	742	647	037	782	40
30	.2672	.9636	.2773	3.606	1.038	3.742	30
40	700	628	805	566	039	703	20
50	728	621	836	526	039	665	10
16°00'	.2756	.9613	.2867	3.487	1.040	3.628	74°00'
10	784	605	899	450	041	592	50
20	812	596	931	412	042	556	40
30	.2840	.9588	.2962	3.376	1.043	3.521	30
40	868	580	994	340	044	487	20
50	896	572	3026	305	045	453	10
17°00'	.2924	.9563	.3057	3.271	1.046	3.420	73°00'
10	952	555	089	237	047	388	50
20	979	546	121	204	048	357	40
30	.3007	.9537	.3153	3.172	1.048	3.326	30
40	035	528	185	140	049	295	20
50	062	520	217	108	050	265	10
18°00'	.3090	.9511	.3249	3.078	1.051	3.236	72°00'
	Cos	Sin	Cot	Tan	Csc	Sec	Degrees

TABLE 18-2 *(CONTD.)*

Degrees	Sin	Cos	Tan	Cot	Sec	Csc		
18°00'	.3090	.9511	.3249	3.078	1.051	3.236	72°00'	
10	118	502	281	0.47	052	207	50	
20	145	492	314	0.18	053	179	40	
30	.3173	.9483	.3346	2.989	1.054	3.152	30	
40	201	474	378	960	056	124	20	
50	228	465	411	932	057	098	10	
19°00'	.3256	.9455	.3443	2.904	1.058	3.072	71°00'	
10	283	446	476	877	059	046	50	
20	311	436	508	850	060	021	40	
30	.3338	.9426	.3541	2.824	1.061	2.996	30	
40	365	417	574	798	062	971	20	
50	393	407	607	773	063	947	10	
20°00'	.3420	.9397	.3640	2.747	1.064	2.924	70°00'	
10	448	387	673	723	065	901	50	
20	475	377	706	699	066	878	40	
30	.3502	.9367	.3739	2.675	1.068	2.855	30	
40	529	356	772	651	069	833	20	
50	557	346	805	628	070	812	10	
21°00'	.3584	.9336	.3839	2.605	1.071	2.790	69°00'	
10	611	325	872	583	072	769	50	
20	638	315	906	560	074	749	40	
30	.3665	.9304	.3939	2.539	1.075	2.729	30	
40	692	293	973	517	076	709	20	
50	719	283	.4006	496	077	689	10	
22°00'	.3746	.9272	.4040	2.475	1.079	2.669	68°00'	
10	773	261	074	455	080	650	50	
20	800	250	108	434	081	632	40	
30	.3827	.9239	.4142	2.414	1.082	2.613	30	
40	854	228	176	394	084	595	20	
50	881	216	210	375	085	577	10	
23°00'	.3907	.9205	.4245	2.356	1.086	2.559	67°00'	
10	934	194	279	337	088	542	50	
20	961	182	314	318	089	525	40	
30	.3987	.9171	.4348	2.300	1.090	2.508	30	
40	.4014	159	383	282	092	491	20	
50	041	147	417	264	093	475	10	
24°00'	.4067	.9135	.4452	2.246	1.095	2.459	66°00'	
10	094	124	487	229	096	443	50	
20	120	112	522	211	097	427	40	
30	.4147	.9100	.4557	2.194	1.099	2.411	30	
40	173	088	592	177	100	396	20	
50	200	075	628	161	102	381	10	
25°00'	.4226	.9063	.4663	2.145	1.103	2.366	65°00'	
10	253	051	699	128	105	352	50	
20	279	038	734	112	106	337	40	
30	.4305	.9026	.4770	2.097	1.108	2.323	30	
40	331	013	806	081	109	309	20	
50	358	001	841	066	111	295	10	
26°00'	.4384	.8988	.4877	2.050	1.113	2.281	64°00'	
10	410	975	913	035	114	268	50	
20	436	962	950	020	116	254	40	
30	.4462	.8949	.4986	2.006	1.117	2.241	30	
40	488	936	.5022	1.991	119	228	20	
50	514	923	059	977	121	215	10	
27°00'	.4540	.8910	.5095	1.963	1.122	2.203	63°00'	
	Cos	Sin	Cot	Tan	Csc	Sec	Degrees	

Degrees	Sin	Cos	Tan	Cot	Sec	Csc		
27°00'	.4540	.8910	.5095	1.963	1.122	2.203	63°00'	
10	566	897	132	949	124	190	50	
20	592	884	169	935	126	178	40	
30	.4617	.8870	.5206	1.921	1.127	2.166	30	
40	643	857	243	907	129	154	20	
50	669	843	280	894	131	142	10	
28°00'	.4695	.8829	.5317	1.881	1.133	2.130	62°00'	
10	720	816	354	868	134	118	50	
20	746	802	392	855	136	107	40	
30	.4772	.8788	.5430	1.842	1.138	2.096	30	
40	797	774	467	829	140	085	20	
50	823	760	505	816	142	074	10	
29°00'	.4848	.8746	.5543	1.804	1.143	2.063	61°00'	
10	874	732	581	792	145	052	50	
20	899	718	619	780	147	041	40	
30	.4924	.8704	.5658	1.767	1.149	2.031	30	
40	950	689	696	756	151	020	20	
50	975	675	735	744	153	010	10	
30°00'	.5000	.8660	.5774	1.732	1.155	2.000	60°00'	
10	025	646	812	720	157	1.990	50	
20	050	631	851	709	159	980	40	
30	.5075	.8616	.5890	1.698	1.161	1.970	30	
40	100	601	930	686	163	961	20	
50	125	587	969	675	165	951	10	
31°00'	.5150	.8572	.6009	1.664	1.167	1.942	59°00'	
10	175	557	048	653	169	932	50	
20	200	542	088	643	171	923	40	
30	.5225	.8526	.6128	1.632	1.173	1.914	30	
40	250	511	168	621	175	905	20	
50	275	496	208	611	177	896	10	
32°00'	.5299	.8480	.6249	1.600	1.179	1.887	58°00'	
10	324	465	289	590	181	878	50	
20	348	450	330	580	184	870	40	
30	.5373	.8434	.6371	1.570	1.186	1.861	30	
40	398	418	412	560	188	853	20	
50	422	403	453	550	190	844	10	
33°00'	.5446	.8387	.6494	1.540	1.192	1.836	57°00'	
10	471	371	536	530	195	828	50	
20	495	355	577	520	197	820	40	
30	.5519	.8339	.6619	1.511	1.199	1.812	30	
40	544	323	661	501	202	804	20	
50	568	307	703	492	204	796	10	
34°00'	.5592	.8290	.6745	1.483	1.206	1.788	56°00'	
10	616	274	787	473	209	781	50	
20	640	258	830	464	211	773	40	
30	.5664	.8241	.6873	1.455	1.213	1.766	30	
40	688	225	916	446	216	758	20	
50	712	208	959	437	218	751	10	
35°00'	.5736	.8192	.7002	1.428	1.221	1.743	55°00'	
10	760	175	046	419	223	736	50	
20	783	158	089	411	226	729	40	
30	.5807	.8141	.7133	1.402	1.228	1.722	30	
40	831	124	177	393	231	715	20	
50	854	107	221	385	233	708	10	
36°00'	.5878	.8090	.7265	1.376	1.236	1.701	54°00'	
	Cos	Sin	Cot	Tan	Csc	Sec	Degrees	

TABLE 18-2 (CONTD.)

Degrees	Sin	Cos	Tan	Cot	Sec	Csc	
36°00'	.5878	.8090	.7265	1.376	1.236	1.701	54°00'
10	901	073	310	368	239	695	50
20	925	056	355	360	241	688	40
30	.5948	.8039	.7400	1.351	1.244	1.681	30
40	972	021	445	343	247	675	20
50	995	004	490	335	249	668	10
37°00'	.6018	.7986	.7536	1.327	1.252	1.662	53°00'
10	041	969	581	319	255	655	50
20	065	951	627	311	258	649	40
30	.6088	.7934	.7673	1.303	1.260	1.643	30
40	111	916	720	295	263	636	20
50	134	898	766	288	266	630	10
38°00'	.6157	.7880	.7813	1.280	1.269	1.624	52°00'
10	180	862	860	272	272	618	50
20	202	844	907	265	275	612	40
30	.6225	.7826	.7954	1.257	1.278	1.606	30
40	248	808	.8002	250	281	601	20
50	271	790	050	242	284	595	10
39°00'	.6293	.7771	.8098	1.235	1.287	1.589	51°00'
10	316	753	146	228	290	583	50
20	338	735	195	220	293	578	40
30	.6361	.7716	.8243	1.213	1.296	1.572	30
40	383	698	292	206	299	567	20
50	406	679	342	199	302	561	10
40°00'	.6428	.7660	.8391	1.192	1.305	1.556	50°00'
10	450	642	441	185	309	550	50
20	472	623	491	178	312	545	40
30	.6494	.7604	.8541	1.171	1.315	1.540	30
40	517	585	591	164	318	535	20
50	539	566	642	157	322	529	10
41°00'	.6561	.7547	.8693	1.150	1.325	1.524	49°00'
10	583	528	744	144	328	519	50
20	604	509	796	137	332	514	40
30	.6626	.7490	.8847	1.130	1.335	1.509	30
40	648	470	899	124	339	504	20
50	670	451	952	117	342	499	10
42°00'	.6691	.7431	.9004	1.111	1.346	1.494	48°00'
10	713	412	057	104	349	490	50
20	734	392	110	098	353	485	40
30	.6756	.7373	.9163	1.091	1.356	1.480	30
40	777	353	217	085	360	476	20
50	799	333	271	079	364	471	10
43°00'	.6820	.7314	.9325	1.072	1.367	1.466	47°00'
10	841	294	380	066	371	462	50
20	862	274	435	060	375	457	40
30	.6884	.7254	.9490	1.054	1.379	1.453	30
40	905	234	545	048	382	448	20
50	926	214	601	042	386	444	10
44°00'	.6947	.7193	.9657	1.036	1.390	1.440	46°00'
10	967	173	713	030	394	435	50
20	988	153	770	024	398	431	40
30	.7009	.7133	.9827	1.018	1.402	1.427	30
40	030	112	884	012	406	423	20
50	050	092	942	006	410	418	10
45°00'	.7071	.7071	1.000	1.000	1.414	1.414	45°00'
	Cos	Sin	Cot	Tan	Csc	Sec	Degrees

19
CONVERSIONS

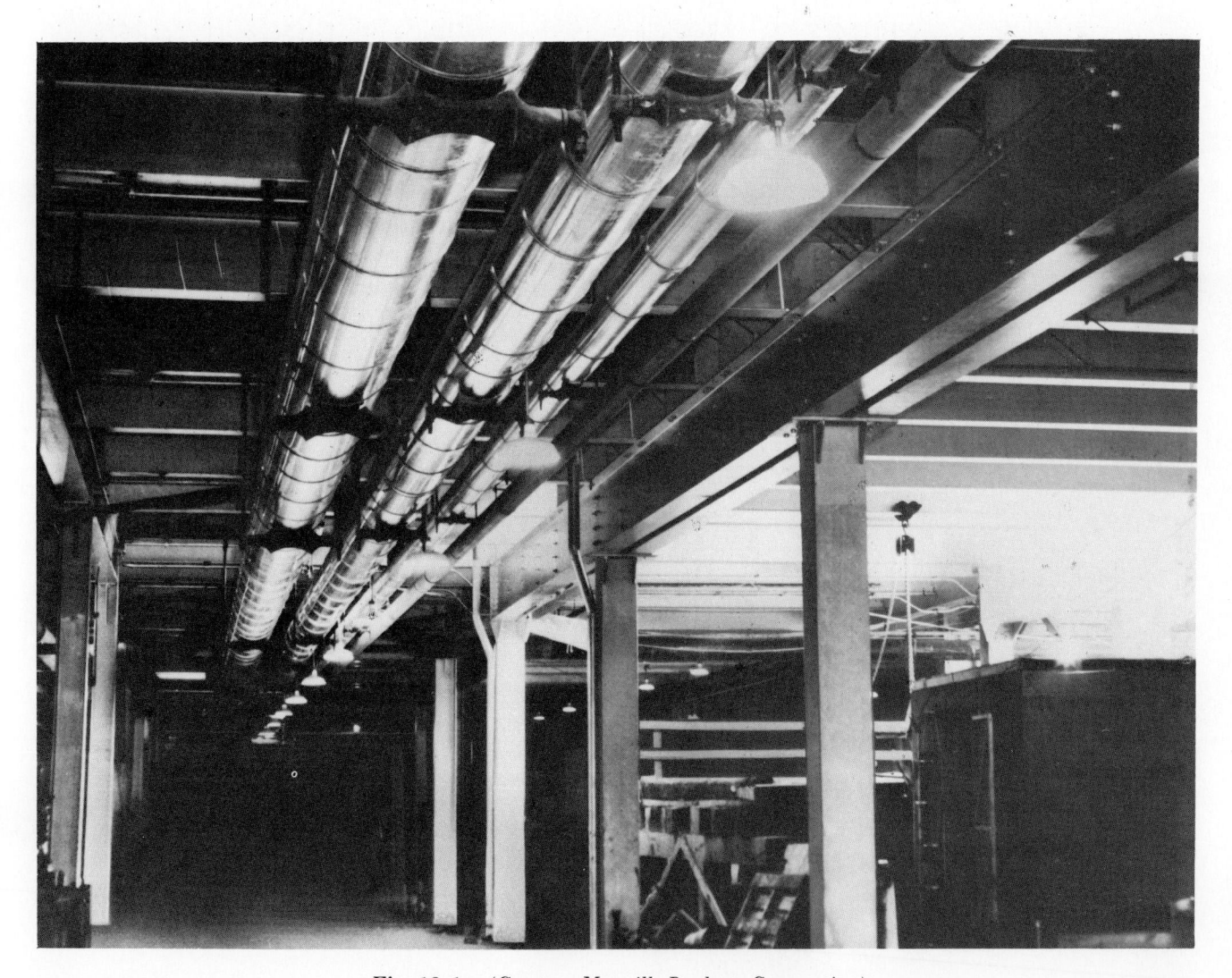

Fig. 19-1 *(Courtesy Manville Products Corporation.)*

TABLE 19-1 DECIMAL AND METRIC EQUIVALENTS OF COMMON FRACTIONS OF AN INCH (*COURTESY LUNKENHEIMER COMPANY, CINCINNATI, OH 45214.*)

Fraction	Decimal	Millimeter	Fraction	Decimal	Millimeter
1/64	.015625	0.39688	33/64	.515625	13.09690
1/32	.03125	0.79375	17/32	.53125	13.49378
3/64	.046875	1.19063	35/64	.546875	13.89065
1/16	.0625	1.58750	9/16	.5625	14.28753
5/64	.078125	1.98438	37/64	.578125	14.68440
3/32	.09375	2.38125	19/32	.59375	15.08128
7/64	.109375	2.77813	39/64	.609375	15.47816
1/8	.125	3.17501	5/8	.625	15.87503
9/64	.140625	3.57188	41/64	.640625	16.27191
5/32	.15625	3.96876	21/32	.65625	16.66878
11/64	.171875	4.36563	43/64	.671875	17.06566
3/16	.1875	4.76251	11/16	.6875	17.46253
13/64	.203125	5.15939	45/64	.703125	17.85941
7/32	.21875	5.55626	23/32	.71875	18.25629
15/64	.234375	5.95314	47/64	.734375	18.65316
1/4	.25	6.35001	3/4	.75	19.05004
17/64	.265625	6.74689	49/64	.765625	19.44691
9/32	.28125	7.14376	25/32	.78125	19.84379
19/64	.296875	7.54064	51/64	.796875	20.24067
5/16	.3125	7.93752	13/16	.8125	20.63754
21/64	.328125	8.33439	53/64	.828125	21.03442
11/32	.34375	8.73127	27/32	.84375	21.43129
23/64	.359375	9.12814	55/64	.859375	21.82817
3/8	.375	9.52502	7/8	.875	22.22504
25/64	.390625	9.92189	57/64	.890625	22.62192
13/32	.40625	10.31877	29/32	.90625	23.01880
27/64	.421875	10.71565	59/64	.921875	23.41567
7/16	.4375	11.11252	15/16	.9375	23.81255
29/64	.453125	11.50940	61/64	.953125	24.20942
15/32	.46875	11.90627	31/32	.96875	24.60630
31/64	.484375	12.30315	63/64	.984375	25.00318
1/2	.5	12.70003	1	1.0	25.40005

TABLE 19-2 UNIT CONVERSIONS (*COURTESY ITT GRINNELL CORPORATION.*)

TEMPERATURE

°C = (°F − 32) × 5/9

VOLUME

1 gal. (U.S.) = 128 fl. oz. (U.S.)
 = 231 cu. in.
 = 0.833 gal. (Brit.)
1 cu. ft. = 7.48 gal. (U.S.)

WEIGHT OF WATER

1 cu. ft. at 50°F. weighs 62.41 lb.
1 gal. at 50°F. weighs 8.34 lb.
1 cu. ft. of ice weighs 57.2 lb.
Water is at its greatest density at 39.2°F.
1 cu. ft. at 39.2°F. weighs 62.43 lb.

WEIGHT OF LIQUID

1 gal. (U.S.) = 8.34 lb. × sp. gr.
1 cu. ft. = 62.4 lb. × sp. gr.
1 lb. = 0.12 U.S. gal. ÷ sp. gr.
 = 0.016 cu. ft. ÷ sp. gr.

WORK

1 Btu (mean) = 778 ft. lb.
 = 0.293 watt hr.
 = 1/180 of heat required to change temp of 1 lb. water from 32°F to 212°F
1 hp-hr = 2545 Btu (mean)
 = 0.746 kwhr
1 Kwhr = 3413 Btu (mean)
 = 1.34 hp-hr

FLOW

1 gpm = 0.134 cu. ft. per min.
 = 500 lb. per hr. × sp. gr.
500 lb. per hr. = 1 gpm ÷ sp. gr.
1 cu. ft. per min. (cfm) = 448.8 gal. per hr. (ghp)

POWER

1 Btu per hr. = 0.293 watt
 = 12.96 ft. lb. per min.
 = 0.00039 hp
1 ton refrigeration (U.S.) = 288,000 Btu per 24 hr.
 = 12,000 Btu per hr.
 = 200 Btu per min.
 = 83.33 lb. ice melted per hr. from and at 32°F.
 = 2000 lb. ice melted per 24 hr. from and at 32°F.
1 hp = 550 ft. lb. per sec.
 = 746 watt
 = 2545 Btu per hr.
1 boiler hp = 33,480 Btu per hr.
 = 34.5 lb. water evap. per hr. from and at 212°F.

1 kw. = 9.8 kw.
 = 3413 Btu per hr.

MASS

1 lb. (avoir.) = 16 oz. (avoir.)
 = 7000 grain
1 ton (short) = 2000 lb.
1 ton (long) = 2240 lb.

PRESSURE

1 lb. per sp. in. = 2.31 ft. water at 60°F
 = 2.04 in. hg at 60°F.
1 ft. water at 60°F = 0.433 lb. per sq. in.
 = 0.884 in. hg at 60°F
1 in. Hg at 60°F = 0.49 lb. per sq. in.
 = 1.13 ft. water at 60°F
lb. per sp. in. Absolute (psia) = lb. per sq. in. gauge (psig) + 14.7

TABLE 19-3 HARDNESS CONVERSION NUMBERS (*COURTESY ITT GRINNELL CORPORATION.*)

Brinell Indentation Diameter, mm.	Brinell Hardness No.—10-mm. Ball Standard or Tungsten Carbide Ball 3000-kg. Load	Diamond Pyramid Hardness Number. 50-kg. Load	Rockwell Hardness Number B-Scale 100-kg. Load 1/16 in. Dia. Ball	Rockwell Hardness Number C-Scale 150-kg. Load Brale Penetrator	Rockwell Superficial Hardness Number Superficial Brale Penetrator 15-N Scale 15-kg. Load	30-N Scale 30-kg. Load	45-N Scale 45-kg. Load	Shore Scleroscope Hardness Number	Tensile Strength (Approx.) 1000 PSI.
2.95	429	455	—	45.7	83.4	64.6	49.9	61	217
3.00	415	440	—	44.5	82.8	63.5	48.4	59	210
3.05	401	425	—	43.1	82.0	62.3	46.9	58	202
3.10	388	410	—	41.8	81.4	61.1	45.3	56	195
3.15	375	396	—	40.4	80.6	59.9	43.6	54	188
3.20	363	383	—	39.1	80.0	58.7	42.0	52	182
3.25	352	372	(110.0)	37.9	79.3	57.6	40.5	51	176
3.30	341	360	(109.0)	36.9	78.6	56.4	39.1	50	170
3.35	331	350	(108.5)	35.5	78.0	55.4	37.8	48	166
3.40	321	339	(108.0)	34.3	77.3	54.3	36.4	47	160
3.45	311	328	(107.5)	33.1	76.7	53.3	34.4	46	155
3.50	302	319	(107.0)	32.1	76.1	52.2	33.8	45	150
3.55	293	309	(106.0)	30.9	75.5	51.2	32.4	43	145
3.60	285	301	(105.5)	29.9	75.0	50.3	31.2	—	141
3.65	277	292	(104.5)	28.8	74.4	49.3	29.9	41	137
3.70	269	284	(104.0)	27.6	73.7	48.3	28.5	40	133
3.75	262	276	(103.0)	26.6	73.1	47.3	27.3	39	129
3.80	255	269	(102.0)	25.4	72.5	46.2	26.0	38	126
3.85	248	261	(101.0)	24.2	71.7	45.1	24.5	37	122
3.90	241	253	100.0	22.8	70.9	43.9	22.8	36	118
3.95	235	247	99.0	21.7	70.3	42.9	21.5	35	115
4.00	229	241	98.2	20.5	69.7	41.9	20.1	34	111
4.05	223	234	97.3	(18.8)	—	—	—	—	—
4.10	217	228	96.4	(17.5)	—	—	—	33	105
4.15	212	222	95.5	(16.0)	—	—	—	—	102
4.20	207	218	94.6	(15.2)	—	—	—	32	100
4.25	201	212	93.8	(13.8)	—	—	—	31	98
4.30	197	207	92.8	(12.7)	—	—	—	30	95
4.35	192	202	91.9	(11.5)	—	—	—	29	93
4.40	187	196	90.7	(10.0)	—	—	—	—	90
4.45	183	192	90.0	(9.0)	—	—	—	28	89
4.50	179	188	89.0	(8.0)	—	—	—	27	87
4.55	174	182	87.8	(6.4)	—	—	—	—	85
4.60	170	178	86.8	(5.4)	—	—	—	26	83
4.65	167	175	86.0	(4.4)	—	—	—	—	81
4.70	163	171	85.0	(3.3)	—	—	—	25	79
4.80	156	163	82.9	(0.9)	—	—	—	—	76
4.90	149	156	80.8	—	—	—	—	23	73
5.00	143	150	78.7	—	—	—	—	22	71
5.10	137	143	76.4	—	—	—	—	21	67
5.20	131	137	74.0	—	—	—	—	—	65
5.30	126	132	72.0	—	—	—	—	20	63
5.40	121	127	69.8	—	—	—	—	19	60
5.50	116	122	67.6	—	—	—	—	18	58
5.60	111	117	65.7	—	—	—	—	15	56

NOTE: Values in ( ) are beyond normal range; given for information only.

CONVERSIONS

TABLE 19-4 POWER REQUIRED FOR PUMPING *(COURTESY CRANE CORPORATION.)*

Gals. per Min.	Theoretical Horsepower Required to Raise Water (at 60 F) To Different Heights														
	5 feet	10 feet	15 feet	20 feet	25 feet	30 feet	35 feet	40 feet	45 feet	50 feet	60 feet	70 feet	80 feet	90 feet	100 feet
5	0.006	0.013	0.019	0.025	0.032	0.038	0.044	0.051	0.057	0.063	0.076	0.088	0.101	0.114	0.126
10	0.013	0.025	0.038	0.051	0.063	0.076	0.088	0.101	0.114	0.126	0.152	0.177	0.202	0.227	0.253
15	0.019	0.038	0.057	0.076	0.095	0.114	0.133	0.152	0.171	0.190	0.227	0.265	0.303	0.341	0.379
20	0.025	0.051	0.076	0.101	0.126	0.152	0.177	0.202	0.227	0.253	0.303	0.354	0.404	0.455	0.505
25	0.032	0.063	0.095	0.126	0.158	0.190	0.221	0.253	0.284	0.316	0.379	0.442	0.505	0.568	0.632
30	0.038	0.076	0.114	0.152	0.190	0.227	0.265	0.303	0.341	0.379	0.455	0.531	0.606	0.682	0.758
35	0.044	0.088	0.133	0.177	0.221	0.265	0.310	0.354	0.398	0.442	0.531	0.619	0.707	0.796	0.884
40	0.051	0.101	0.152	0.202	0.253	0.303	0.354	0.404	0.455	0.505	0.606	0.707	0.808	0.910	1.011
45	0.057	0.114	0.171	0.227	0.284	0.341	0.398	0.455	0.512	0.568	0.682	0.796	0.910	1.023	1.137
50	0.063	0.126	0.190	0.253	0.316	0.379	0.442	0.505	0.568	0.632	0.758	0.884	1.011	1.137	1.263
60	0.076	0.152	0.227	0.303	0.379	0.455	0.531	0.606	0.682	0.758	0.910	1.061	1.213	1.364	1.516
70	0.088	0.177	0.265	0.354	0.442	0.531	0.619	0.707	0.796	0.884	1.061	1.238	1.415	1.592	1.768
80	0.101	0.202	0.303	0.404	0.505	0.606	0.707	0.808	0.910	1.011	1.213	1.415	1.617	1.819	2.021
90	0.114	0.227	0.341	0.455	0.568	0.682	0.796	0.910	1.023	1.137	1.364	1.592	1.819	2.046	2.274
100	0.126	0.253	0.379	0.505	0.632	0.758	0.884	1.011	1.137	1.263	1.516	1.768	2.021	2.274	2.526
125	0.158	0.316	0.474	0.632	0.790	0.947	1.105	1.263	1.421	1.579	1.895	2.211	2.526	2.842	3.158
150	0.190	0.379	0.568	0.758	0.947	1.137	1.326	1.516	1.705	1.895	2.274	2.653	3.032	3.411	3.790
175	0.221	0.442	0.663	0.884	1.105	1.326	1.547	1.768	1.990	2.211	2.653	3.095	3.537	3.979	4.421
200	0.253	0.505	0.758	1.011	1.263	1.516	1.768	2.021	2.274	2.526	3.032	3.537	4.042	4.548	5.053
250	0.316	0.632	0.947	1.263	1.579	1.895	2.211	2.526	2.842	3.158	3.790	4.421	5.053	5.684	6.316
300	0.379	0.758	1.137	1.516	1.895	2.274	2.653	3.032	3.411	3.790	4.548	5.305	6.063	6.821	7.579
350	0.442	0.884	1.326	1.768	2.211	2.653	3.095	3.537	3.979	4.421	5.305	6.190	7.074	7.958	8.842
400	0.505	1.011	1.516	2.021	2.526	3.032	3.537	4.042	4.548	5.053	6.063	7.074	8.084	9.095	10.11
500	0.632	1.263	1.895	2.526	3.158	3.790	4.421	5.053	5.684	6.316	7.579	8.842	10.11	11.37	12.63

Gals. per Min.	125 feet	150 feet	175 feet	200 feet	250 feet	300 feet	350 feet	400 feet
5	0.158	0.190	0.221	0.253	0.316	0.379	0.442	0.505
10	0.316	0.379	0.442	0.505	0.632	0.758	0.884	1.011
15	0.474	0.568	0.663	0.758	0.947	1.137	1.326	1.516
20	0.632	0.758	0.884	1.011	1.263	1.516	1.768	2.021
25	0.790	0.947	1.105	1.263	1.579	1.895	2.211	2.526
30	0.947	1.137	1.326	1.516	1.895	2.274	2.653	3.032
35	1.105	1.326	1.547	1.768	2.211	2.653	3.095	3.537
40	1.263	1.516	1.768	2.021	2.526	3.032	3.537	4.042
45	1.421	1.705	1.990	2.274	2.842	3.411	3.979	4.548
50	1.579	1.895	2.211	2.526	3.158	3.790	4.421	5.053
60	1.895	2.274	2.653	3.032	3.790	4.548	5.305	6.063
70	2.211	2.653	3.095	3.537	4.421	5.305	6.190	7.074
80	2.526	3.032	3.537	4.042	5.053	6.063	7.074	8.084
90	2.842	3.411	3.979	4.548	5.684	6.821	7.958	9.095
100	3.158	3.790	4.421	5.053	6.316	7.579	8.842	10.11
125	3.948	4.737	5.527	6.316	7.895	9.474	11.05	12.63
150	4.737	5.684	6.632	7.579	9.474	11.37	13.26	15.16
175	5.527	6.632	7.737	8.842	11.05	13.26	15.47	17.68
200	6.316	7.579	8.842	10.11	12.63	15.16	17.68	20.21
250	7.895	9.474	11.05	12.63	15.79	18.95	22.11	25.26
300	9.474	11.37	13.26	15.16	18.95	22.74	26.53	30.32
350	11.05	13.26	15.47	17.68	22.11	26.53	30.95	35.37
400	12.63	15.16	17.68	20.21	25.26	30.32	35.37	40.42
500	15.79	18.95	22.11	25.26	31.58	37.90	44.21	50.53

Specific gravity of water................page A-6
Specific gravity of liquids other than water....page A-7

HORSEPOWER = 33 000 ...ft-lb/min
= 550 ...ft-lb/sec
= 2544.48 ...Btu/hr
= 745.7 ...watts

$(whp) = QH\rho \div 247\,000 = QP \div 1714$
$(bhp) = (whp) \div e_p = QH\rho \div 247\,000\, e_p$
$(e_p) = QH\rho \div 247\,000\, (bhp)$

where: (whp) = water horsepower
H = pump head in feet
(bhp) = brake horsepower
e_p = pump efficiency

Overall efficiency (e_o) takes into account all losses in the pump and driver.

$$e_o = e_p\, e_D\, e_T$$

where: e_D = driver efficiency
e_T = transmission efficiency
e_V = volumetric efficiency

$$e_V(\%) = \frac{\text{actual pump displacement } (Q)\,(100)}{\text{theoretical pump displacement } (Q)}$$

Note: For fluids other than water, multiply table values by specific gravity. In pumping liquids with a viscosity considerably higher than that of water, the pump capacity and head are reduced. To calculate the horsepower for such fluids, pipe friction head must be added to the elevation head to obtain the total head; this value is inserted in the first horsepower equation given above.

TABLE 19-5   CONVERSION FACTORS (*COURTESY ITT GRINNELL CORPORATION.*)

Multiply	by	To Obtain	Multiply	by	To Obtain
Absolute viscosity (poise)	1	Gram/second centimeter	BTU/minute	17.57	Watts
Absolute viscosity (centipoise)	0.01	Poise	BTU/pound	0.556	Calories (Kg)/Kilogram
Acceleration due to gravity (g)	32.174	Feet/second^2	Bushels	2150.4	Cubic inches
	980.6	Centimeters/second^2		35.24	Liters
Acre	0.4047	Hectares		4	Pecks
	10	Square Chains		32	Quarts (dry)
	43,560	Square Feet	Cables	120	Fathoms
	4047	Square Meters	Calories (gm)	0.003968	BTU
	0.001562	Square Miles		0.001	Calories (Kg)
	4840	Square Yards		3.088	Foot pounds
	160	Square Rods		1.558 × 10^{-6}	Horse power hours
Acre-feet	43,560	Cubic Feet		4.185	Joules
	325,851	Gallons (US)		0.4265	Kilogram meters
	1233.49	Cubic Meters		1.1628 × 10^{-6}	Kilowatt hours
	1,233,490	Liters		0.0011628	Watt hours
Acre-feet/hour	726	Cubic feet/Minute	Cal (gm)/sec/cm^2/°C/cm	242.13	BTU/Hr/ft^2/°F/ft
	5430.86	Gallons/Minute	Calories (Kg)	3.968	BTU
Angstroms	10^{-10}	Meters		1000	Calories (gm)
Ares	0.01	Hectares		3088	Foot pounds
	1076.39	Square Feet		0.001558	Horse power hours
	0.02471	Acres		4185	Joules
Atmospheres	76.0	Cms of Hg at 32° F		426.5	Kilogram meters
	29.921	Inches of Hg at 32° F		0.0011628	Kilowatt hours
	33.94	Feet of Water at 62° F		1.1628	Watt hours
	10,333	Kgs/Square meter	Calories (Kg)/Cu meter	0.1124	BTU/Cu foot at 0° C
	14.6963	Pounds/Square inch	Cal (Kg)/Hr/M^2/°C/M	0.671	BTU/Hr/ft^2/°F/foot
	1.058	Tons/Square foot	Calories (Kg)/Kg	1.8	BTU/pound
	1013.15	Millibars	Calories (Kg)/minute	51.43	Foot pounds/second
	235.1408	Ounces/Square inch		0.09351	Horse power
Bags of cement	94	Pounds of cement		0.06972	Kilowatts
Barrels of oil	42	Gallons of oil (US)	Carats (diamond)	200	Milligram
Barrels of cement	376	Pounds of cement	Centares (Centiares)	1	Square meters
Barrels (not legal)	31	Gallons (US)	Centigram	0.01	Grams
or	31.5	Gallons (US)	Centiliters	0.01	Liters
Board feet	144 × 1 in.*	Cubic inches	Centimeters	0.3937	Inches
Boiler horse power	33,479	BTU/hour		0.032808	Feet
	9.803	Kilowatts		0.01	Meters
	34.5	Pounds of water evaporated/hour at 212° F		10	Millimeters
BTU	252.016	Calories (gm)	Centimeters of Hg at 32°F	0.01316	Atmospheres
	0.252	Calories (Kg)		0.4461	Feet of water at 62° F
	777.54	Foot pounds		136	Kgs/Square meter
	0.0003927	Horse power hours		27.85	Pounds/Square foot
	1054.2	Joules		0.1934	Pounds/Square inch
	107.5	Kilogram meters	Centimeters/second	1.969	Feet/minute
	0.0002928	Kilowatt hours		0.03281	Feet/second
BTU/Cu foot	8.89	Calories (Kg)/Cu meter at 32° F		0.036	Kilometers/hour
				0.6	Meters/minute
BTU/Hr/ft^2/°F/ft	0.00413	Cal (gm)/Sec/cm^2/°C/cm		0.02237	Miles/hour
	1.49	Cal (Kg)/Hr/M^2/°C/Meter		0.0003728	Miles/minute
			Centimeters/second^2	0.03281	Feet/second^2
BTU/minute	12.96	Foot pounds/second	Centipoise	0.000672	Pounds/sec foot
	0.02356	Horse power		2.42	Pounds/hour foot
	0.01757	Kilowatts		0.01	Poise
			Chains (Gunter's)	4	Rods
				66	Feet
				100	Links

* For thickness less than 1 in. use actual thickness in decimals of an inch.

CONVERSIONS

181

TABLE 19-5 *(CONTD.)*

Multiply	by	To Obtain
Cheval-vapeur	1	Metric horse power
	75	Kilogram meters/second
	0.98632	Horse power
Circular inches	10^6	Circular mils
	0.7854	Square inches
	785,400	Square mils
Circular mils	0.7854	Square mils
	10^{-6}	Circular inches
	7.854×10^{-5}	Square inches
Cubic centimeters	3.531×10^{-5}	Cubic feet
	0.06102	Cubic inches
	10^{-6}	Cubic meters
	1.308×10^{-6}	Cubic yards
	0.0002642	Gallons (US)
	0.001	Liters
	0.002113	Pints (liq. US)
	0.001057	Quarts (liq. US)
	0.0391	Ounces (fluid)
Cubic feet	28,320	Cubic centimeters
	1728	Cubic inches
	0.02832	Cubic meters
	0.03704	Cubic yards
	7.48052	Gallons (US)
	28.32	Liters
	59.84	Pints (liq. US)
	29.92	Quarts (liq. US)
	2.296×10^{-5}	Acre feet
	0.803564	Bushels
Cubic feet of water	62.4266	Pounds at 39.2° F
	62.3554	Pounds at 62° F
Cubic feet/minute	472	Cubic centimeters/sec
	0.1247	Gallons (US)/second
	0.472	Liters/second
	62.36	Pounds water/min at 62°F
	7.4805	Gallons (US)/minute
	10,772	Gallons/24 hours
	0.033058	Acre feet/24 hours
Cubic feet/second	646,317	Gallons (US)/24 hours
	448.831	Gallons/minute
	1.98347	Acre feet/24 hours
Cubic inches	16.387	Cubic centimeters
	0.0005787	Cubic feet
	1.639×10^{-5}	Cubic meters
	2.143×10^{-5}	Cubic yards
	0.004329	Gallons (US)
	0.01639	Liters
	0.03463	Pints (liq. US)
	0.01732	Quarts (liq. US)
Cubic meters	10^6	Cubic centimeters
	35.31	Cubic feet
	61,023	Cubic inches
	1.308	Cubic yards
	264.2	Gallons (US)
	1000	Liters
	2113	Pints (liq. US)
	1057	Quarts (liq. US)
Cubic yards	764,600	Cubic centimeters
	27	Cubic feet
	46,656	Cubic inches
	0.7646	Cubic meters
	202	Gallons (US)
	764.6	Liters
	1616	Pints (liq. US)
	807.9	Quarts (liq. US)
Cubic yards/minute	0.45	Cubic feet/second
	3.367	Gallons (US)/second
	12.74	Liters/second

Multiply	by	To Obtain
Cubit	18	Inches
Days (mean)	1440	Minutes
	24	Hours
	86,400	Seconds
Days (sidereal)	86,164.1	Solar seconds
Decigrams	0.1	Grams
Deciliters	0.1	Liters
Decimeters	0.1	Meters
Degrees (angle)	60	Minutes
	0.01745	Radians
	3600	Seconds
Degrees F [less 32]	0.5556	Degrees C
Degrees F	1 [plus 460]	Degrees F above absolute 0
Degrees C	1.8 [plus 32]	Degrees F
	1 [plus 273]	Degrees C above absolute 0
Degrees/second	0.01745	Radians/second
	0.1667	Revolutions/minute
	0.002778	Revolutions/second
Dekagrams	10	Grams
Dekaliters	10	Liters
Dekameters	10	Meters
Diameter (circle)	3.14159265359	Circumference
(approx)	3.1416	
(approx)	3.14	
(approx)	$\frac{22}{7}$	
Diameter (circle)	0.88623	Side of equal square
	0.7071	Side of inscribed square
Diameter³ (sphere)	0.5236	Volume (sphere)
Diam (major) × diam (minor)	0.7854	Area of ellipse
Diameter² (circle)	0.7854	Area (circle)
Diameter² (sphere)	3.1416	Surface (sphere)
Diam (inches) × RPM	0.262	Belt speed ft/minute
Digits	0.75	Inches
Drams (avoirdupois)	27.34375	Grains
	0.0625	Ounces (avoir.)
	1.771845	Grams
Fathoms		Feet
Feet	30.48	Centimeters
	12	Inches
	0.3048	Meters
	⅓	Yards
	0.06061	Rods
Feet of water at 62	0.029465	Atmospheres
	0.88162	Inches of Hg at 32° F
	62.3554	Pounds/square foot
	0.43302	Pounds/square inch
	304.44	Kilogram/sq meter
Feet/minute	0.5080	Centimeters/second
	0.01667	Feet/second
	0.01829	Kilometers/hour

TABLE 19-5 (CONTD.)

Multiply	by	To Obtain
Feet/minute	0.3048	Meters/minute
	0.01136	Miles/hour
Feet/second	30.48	Centimeters/second
	1.097	Kilometers/hour
	0.5921	Knots
	18.29	Meters/minute
	0.6818	Miles/hour
	0.01136	Miles/minute
Feet/second^2	30.48	Centimeters/second^2
	0.3048	Meters/second^2
Flat of a hexagon	1.155	Distance across corners
Flat of a square	1.414	Distance across corners
Foot pounds	0.0012861	BTU
	0.32412	Calories (gm)
	0.0003241	Calories (Kg)
	5.05×10^{-7}	Horse power hours
	1.3558	Joules
	0.13826	Kilogram meters
	3.766×10^{-7}	Kilowatt hours
	0.0003766	Watt hours
Foot pounds/minute	0.001286	BTU/minute
	0.01667	Foot pounds/second
	3.03×10^{-5}	Horse power
	0.0003241	Calories (Kg)/minute
	2.26×10^{-5}	Kilowatts
Foot pounds/second	0.07717	BTU/minute
	0.001818	Horse power
	0.01945	Calories (Kg)/minute
	0.001356	Kilowatts
Furlong	40	Rods
	220	Yards
	660	Feet
	0.125	Miles
	0.2042	Kilometers
Gallons (Imperial)	277.42	Cubic inches
	4.543	Liters
	1.20095	Gallons (US)
Gallons (US)	3785	Cubic centimeters
	0.13368	Cubic feet
	231	Cubic inches
	0.003785	Cubic meters
	0.004951	Cubic yards
	3.785	Liters
	8	Pints (liq. US)
	4	Quarts (liq. US)
	0.83267	Gallons (Imperial)
	3.069×10^{-6}	Acre feet
Gallons (US) of water at 62° F	8.3357	Pounds of water
Gallons (US) of water/minute	6.0086	Tons of water/24 hours
Gallons (US)/minute	0.002228	Cubic feet/second
	0.13368	Cubic feet/minute
	8.0208	Cubic feet/hour
	0.06309	Liters/second
	3.78533	Liters/minute
	0.0044192	Acre feet/24 hours
Grains	1	Grains (avoirdupois)
	1	Grains (apothecary)
	1	Grains (troy)
	0.0648	Grams
	0.0020833	Ounces (troy)
	0.0022857	Ounces (avoir.)

Multiply	by	To Obtain
Grains/gallon (US)	17.118	Parts/million
	142.86	Pounds/million gallons (US)
Grams	980.7	Dynes
	15.43	Grains
	0.001	Kilograms
	1000	Milligrams
	0.03527	Ounces (avoir.)
	0.03215	Ounces (troy)
	0.002205	Pounds
Grams/centimeter	0.0056	Pounds/inch
Grams/cubic centimeter	62.43	Pounds/cubic foot
	0.03613	Pounds/cubic inch
	4.37	Grains/100 cubic ft
Grams/liter	58.417	Grains/gallon (US)
	8.345	Pounds/100 gallons (US)
	0.062427	Pounds/cubic foot
	1000	Parts/million
Gravity (g)	32.174	Feet/second^2
	980.6	Centimeters/second^2
Hand	4	Inches
	10.16	Centimeters
Hectares	2.471	Acres
	107,639	Square feet
	100	Ares
Hectograms	100	Grams
Hectoliters	100	Liters
Hectometers	100	Meters
Hectowatts	100	Watts
Hogshead	63	Gallons (US)
	238.4759	Liters
Horse power	42.44	BTU/minute
	33,000	Foot pounds/minute
	550	Foot pounds/second
	1.014	Metric horse power (Cheval vapeur)
	10.7	Calories (Kg)/min
	0.7457	Kilowatts
	745.7	Watts
Horse power (boiler)	33,479	BTU/hour
	9.803	Kilowatts
	34.5	Pounds of water evaporated/hour at 212° F
Horse power hours	2546.5	BTU
	641,700	Calories (gm)
	641.7	Calories (Kg)
	1,980,000	Foot pounds
	2,684,500	Joules
	273,740	Kilogram meters
	0.7455	Kilowatt hours
	745.5	Watt hours
Inches	2.54	Centimeters
	0.08333	Feet
	1000	Mils
	12	Lines
	72	Points
Inches of Hg at 32° F	0.03342	Atmospheres
	345.3	Kilograms/square meter
	70.73	Pounds/square foot
	0.49117	Pounds/square inch
	1.1343	Feet of water at 62° F

CONVERSIONS

TABLE 19-5   *(CONTD.)*

Multiply	by	To Obtain	Multiply	by	To Obtain
Inches of Hg at 32° F	13.6114	Inches of water at 62° F	Kilowatt hours	860,500	Calories (gm)
	7.85872	Ounces/square inch		860.5	Calories (Kg)
Inches of water at 62° F	0.002455	Atmospheres		2,655,200	Foot pounds
	25.37	Kilograms/square meter		1.341	Horse power hours
	0.5771	Ounces/square inch		3,600,000	Joules
	5.1963	Pounds/square foot		367,100	Kilogram meters
	0.03609	Pounds/square inch		1000	Watt hours
	0.07347	Inches of Hg at 32° F	Knots	1	Nautical miles/hour
Joules	0.00094869	BTU		1.1516	Miles/hour
	0.239	Calories (gm)		1.8532	Kilometers/hour
	0.000239	Calories (Kg)	Leagues	3	Miles
	0.73756	Foot pounds	Lines	0.08333	Inches
	3.72×10^{-7}	Horse power hours	Links	7.92	Inches
	0.10197	Kilogram meters	Liters	1000	Cubic centimeters
	2.778×10^{-7}	Kilowatt hours		0.03531	Cubic feet
	0.0002778	Watt hours		61.02	Cubic inches
	1	Watt second		0.001	Cubic meters
Kilograms	980,665	Dynes		0.001308	Cubic yards
	2.205	Pounds		0.2642	Gallons (US)
	0.001102	Tons (short)		0.22	Gallons (Imp)
	1000	Grams		2.113	Pints (liq. US)
	35.274	Ounces (avoir.)		1.057	Quarts (liq. US)
	32.1507	Ounces (troy)		8.107×10^{-7}	Acre Feet
Kilogram meters	0.009302	BTU		2.2018	Pounds of water at 62° F
	2.344	Calories (gm)	Liters/minute	0.0005886	Cubic feet/second
	0.002344	Calories (Kg)		0.004403	Gallons (US)/second
	7.233	Foot pounds		0.26418	Gallons (US)/minute
	3.653×10^{-6}	Horse power hours	Meters	100	Centimeters
	9.806	Joules		3.281	Feet
	2.724×10^{-6}	Kilowatt hours		39.37	Inches
	0.002724	Watt hours		1.094	Yards
Kilograms/cubic meter	0.06243	Pounds/cubic foot		0.001	Kilometers
Kilograms/meter	0.6720	Pounds/foot		1000	Millimeters
Kilograms/sq centimeter	14.223	Pounds/sq inch	Meters/minute	1.667	Centimeters/second
	1	Metric atmosphere		3.281	Feet/minute
Kilogram/sq meter	9.678×10^{-5}	Atmospheres		0.05468	Feet/second
	0.003285	Feet of water at 62° F		0.06	Kilometers/hour
	0.002896	Inches of Hg at 32° F		0.03728	Miles/hour
	0.2048	Pounds/square foot	Meters/second	196.8	Feet/minute
	0.001422	Pounds/square inch		3.281	Feet/second
	0.007356	Centimeters of Hg at 32° F		3.6	Kilometers/hour
Kiloliters	1000	Liters		0.06	Kilometers/minute
Kilometers	100,000	Centimeters		2.237	Miles/hour
	1000	Meters		0.03728	Miles/minute
	3281	Feet	Microns	10^{-6}	Meters
	0.6214	Miles		0.001	Millimeters
	1094	Yards		0.03937	Mils
Kilometers/hour	27.78	Centimeters/second	Mils	0.001	Inches
	54.68	Feet/minute		0.0254	Millimeters
	0.9113	Feet/second		25.4	Microns
	16.67	Meters/minute	Miles	160,934	Centimeters
	0.6214	Miles/hour		5280	Feet
	0.5396	Knots		63,360	Inches
Kilometers/hr/sec	27.78	Centimeters/sec/sec		1.609	Kilometers
	0.9113	Feet/sec/sec		1760	Yards
	0.2778	Meters/sec/sec		80	Chains
Kilowatts	56.92	BTU/minute		320	Rods
	44,250	Foot pounds/minute		0.8684	Nautical miles
	737.6	Foot pounds/second	Miles/hour	44.70	Centimeters/second
	1.341	Horse power		88	Feet/minute
	14.34	Calories (Kg)/min		1.467	Feet/second
	1000	Watts		1.609	Kilometers/hour
Kilowatt hours	3413	BTU		0.8684	Knots
				26.82	Meters/minute

TABLE 19-5 *(CONTD.)*

Multiply	by	To Obtain
Miles/minute	2682	Centimeters/second
	88	Feet/second
	1.609	Kilometers/minute
	60	Miles/hour
Millibars	0.000987	Atmosphere
Milliers	1000	Kilograms
Milligrams	0.001	Grams
	0.01543	Grains
Milligrams/liter	1	Parts/million
Milliliters	0.001	Liters
Million gals/24 hours	1.54723	Cubic feet/second
Millimeters	0.1	Centimeters
	0.03937	Inches
	39.37	Mils
	1000	Microns
Miner's inches	1.5	Cubic feet/minute
Minutes (angle)	0.0002909	Radians
Nautical miles	6080.2	Feet
	1.1516	Miles
Ounces (avoirdupois)	16	Drams (avoir.)
	437.5	Grains
	0.0625	Pounds (avoir.)
	28.349527	Grams
	0.9115	Ounces (troy)
Ounces (fluid)	1.805	Cubic inches
	0.02957	Liters
	29.57	Cubic centimeters
	0.25	Gills
Ounces (troy)	480	Grains
	20	Pennyweights (troy)
	0.08333	Pounds (troy)
	31.103481	Grams
	1.09714	Ounces (avoir.)
Ounces/square inch	0.0625	Pounds/square inch
	1.732	Inches of water at 62° F
	4.39	Centimeters of water at 62° F
	0.12725	Inches of Hg at 32° F
	0.004253	Atmospheres
Palms	3	Inches
Parts/million	0.0584	Grains/gallon (US)
	0.07016	Grains/gallon (Imp)
	8.345	Pounds/million gal (US)
Pennyweights (troy)	24	Grains
	1.55517	Grams
	0.05	Ounces (troy)
	0.0041667	Pounds (troy)
Pints (liq. US)	4	Gills
	16	Ounces (fluid)
	0.5	Quarts (liq. US)
	28.875	Cubic inches
	473.1	Cubic centimeters
Pipe	126	Gallons (US)
Points	0.01389	Inches
Poise	0.0672	Pounds/sec foot
	242	Pounds/hour foot
	100	Centipoise

Multiply	by	To Obtain
Poncelots	100	Kilogram meters/second
	1.315	Horse power
Pounds (avoirdupois)	16	Ounces (avoir.)
	256	Drams (avoir.)
	7000	Grains
	0.0005	Tons (short)
	453.5924	Grams
	1.21528	Pounds (troy)
	14.5833	Ounces (troy)
Pounds (troy)	5760	Grains
	240	Pennyweights (troy)
	12	Ounces (troy)
	373.24177	Grams
	0.822857	Pounds (avoir.)
	13.1657	Ounces (avoir.)
	0.00036735	Tons (long)
	0.00041143	Tons (short)
	0.00037324	Tons (metric)
Pounds of water at 62° F	0.01604	Cubic feet
	27.72	Cubic inches
	0.120	Gallons (US)
Pounds of water/min at 62° F	0.0002673	Cubic feet/second
Pounds/cubic foot	0.01602	Grams/cubic centimeter
	16.02	Kilograms/cubic meter
	0.0005787	Pounds/cubic inch
Pounds/cubic inch	27.68	Grams/cubic centimeter
	27,680	Kilograms/cubic meter
	1728	Pounds/cubic foot
Pounds/foot	1.488	Kilograms/meter
Pounds/inch	178.6	Grams/centimeter
Pounds/hour foot	0.4132	Centipoise
	0.004132	Poise grams/sec cm
Pounds/sec foot	14.881	Poise grams/sec cm
	1488.1	Centipoise
Pounds/square foot	0.016037	Feet of water at 62° F
	4.882	Kilograms/square meter
	0.006944	Pounds/square inch
	0.014139	Inches of Hg at 32° F
	0.0004725	Atmospheres
Pounds/square inch	0.068044	Atmospheres
	2.30934	Feet of water at 62° F
	2.0360	Inches of Hg at 32° F
	703.067	Kilograms/square meter
	27.912	Inches of water at 62° F
Quadrants (angular)	90	Degrees
	5400	Minutes
	324,000	Seconds
	1.751	Radians
Quarts (dry)	67.20	Cubic inches
Quarts (liq. US)	2	Pints (liq. US)
	0.9463	Liters
	32	Ounces (fluid)
	57.75	Cubic inches
	946.3	Cubic centimeters
Quintal, Argentine	101.28	Pounds
Brazil	129.54	Pounds
Castile, Peru	101.43	Pounds
Chile	101.41	Pounds
Metric	220.46	Pounds
Mexico	101.47	Pounds

CONVERSIONS

TABLE 19-5 *(CONTD.)*

Multiply	by	To Obtain	Multiply	by	To Obtain
Quires	25	Sheets	Square miles	27,878,400	Square feet
Radians	57.30	Degrees		2.590	Square kilometers
	3438	Minutes		259	Hectares
	206,625	Seconds		3,097,600	Square yards
	0.637	Quadrants		102,400	Square rods
				1	Sections
Radians/second	57.30	Degrees/second	Square millimeters	0.01	Square centimeters
	0.1592	Revolutions/second		0.00155	Square inches
	9.549	Revolutions/minute		1550	Square mils
				1973	Circular mils
Radians/second^2	573.0	Revolutions/minute^2	Square mils	1.27324	Circular mils
	0.1592	Revolutions/second^2		0.0006452	Square millimeters
Reams	500	Sheets		10^{-6}	Square inches
Revolutions	360	Degrees	Square yards	0.0002066	Acres
	4	Quadrants		9	Square feet
	6.283	Radians		0.8361	Square meters
Revolutions/minute	6	Degrees/second		3.228×10^{-7}	Square miles
	0.1047	Radians/second	Stere	1	Cubic meters
	0.01667	Revolutions/second			
Revolutions/minute^2	0.001745	Radians/second^2	Stone	14	Pounds
	0.0002778	Revolutions/second^2		6.35029	Kilograms
Revolutions/second	360	Degrees/second	Tons (long)	1016	Kilograms
	6.283	Radians/second		2240	Pounds
	60	Revolutions/minute		1.12	Tons (short)
Revolutions/second^2	6.283	Radians/second^2	Tons (metric)	1000	Kilograms
	3600	Revolutions/minute^2		2205	Pounds
Rods	16.5	Feet		1.1023	Tons (short)
	5.5	Yards	Tons (short)	2000	Pounds
Seconds (angle)	4.848×10^{-6}	Radians		32,000	Ounces
Sections	1	Square miles		907.185	Kilograms
Side of a square	1.4142	Diameter of inscribed circle		0.90718	Tons (metric)
				0.89286	Tons (long)
	1.1284	Diameter of circle with equal area	Tons of refrigeration	12,000	BTU/hour
				288,000	BTU/24 hours
Span	9	Inches	Tons of water/24 hours at 62° F	83.33	Pounds of water/hour
Square centimeters	0.001076	Square feet		0.16510	Gallons (US)/minute
	0.1550	Square inches		1.3263	Cubic feet/hour
	0.0001	Square meters	Watts	0.05692	BTU/minute
	100	Square millimeters		44.26	Foot pounds/minute
Square feet	2.296×10^{-5}	Acres		0.7376	Foot pounds/second
	929.0	Square centimeters		0.001341	Horse power
	144	Square inches		0.01434	Calories (Kg)/minute
	0.0929	Square meters		0.001	Kilowatts
	3.587×10^{-8}	Square miles		1	Joule/second
	0.1111	Square yards	Watt hours	3.413	BTU
Square inches	6.452	Square centimeters		860.5	Calories (gm)
	0.006944	Square feet		0.8605	Calories (Kg)
	645.2	Square millimeters		2655	Foot pounds
	1.27324	Circular inches		0.001341	Horse power hours
	1,273,239	Circular mils		3600	Joules
	1,000,000	Square mils		367.1	Kilogram meters
Square kilometers	247.1	Acres		0.001	Kilowatt hours
	10,760,000	Square feet	Watts/square inch	8.2	BTU/square foot/minute
	1,000,000	Square meters		6373	Foot pounds/sq ft/minute
	0.3861	Square miles		0.1931	Horse power/square foot
	1,196,000	Square yards	Yards	91.44	Centimeters
Square meters	0.0002471	Acres		3	Feet
	10.764	Square feet		36	Inches
	1.196	Square yards		0.9144	Meters
	1	Centares		0.1818	Rods
Square miles	640	Acres	Year (365 days)	8760	Hours

20
ABBREVIATIONS

Fig. 20-1 (*Courtesy Stockham Valves and Fittings.*)

A

A	Absolute
A	Air
A	Anchor
ABS	Absolute
AC	Air closes
AC	Combustion air
ACCUM	Accumulator
AFD	Auxiliary feedwater
AGA	American Gas Association
AI	All iron
AI	Instrument air
AISC	American Institute of Steel Construction
AL	Aluminum
ALY	Alloy
AMER STD	American Standard
ANSI	American National Standards Institute
AO	Air opens
AP	Plant air
API	American Petroleum Institute
APPROX	Approximate
ARCH	Architectural
AS	Starting air
ASA	American Standards Association
ASB	Asbestos
ASHVE	American Society of Heating and Ventilating Engineers
ASME	American Society of Mechanical Engineers
ASNT	American Society for Nondestructive Testing
ASSY	Assembly
ASTE	American Society of Testing Engineers
ASTM	American Society for Testing and Materials
AUT	Automatic vent trap
AUX	Auxiliary
AV	Average
AV	Ventilation or cooling air
AWS	American Welding Society
AWWA	American Waterworks Association
AZ	Azimuth

B

B	Beveled
B&B	Bell and bell
BB	Bolted bonnet
BBE	Bevel both ends
BBL	Barrel
BC	Between centers
BC	Bolt cap
BC	Bolt circle
BD	Blowdown
BE	Beveled end(s)
BF	Blind flange
BF	Bottom flat
BHN	Brinell hardness number
BL	Bottom level
BLDG	Building
BLE	Beveled large end
BLK	Black
BLVD	Beveled
BM	Beam
B/M	Bill of material
BOC	Bottom of concrete
BOM	Bill of materials
BOP	Bottom of pipe
BOS	Bottom of steel
BR	Bronze
BRS	Brass
B&S	Bell and spigot
BTM	Bottom
BTU	British thermal unit
BUSH	Bushing
BW	Butt weld
BW	Butt-welded
B/W	Butt-weld pipe
BWG	Birmingham wire gauge

C

C	Center line
C	Centigrade
C	Channel or channel steel
C	Condensate
°C	Degrees centigrade
CAS	Cast alloy steel
CBD	Continuous blowdown
CtoC	Center to center
CCW	Component cooling water
CD	Closed drain
CENT	Centigrade
CtoF	Center to face
CFM	Cubic feet per minute
CFS	Cubic feet per second
CHG	Change
CHKD	Checked
CHO	Chain-operated
CHO	Chain operator
CHU	Centigrade heat unit
CI	Cast iron

CL   Center line
CL   Clearance
CM   Centimeter
CO   Chain operator
CO   Clean out
CO   Company
CO_2 Carbon dioxide
COL   Column
COMP   Compressor
CON   Concentric
CONC   Concentric
CONC   Concrete
COND   Condensate
CONN   Connection
CONSTR Construction
CONT   Continuation
CORR   Corrosion
CP   Chemicals
CPLG   Coupling
CR   Conductivity recorder
Cr   Chromium
Cr13   Type 410 stainless steel
CS   Carbon steel
CS   Cast steel
CS   Cold spring
CSO   Car seal open
CTMT   Containment
CTR   Center
CTS   Containment spray
CU   Cubic
Cu   Copper
CVC   Chemical and volume control
CWP   Cold working pressure

D

DC   Density recorder
DC   Drain closed
DC   Drain connection
DD   Double disk
DEG   Degree
DEG(°) Degrees
DET   Detail
DF   Drain funnel
DI   Ductile iron
DIA   Diameter
DIM   Dimension
DIN   Deutsche Industrie Norm (German Standard)
DISCH Discharge
DO   Drain open

DP   Process sewer
DPI   Differential pressure indicator
DRG   Drawing (not preferred)
DS   Sanitary sewer
DW   Storm sewer
DWG   Drawing
DWN   Drawn
DXS   Double extra strength

E

E   East
EA   Each
EBD   Emergency blowdown valve
ECC   Eccentric
ECN   Engineering change number
EF   Electric furnace
EFW   Electric fusion welded
EJMA   Expansion Joint Manufacturers Association
EL   Elevation
ELB   Elbowlet
ELEV   Elevation
ELL   Elbow
EMBED Embedment
ENGR   Engineer
EP   Equipment piece
EQUIP Equipment
ERW   Electric resistance welded
ESD   Emergency shutdown valve
EXCH   Exchanger
EXH   Exhaust
EX-HY Extra heavy
EXIST Existing
EXPJT Expansion joint

F

°F   Fahrenheit
F   Furnished by others
FA   Flow alarm
FAB   Fabricate
FAHR   Fahrenheit
FBW   Furnace butt-welded
FCN   Field change number
FCV   Flow control valve
FD   Feedwater
F&D   Faced and drilled
FDS&F Faced, drilled, and spot-faced
FE   Flanged end
FE   Flow element

ABBREVIATIONS

F-F   Face to face
FtoF   Face to face
FF   Flange face
FF   Flat face(d)
FF   Full face(of gasket)
FH   Fixed hanger
FI   Flow indicator
FIC   Flow indicating controller
FICV   Flow indicating control valve
FIG   Figure or figure number
FL   Floor
FLD   Field
FLG(Flg.)   Flange or flanges
FLGD   Flanged
FmI   Displacement flowmeter
FO   Flow orifice
FOB   Flat on bottom
FOT   Flat on top
FPS   Feet per second
FR   Flow recorder
FR   From
FRC   Flow recording controller
FRCV   Flow recording control valve
FRP   Fiberglass-reinforced pipe
FS   Far side
FS   Flow switch
FS   Forged steel
FSD   Flat side down
FSS   Forged stainless steel
FSU   Flat side up
FT(')   Foot or feet
FTG   Fitting
FW   Field weld

G

G   Gage or gauge
G   Gas
G   Grade
G   Gram
GA   Gage or gauge
GAL   Gallon
GALV   Galvanized
GEN   General
GEN   Generator
GF   Fuel gas
GG   Gauge glass
GJ   Ground joint
GN   Nitrogen or inert gas

GPH   Gallons per hour
GPM   Gallons per minute
GR   Grade
GRD   Ground
GRJT   Ground joint
GRV   Groove
GU   Guide
GV   Gate valve
GW   Waste gas

H

H   Horizontal
H   Hour
H   Hydrogen
HC   Hand (manual) controller
HC   Hose connection
HC   Hydrocarbon
HCV   Hand-operated control valve
HDR   Header
HEX   Hexagon
HF   Hard (stellite) face
Hg   Mercury
HGR   Hanger
HIC   Hand-actuated pneumatic controller
HOR   (HORIZ)-Horizontal
HP   High point
HPT   Hose-pipe thread
HR   Hour
HR   Hanger (rod)
HS   Hanger (spring)
HTR   Heater
HVAC   Heating, ventilating, and air conditioning
HVY   Heavy
HYD   Hydraulic

I

I   Iron
IBBM   Iron body bronze (or brass) mounted
IBD   Intermittent blowdown
ID   Inside diameter
ID   Inside depth of dish
IE   Invert elevation
IMP   Imperial (British unit)
IN (")   Inch or inches
INS   Insulate or insulation
INT   Integral
INV   Invert (inside bottom of pipe)
IPS   Iron pipe size

IS Inside screw (of valve stem)
ISA Instrument Society of America
ISO Isometric drawing
ISRS Inside screw rising stem
IS&Y Inside screw and yoke

J

JCT Junction
JT Joint

K

K Kilo, times one thousand, ×1000
KG Kilogram
KW Kilowatt(s)

L

L Angle (structural 4-in. angle shape)
L Liquid
LA Level alarm
LB, lb Pound weight
LBS Pounds
LC Level controller
LC Lock closed
LCR Level control recorder
LCV Level control valve
LG Level glass
LI Level indicator
LIC Level indicating controller
LICV Level indicating control valve
LLA Liquid level alarm
LLC Liquid level controller
LLI Liquid level indicator
LLR Liquid level recorder
LO Lock opened
LOC Location
L-O-L Latrolet
LP Low point
LR Level recorder
LR Long radius
LRC Level recording controller
LS Level switch
LW Lap weld
L/W Lapweld pipe

M

M Mega, times one million
M Meter

M Miscellaneous shapes (steel)
M Monel metal
MACH Machined
MATL Material
MAWP Maximum allowable working pressure
MAX Maximum
MB Machine bolts
M/C Machine
MCC Motor control center
MECH Mechanical
M&F Male and female
MFG Manufacturing
MFR Manufacturer
MI Malleable iron
MIN Minimum
MIN Minute (time)
MISC Miscellaneous
MM Millimeter
Mo Molybdenum
MS Mild steel
MSS Manufacturers Standardization Society (valve and fittings industry)
MTD Mounted
MW Miter weld

N

N North
NC Normally closed
NEC National Electric Code
NEG(−) Negative
NEMA National Electrical Manufacturers Association
NEWWA New England Water Works Association
Ni Nickel
NICU Nickel-copper alloy
NIP Nipple
NO Normally opened
NO(#) Number
NOM Nominal
NOZ Nozzle
NPS National pipe size
NPS Nominal pipe size
NPSH Net positive suction head
NPT National pipe thread
NPTF National pipe thread female
NPTM National pipe thread male
NS Near side
NTS Not to scale

O

OD   Open drain
OD   Outside diameter
OH   Heating oil or dowtherm
OH   Open hearth
OL   Lube oil
OPP Opposite
ORIG Original
OS   Seal oil
OS&Y Outside screw and yoke
OV   Hydraulic oil
OWG Oil, water, gas

P

P   Personnel protection
PA   Pressure alarm
PC   Pressure controller
PCV   Presure control valve
PdC   Pressure differential controller
PdI   Pressure differential indicator
PdRC Pressure differential recording controller
PE   Plain end (not beveled)
PE   Pressure test connection
PERP Perpendicular
PF   Process fluid (no distinction)
PFI   Pipe Fabrication Institute
PI   Point of intersection
PI   Pressure indicator
PIC   Pressure indicator controller
PICV Pressure indicator control valve
P&ID Piping and instrument diagram
PIM   Pressure indicating manometer
PL   Plate
PL   Process liquid
PO   Pump out
POE   Plain one end
POS(+) Positive
PR   Pair
PR   Pump
PR   Pressure recorder
PR   Pressure regulator
PRC   Pressure recording controller
PRESS Pressure
PRI   Primary
PRV   Pressure-reducing valve
PS   Pipe support
PS   Pressure switch
PSD   Rupture disk
PSI(psi) Pounds per square inch
PSIA Pounds per square inch absolute
PSIG Pounds per square inch gauge
PSV   Pressure safety valve or relief valve
PT   Point
PV   Process vapor
PVC   Polyvinyl chloride
PX   Process fluid special hazard

Q

QO   Quick opening
QTY   Quantity
QUAD Quadrant

R

R   Radius
RA   Refrigerant ammonia
RB   Reactor building
RC   Reactor coolant
RE   Refrigerant (ethylene)
REAC Reactor
RECD Received
RED   Reducer (or reducing)
REF   Reference
REINF Reinforce
REQ (REQ'D) Required
REV   Revision
RF   Raised face
RF   Refrigerant (Freon)
RFC   Ratio flow controller
RFI   Ratio flow indicator
RHR   Residual heat removal
RJ(RTJ) Ring-type joint
R/L   Random length
RP   Refrigerant (propane or propylene)
RPM(rpm) Revolutions per minute
RR   Refrigerant (no distinction)
RS   Rising stem

S

S   South
S   Steam pressure
SA   Sludge acid
SA   Sulfuric acid
SAE   Society of Automotive Engineers
SC   Sample connection
SC   Steam condensate

SCD Screwed
SCFH Standard cubic feet per hour
SCFM Standard cubic feet per minute
SCH Schedule
SCHED Schedule
SCR Screwed ends
SCRD Screwed
SD Storm drain
SE Steam exhaust
SECT Section
SF Semifinished
SG Sight glass
SGA Special gravity alarm
SGC Special gravity controller
SGI Special gravity indicator
SGR Special gravity recorder
SH(SHT) Sheet
SH Steam-high pressure
SI Safety injection
SJ Solder joints
SK Sketch
SL Slip-on
SL Steam-low pressure
SL FLG Slip-on flange
SLOT Slotted
SLV Sleeve
SM Steam-medium pressure
SMLS Seamless
SNUB Snubber
SO Steam out
SOL Sockolet
SP Steam pressure
SPEC Specification
SPG Spring
SPI Special
SQ Square
SR Short radius
SR Speed recorder
SS Stainless steel
SSP Steam service pressure
STD Standard
STIFF Stiffener
STL Steel
STM Steam
STR(STRUCT) Structure
SUCT Suction
SUPT Support
SW Socket weld
SW Socket-welded ends
SWG Swag

SWP Standard working pressure
SYS System

T

T Threaded
T Steam trap
TA Temperature alarm
TAN Tangent
TBE Threaded both ends
TC Temperature controller
TC Test connection
T&C Threaded and coupled
TCV Temperature control valve
TdC Temperature differential controller
TdI Temperature differential indicator
TdR Temperature differential recorder
TE Thread end
TE Threaded end
TECH Technical
TEF Teflon
TEMP Temperature
TENS Tension
T&G Tongue and groove
THD Threaded
THRU Through
TI Temperature indicator
TIC Temperature controller
TICV Temperature-control valve
TK Tank
TL Top level
TLE Thread large end
TOC Top of concrete
TOE Thread one end
TOG Top of grating
T-O-L Thread-o-let
TOP Top of pipe
TOS (T/S) Top of steel
TP Equipment trim pumps
TP Type
TPD Tons per day
TR Temperature recorder
TRANS Transactions
TRC Temperature recorder controller
TS Temperature switch
TSE Thread small end
TT Equipment-exchanger tubes
T-T Tangent to tangent
TURB Turbine
TV Temperature valve

ABBREVIATIONS

TW	Temperature wall	WC	Water column
TYP	Typical	WE	Weld end
		WF	Welded flange
		WF	Wide flange

U

UA	Unit alarm
UB	Union bonnet

V

VA	Valve
VB	Vortex breaker
VC	Vitrified clay
VERT	Vertical
VF	Vent to flare
VOL	Volume
VR	Viscosity recorder
VS	Vent to stack

W

W	West
W	Wide flange steel shape
W	Width
W/	With
WB	Welded bonnet
WC	Plant cooling water
WC	Water column
WE	Weld end
WF	Welded flange
WF	Wide flange
WH	Weep hole
WJ	Jacket or closed cycle water
WLD	Weld
WN	Weld neck
WN FLG	Weld neck flange
WOG	Water, oil, and gas (pressure)
W-O-L	Weld-o-let
WP	Potable water
WP	Working point
WR	Raw water
WR	Weight recorder
WS	Salt water
WSP	Working steam pressure
WT	Treated water
WT	Weight
WWP	Working water pressure

X

XH	Extra heavy
XS	Extra strong
XXH	Double extra heavy
XXS	Double extra strong

21

GLOSSARY

Fig. 21-1

A

ABS (acrylonitrile-butadiene-styrene) Plastic used in manufacturing drainage pipe and fittings.

Accumulator Container in which fluids or gases are stored under pressure. The term also refers to a holding tank used for temporary storage.

Actuator Any device that will operate a valve by remote control (fluid motor, air cylinder, hydraulic cylinder, electric motor).

Adjustability Ability of a pipe system and its support system to enable field installation and adequate functioning when the installation position differs from the design.

Air, compressed Air having pressure greater than atmospheric pressure.

Air, free Air subject to only atmospheric conditions and not contaminated.

Alley, pipe Main bank of pipe headers located inside the limits of a structure; known as a pipe bridge when outside the limits.

Alloy steel Steel which owes its distinctive properties to elements other than carbon.

Anchor Rigid support that keeps the pipe from translation or rotation movement at one point along the piping system; also prevents transmission of forces and moments (thermal, shock, vibration) between both sides of the pipe.

Anchor knot Means of fastening a rope to a ring or post.

Angle valve Valve designed so that the inlet and outlet are at a 90 degree angle to each other.

Annulus Doughnut-shaped duct or pipe.

Antihammer device Air chamber, such as a closed length of pipe or a coil, designed to absorb shock caused by a rapidly closed valve.

Autoclave Vessel used to hold material, medium, or reactants under prescribed conditions, including temperature, pressure, and movement.

Automatic valve Any valve whose position, as to degree of opening, is controlled by means other than manual.

B

Backflow Flow of a fluid in a pipe in the opposite direction to which it normally flows.

Backing ring Metal strip used to prevent melted metal from the welding process from entering a pipe when completing a butt-welded joint.

Back splice Stopper to prevent unlaying of strands at the end of a rope. It starts as a crown knot, then has several additional tucks.

Baffle Series of obstructions used in petrochemical vessels to guide or mix process materials.

Balancing line Vent line from the top of a vessel receiving the liquid to a point above the surface of the liquid in the vessel from which the liquid is being drained.

Ball valve Regulating valve that uses a ball with a hole through it to control the flow of fluid; normally has 90-degree open-and-shut operation.

Barometric leg Vertical pipe which is sealed off from the atmosphere. A vacuum is thereby produced above the liquid standing in the pipe similar to a barometer.

Battery Group of similar reaction vessels or tanks.

Battery limit Lines used on a plot plan to determine the outside limits of a unit, usually established on the piping index drawing, plot plan, and site plan.

Bay line Line of structural steel columns that spans the width of a building; the "bay" is the space between two bay lines.

Bay of steel Steel surrounding a space bounded by the four nearest columns and the floor above.

Bell or hub Enlarged end of some types of pipe which fit over the next pipe section.

Bell and spigot joints Type of joint usually found in waste piping. The joint is sealed by some sort of packing.

Bend Vertical plane of steel structure that extends along a column row in a building; consists of two columns with horizontal connecting members. In a pipe rack, the bent is two columns and one or more horizontal connecting members which form a U shape.

Bend angle (pipe) Angle at the center of the bend between radial lines from the beginning and end of the bend to the center.

Bight Simple loop; a part of all knots.

Black pipe Steel pipe that is not galvanized.

Blank flange Flange in which the bolt holes have not been drilled.

Blind flange Solid platelike fitting used to seal the end of a flanged-end pipeline; also known as dead end.

Blowdown Discharged material from relief valves.

Blowdown-blowback Difference between relieving and closing pressures in a relief or safety valve. Closing pressure is always less than the pressure at which the safety valve begins to open (relieve).

Blowdown system Discharge system which removes excess or relieved medium from vessels, relief valves, or other equipment commonly associated with a piping system.

Blowdown tank Vessel into which line material or contents of another vessel can be emptied immediately in an emergency.

Blowoff Controlled discharge of excess pressure.

Boiler Vessel in which water is heated to generate steam under pressure.

Boiler, fire-tube Boiler system with heating tubes located within the shell. The use of tubes submerged in boiling water increases the area of contact between the water and the fire gases passing through the tubes.

Boiler, water-tube Reverse of the fire-tube boiler. Hot gases are in contact with the outside surface of the tubes and the water is inside.

Boiler feed Water fed into a boiler.

Bonnet Upper portion of the gate valve body into which the disk of a gate valve rises when it is opened.

Boom Long part of a crane which makes it possible for the load sheaves to be maneuvered directly over the load to be lifted.

Boom (crane) Member hinged to the front of the rotating superstructure with the outer end supported by ropes leading to a gantry, or A frame. Used to support hoisting tackle.

Boom angle Angle between the longitudinal centerline of the boom and the horizontal.

Boom hoist Hoist drum and rope reeving system used to raise and lower the boom.

Bowline Nonslipping eye knot.

Braided wire rope Rope formed by braiding component wire ropes.

Branch Pipe, usually of small diameter, that enters into or exits from the main run pipe.

Branch tee Tee having multiple branches.

Braze weld or brazing Process of joining metals using a nonferrous filler metal or alloy, the melting point of which is higher than 800°F but lower than that of the metals to be joined.

Breaking strength Load or tension required to break a fiber or wire rope.

Break out flange Flanges placed in pipeline in order that that part of the line may be removed to facilitate maintenance work.

Bridge crane Fabricated structural crane operating on elevated tracks and bridged over the area of lifting.

Bubble cap Part of a tray of a distillation column.

Bucket trap Device which permits the flow of condensate through it, but will not permit the flow of steam.

Bushing Tapped or threaded fitting (male outside, female inside) used to reduce the size of the end opening of a fitting or valve.

Butt weld Circumferential weld in pipe fusing the abutting pipe walls completely from inside wall to outside wall.

Butt-welded pipe Pipe welded along a seam, edge to edge and not scarfed or lapped.

Bypass Pipe loop which provides partial or full flow of material around a piece of equipment or valve station.

C

Cab Housing which covers the rotating superstructure machinery and/or operator's station. On truck-crane trucks a separate cab covers the driver's station.

Cages Enclosures around ladders at a certain elevation to prevent personnel from falling.

Cap Fitting that seals the end of a pipe permanently.

Cap end End enclosure of a vessel. Can be dished elliptical or flanged and bolted.

Carbon steel Steel owing its distinctive properties chiefly to the various percentages of carbon (as distinguished from the other elements) which it contains.

Caulk Material used to seal joints in cast iron drainage pipe.

Celsius Temperature scale where the boiling point of water is 100° and the freezing point of water is 0°.

Centigrade Temperature scale where the freezing point of water is 0° and the boiling point is 100°.

Centrifugal pump Impellor-type pump with suction at the center that discharges tangentially by the use of volutes.

Chain wheel Manual operator for use on gate valves located above normal reach.

Chain wrench Adjustable tool for holding large pipe up to 4 in. in diameter. A flexible chain replaces usual jaws.

Charge In a batch process, the original material which goes into a system for processing (usually crude stock).

Chase (pipe) Slot in the floor through which a group of vertical piping passes.

Check valve Valve that permits flow in one direction only.

Chemical plant Plant that utilizes various hydrocarbon products from refineries to produce other products for consumer and industrial use.

Choker Hitch made by using a sling in a manner so that the heavier the load, the tighter the sling will hold it.

Clevis U-shaped or stirrup-shaped device used to connect two or more lifting members. Usually referred to as a shackle.

Close nipple Shortest length of a given size pipe which can be threaded externally from both ends; used to connect closely two internally threaded pipe fittings.

Clove hitch Temporary method of fastening a rope at a right angle to a post. Consists of two half hitches.

Codes Standards that provide industry with recognized specifications for convenience, safety, and uniform design.

Coefficient of expansion Number indicating the degree of expansion or contraction of a substance. The coefficient of expansion is not constant and varies with changes in temperature. For linear expansion it is expressed as the change in length of one unit of length of a substance having 1 degree rise in temperature.

Cold bending Bending process for pipes with diameters (up to 42 in.) and walls (up to 3/4 in.). Wrinkling, excessive thinning, and ovality can be avoided with proper equipment and fixtures.

Cold joint Solder joint made with inadequate heat, or the two parts have been moved slightly as the solder is solidifying.

Cold load Spring reading (a function of amount of compression) when pipe suspended from a spring hanger is at room temperature (see Hot load). Cold load is higher than hot load if the thermal movement of the pipe is up, and vice versa.

Column Vertical vessel used for fractional distillation; also called tower or stanchion. Column also refers to vertical steel structural members for buildings, pipe racks, etc.

Column line Straight row of steel columns represented on a model or drawing.

Companion flange Any flange suited to connect with a flanged valve or fitting or with another flange to form a joint.

Compressive stress One that resists a force tending to crush a body.

Compressor Mechanical (piston and cylinder) device used to increase the pressure of air or gas.

Compressor, reciprocating Device similar to the reciprocating pump. Pressurized gas or air is created by the use of cylinders and pistons.

Compressor, rotary Device used to compress air or gas by the rotation of valves.

Compressor, rotary lobe Device used to produce pressure for gas and air by the use of lobed rotors.

Concentric reducer Piping fitting used to reduce the size of pipe so that the pipe may be continued in a smaller size. The reducer's shape is such that the centerlines of the larger and smaller pipes are in a straight line.

Condensate Liquid resulting from condensation of vapor or gas in a line, especially condensed steam.

Condensate utility system Separate piping system which collects the liquid formed from the condensing of steam.

Condenser Device used to condense vapor to liquid by cooling. Condensers are used to lower the turbine's exhaust pressure and to save water that may be lost to the system.

Condensing Process of vapor turning to liquid upon being cooled.

Conduction Transfer of heat or electrical energy through a body without displacement of the particles in the body.

Conduit Structural covering for electrical lines made from a variety of materials (plastics, steel, aluminum). Conduit provides the strength needed to run electrical lines throughout an industrial installation.

Constant-load hanger Device consisting of a lever mechanism in a housing and a coil spring. The movement of piping, within certain limits, will not change the spring forces holding up the piping; thus no additional forces are introduced to the piping being supported.

Continuous blowdown Expulsion of water continuously from steam lines.

Control station Pipe loop arrangement consisting of control valve, various fittings, and isolating (block) valves to regulate the flow, pressure, or quantity of material at a predetermined location.

Control valve Automatic valve operated electrically, pneumatically, or hydraulically and used to regulate flow, temperature, pressure, etc.

Controller Mechanism that operates a valve; it can be manual or automatic.

Convection Transfer of heat between two surfaces—the cold one above, the hot one below—which are separated by a layer of air. The surfaces may be vertical, but the hot one cannot be above the cold one.

Corrosion Gradual destruction or alteration of a metal or alloy caused by direct chemical attack or by electrochemical reaction.

Corrosion allowance Extra metal added to the original calculated thickness to allow for corrosion.

Coupling Pipe fitting containing female threads on both ends. Couplings are used to join two or more lengths of pipe in a straight run or to join a pipe and a fixture.

Coupling, dresser Coupling used for connecting plain-ended pipes.

Cross Fitting the shape of a cross having four openings at 90 degrees to each other.

Cross, straight Fitting in which all outlets are of the same diameter.

D

Davit Small crane at the top of a column (tower) used to lift equipment.

Dead weight Total weight of all the suspended rigging.

Derrick Structure of fabricated member of considerable length or height used to provide a fixed stable point above the load from which a lift can be made.

Development, pipe View on pipe fabrication drawing that shows the unfolding of pipe intersections in one plane. Developments are used for the construction of template drawings for pipe intersections.

Diaphragm Takes the place of a disk in certain valves. Usually made of a rubberlike substance flexible enough to be stretched over the valve bore area.

Diaphragm valve Valve whose port is sealed off by means of a flexible diaphragm.

Dike Wall or enclosure around a tank or holding vessel to provide a backup system in case of ruptures.

The dike must be able to hold the total contents of the enclosed vessel or tank.

Discharge Side of a pump that propels the fluid away.

Dish head End cover for a round vessel which has been formed into a segment of a sphere in order to withstand internal pressure.

Disk (disc) Part of valve which initiates, shuts off, or regulates flow.

Disk, rupture Diaphragm placed between flanges on a vent line. The diaphragm has a specific breaking pressure and will fail before the safety limits of the system have been reached.

Distillation column Device for distilling; also known as distillation tower.

Double acting pump Piston-type pump in which the pumping action takes place regardless of the direction the piston is moving.

Double extra strong Schedule of wrought pipe weights in common use.

Double spring Tandem medium spring used for high deflection with large hot-cold travel variation.

Downcomer Any line or section of equipment (tray) which provides for the downward movement of material.

Drain hub Funnel-shaped arrangement designed into flooring and concrete foundations for drainage. The drain hub is connected directly to the drain line, which removes the collected material.

Drainage system All hubs, piping, fittings, connections, and holding tanks associated with the removal or relocation of unwanted materials that have been discharged from equipment; system includes runoff from rain or cleaning operations.

Dresser coupling Coupling for connecting plain ended pipes.

Drifting Act of moving a suspended load in a horizontal direction using two or more pieces of hoisting equipment.

Drip valve Valve installed on drip legs for the removal of material from pipelines.

Ductility Property of elongation, above the elastic limit but under the tensile strength. A measure of ductility is the percentage of elongation of the fractured piece over its original length.

Dummy leg Piece of pipe or rolled steel section which is welded to the pipe in order to support the line.

E

Eccentric reducer Reducer that is flat on one side; used where trapped air in a line may cause problems.

Economizer Heat exchanger used on a boiler by which exhaust gases are used to preheat the air entering the combustion chamber.

Elastic limit Greatest stress which a material can withstand without a permanent deformation after release of the stress.

Elastomer Resilient material used for seals to prevent leakage and used as a coating to inhibit corrosion.

Elbow Standard fitting that creates a 90-degree or 45-degree bend; can be made for any degree bend. Used to change direction.

Elbow, long-radius Elbow whose radius equals one and one-half pipe diameters.

Elbow, short-radius Elbow whose radius equals 1.

Elbowlet Small fitting that is welded directly to an elbow to create a branch.

Elevation Height of an object above a given reference point.

Ell, street Elbow with male threads on one end and female threads on the other.

Erosion Gradual destruction of metal or other material by the abrasive action of liquids, gases, solids, or mixtures thereof.

Exchanger, heat Device used to transfer heat between two fluids.

Expansion, thermal Expansion of material caused by heating.

Expansion joint Special type of joint in concrete or steel construction to permit expansion due to temperature changes. Also a piping specialty which allows for expansion and/or vibration in piping.

Expansion loop Pipe bend designed in a manner to make the entire pipe more flexible to allow for its thermal expansion.

Expansion roof Special cover on a storage tank which floats on top of the liquid in the tank.

Extension Device attached to a valve to extend the handwheel up to the operating level; used where valves are below normal reach.

Extra strong Pipe sizes corresponding to schedule 80.

Extruded nozzle Nozzle or outlet formed by pulling hemispherically or conically shaped dies through a circular hole from inside the pipe.

Eye splice Splice made at the end of a rope by forming a loop and splicing the running end of the loop into the standing portion of the rope.

F

Fabrication drawing Drawing used to represent pipeline configurations that must be made up in a shop or in the field.

Face Finished contact surface of flanged-end piping or components (valves).

Face-to-face Dimensions from the face of the outlet port to the face of the inlet port of a valve or fitting.

Facing Part of valve body which connects to a companion flange.

Factor of Safety Load or tension at which a piece of equipment will fail divided by the actual load or tension placed on the equipment. The factor of safety for hoisting or rigging equipment is never less than 5.

Fahrenheit Temperature scale where the boiling point of water is 212° and the freezing point of water is 32°.

Feed Primary medium in a process; the charge.

Feed, boiler Water used to produce steam inside a boiler.

Feedwater Provides original and makeup water for the process.

Female thread Internal thread in valves, fittings, and pipes for making screwed connections.

Fiber rope Rope made from nonmetallic materials such as vegetable, animal, or synthetic fibers.

Field weld Weld performed at the construction site.

Fitting Wide range of piping components that enable pipe to change direction, change pipe size, provide branches for auxiliary lines, and provide connections.

Flange Rim on the end of a pipe, valve, or fitting for bolting to another piping element.

Flow, concurrent Two materials passing through a system in the same direction.

Flow, countercurrent Two materials passing through a system in opposite directions.

Flow, laminar Flow in which each particle of the fluid is moving parallel to the walls of the pipe. The velocity of flow is greater at the center of the pipe.

Flow, turbulent Flow in which the velocity of each particle of the fluid changes constantly in magnitude and direction, although the fluid as a whole is moving in the direction parallel to the pipe.

Flow displacement meter Instrument placed directly in the line to indicate the total amount of flow.

Flow element Portable device, consisting of an orifice set with plugged tapped valves and a plate, used for testing flow. Vertical pipe sections are considered part of the measuring device.

Flow indicating controller Control valve and flow indicator used for controlling flow by connections to an orifice through a pneumatic transmitter.

Flow indicator Linear indicator or dial which shows the flow rate. The indicator can be of the differential type or a direct hookup.

Flow rate Volumetric flow per unit time (gpm, cfm).

Flow ratio recording controller Instrument that controls and records the flow ratio of the main line. An orifice must be within close range of the meter. Pneumatic recorders can be either a board-mounted pneumatic type or a direct hookup differential variation.

Flow recording controller Flow recorder which controls medium movement and is connected to an orifice through a pneumatic transmitter.

Flowmeter Any device which indicates and records the quantity of liquid flowing through a pipe.

Flush bottom valve Special valve designed so that the valve seat is flush with the bottom of the tank when inserted in the nozzle.

Foreign matter Dirt, rust, scale, and the like in pipeline.

Forming Method of pipe fabrication, including bending, swagging, lapping, extruding, expanding, and belling. Most forming is done in the shop.

Friction loss Loss of pressure due to friction between the liquid and the pipe wall.

Full-bore port Valve in which diameter of internal bore equals the inside diameter of pipe used on line.

G

Gage Any device used for measuring a quantity or value, such as steam pressure or temperature.

Galvanic corrosion Corrosion caused by the contact of different metals which are improperly isolated in the presence of a liquid that can conduct electricity.

Galvanizing Process which covers the surface of iron or steel with a layer of zinc.

Gantry crane Fabricated structural crane operating on tracks at or near ground level. The traverse beam is elevated and bridged over the area of lifting.

Gas State of matter distinguished from the solid and liquid states by low density and viscosity. A gas always fills its container entirely.

Gasket Thin piece of material, rubber, asbestos, etc., placed between two flanges to prevent leakage.

Gate valve Valve designed for open/shut operation. The controlling disk operates much like a gate.

Gauge level Device for measuring the level of a material in a vessel.

Girder Permanent strength member in building construction from which rigging often is hung.

Glass, sight Glass plate or tube inserted in a vessel or a pipe wall for observation of internal conditions.

Glass lining Glass coating fused to a base material.

Glass pipe Piping fabricated from a special glass.

Globe valve Valve whose design is such that it may be used for throttling. This valve may be used in place of a gate valve, but this is seldom done because of additional cost and greater pressure drop.

Grade beam Beam used to support flooring at grade or ground level.

Gravity system Piping system that is under static head (pressure) only.
Guide Type of hanger support that allows a pipe to move along its length but not sideways.
Guys Rope, usually wire rope, used to hold the top end of a boom, derrick, or gin pole in a fixed position. Must be used in sets of three or more.

H

Half coupling Similar to the coupling, but only one end is threaded; used to create small-diameter branches or to mount instruments.
Half hitch Temporary method of fastening a rope at a right angle to a post. Not recommended unless the end of the rope is strongly seized to the standing part.
Hand hole Small hole in a vessel used for hand access for maintenance and adjustment.
Handwheel Manual operator that sits atop gate valves.
Hanger Device used to suspend pipe from a ceiling or exposed steel.
Head Pressure in a fluid system expressed as equivalent feet of water. Also refers to the end closure of a vessel (horizontal or vertical, elliptical, spherical, or dish-shaped); in some cases large blind flanges may be bolted to the end of a vessel and used as its head.
Head, friction The head required to overcome the friction between the fluid particles in motion and the interior surface of a conductor.
Header Pipe to which two or more pipelines are joined to carry fluid from a common source to various points of use.
Headroom Physical space above a valve, pipe, platform, etc.
Heat, latent Heat used to transform liquid to vapor or vapor to liquid.
Heat exchanger Piece of equipment possessing two separate chambers or sets of coils that is used to transfer heat or cold from one liquid to another.
Heat pump Device used for heating or cooling by transferring heat via a mechanically driven thermodynamic process such as evaporation and condensation.
Hitch Attachment of a rope to a post, pole, ring, hook, etc.
Hoist Hoisting unit mounted overhead in a fixed location or suspended from a small trolley attached to an I-beam track. "Hoist" can also be applied to the power-driven mechanism and drums that are used in conjunction with a derrick, gin pole, or even a material-hoisting elevator.
Hot bending Bending of pipe to a predetermined radius after heating to a suitable temperature.
Hot well Seal pot for barometric legs.
Hub end End connection, caulked or leaded, used on valves, fittings, and pipe (mainly for water supply and sewage lines).

I

Impeller Device that rotates in a centrifugal pump and produces the pumping action.
Impulse trap Steam trap.
Increaser Coupling that has a larger opening at one end (used to increase size of pipe opening). See also Reducer.
Indicator Type of gage or glass that permits visual readings of a certain variable, such as flow or temperature.
Instrument Device designed to sense, transmit, indicate, record, or control any number of variables within a piping system.
Instrument air Separate air system which operates instrumentation in a plant. Instrument air and utility air must be kept separate.
Instrument cable Flexible cable containing one or more plastic tubes which transmit various air pressures to instruments for control purposes. Also known as instrument conduit when cable is rigid.
Instrumentation Application of industrial instruments to a process of manufacturing operation.
Insulation Material used to cover pipelines or vessels to maintain a constant temperature in the line; also used to prevent the transfer of heat between the atmosphere and the line fluid and to protect operators from burns.
Insulation, aluminum-armored Insulation made of calcium silicate and asbestos fiber and covered with a weatherproof aluminum jacket; used for steam and process lines which operate at temperatures up to 1200°F (650°C).
Insulation, antisweat Used to prevent cold water lines from sweating.
Insulation, asbestos-sponge felted Type of insulation applied to pipes, valves, and fittings with temperatures up to 700°F (370°C).
Insulation, cold Insulation applied to pipes, fittings, and valves in refrigeration services or to prevent heat penetration from local steam lines to the fluid in the insulated line.
Insulation, fiberglass Rigid structurally strong insulation for temperatures up to 600°F (315°C).
Insulation, hot Insulation applied to pipes, fittings, and valves in steam services.
Insulation, rock cork Mineral wool product bonded with a waterproof compound.

Insulation, vegetable cork Compressed cork granules baked into a mold and used for low thermal conductivity and high moisture resistance.

Insulation, wool felt Commonly used for both hot and cold water lines. It has a temperature range of 40° to 212°F (5° to 100°C) and is made up of layers of wool felt with an inner waterproof liner.

Insulation, 85 percent magnesia Durable, fireproof, molded insulation used for pipes at temperatures up to 600°F (315°C).

Insulation ring Steel ring used to support insulation attached to the outside of a vessel.

J

Jacketed Pipe or a vessel which is surrounded by another pipe or shell to provide a space to introduce heating or cooling mediums.

Jib Extension attached to the boom point to provide added boom length for lifting loads. The jib may be in line with the boom or offset to a variety of angles.

Jib crane Beam or crane section secured at one end and allowed to rotate around its axis with a hoist or lifting device at the unsupported end.

Joint Point of connection between two piping elements.

L

Lapp joint Type of flanged joint used for joining pipe.

Lateral Fitting that allows 45-degree angle entry into the main pipe run.

Latrolet Small fitting welded directly to the pipe that creates a 45-degree angle entry into the run.

Lead joint Joint made by pouring molten lead into the space between a bell and spigot.

Level gauge Device for measuring level in a vessel.

Level glass Reading device directly connected to a vessel from the low to high points of a level variation. The liquid in the vessel finds its own level and may be observed through glass in the instrument.

Level indicating controller Instrument that indicates a vessel's liquid level and regulates it by pneumatic signal to a control valve.

Level indicator Dial or linear indicator which shows the level of liquid in the vessel.

Level recorder Instrument that makes a permanent record of liquid level in a vessel by pneumatic signal from a displacement-type transmitter on the vessel.

Level recording controller Instrument which has the same type of transmitter as the level recorder with a pneumatic signal to a control valve as well as a recorder.

Lift angle Angle between an imaginary line vertical to the load and the hoist or lifting device.

Line pressure Pressure in a line measured in pounds per square inch and rated in terms of SP or WOG.

Link Piece of metal forged or formed to make endless cord with half circles at each end and straight sides between.

Link, master (gathering ring) Forged or welded steel link used to support all members (legs) of an alloy steel chain or wire rope sling.

Link, master coupling Alloy steel-welded coupling link used as an intermediate link to join alloy steel chain to master links.

Load (dead) Total weight of all the suspended rigging.

Load (live) Weight of the object to be lifted.

Load (total) Sum of dead load plus live load.

Load (working) External load, in pounds, applied to the crane, including the weight of load-attaching equipment such as load blocks, shackles, and slings.

Long radius Elbow whose radius is 1 1/2 times the pipe's diameter.

M

Main Primary piping; section of a piping system which contains the process fluid or a major service.

Male thread External thread on pipes, fittings, and valves for making screwed connections.

Malleable fitting Fitting made for malleable iron.

Malleable iron Cast iron which has been heat-treated in an oven to relieve its brittleness, improve its tensile strength, and enable it to be pounded to a given shape.

Manifold Main line pipe with several branch connections; also referred to as a header.

Manometer Gauge used to measure differential pressure.

Manway Large nozzle and hinged flange on a vessel used to provide access to vessel internals.

N

Nipple Short length of pipe, threaded (male) on both ends, used for joining piping elements.

Nipple, close Nipple which shows no unthreaded pipe between the threaded ends.

Nominal pipe size Commercial designation of a pipe.

Nonmetallic disk (NMD) Composition disk, usually made of asbestos or resins.

Nonrising stem valve Valve whose stem does not rise on opening.

Normalizing Process in which an alloy or carbon steel is heated to a suitable temperature above the transformation range and is subsequently cooled in still air to room temperature.

Notch Small sharp edge in a weld area that may lead to weld failure.

Nozzle Any piece of pipe (stub-in) which is welded to a vessel or piece of equipment and has a flanged end onto which the pipeline with a similar flanged connection can be bolted. It is also possible to use a long-neck welded flange as a nozzle. Nozzles provide for the attachment of a piping system to a vessel, column, or tower. Also any device, fixed or adjustable, which controls a flow rate or discharge pattern from a pipe or line by means of special contour or size of an orifice.

Nozzle orientation Nozzle location in plan.

O

Off plot Area in the general vicinity of the project or on-plot area; battery limits determine what is on or off plot. Also refers to nonprocessing portion of plant, such as storage areas.

On plot Area connected with a project and bounded by the battery limits. Any portion of the project or equipment which is on plot is drawn and constructed on the same set of drawings to provide a complete visual description of the system. Often the plot plan is divided into areas to provide a smaller drawing of individual sections; all are considered to be on plot. Also refers exclusively to processing area of plant.

On site In the field; at the construction site. Any portion of the project which must be designed, constructed, or fabricated at the construction site is referred to as on-site production or construction.

Operator Any device, manual or remote, that activates a valve.

Orifice Device to produce differential pressures for flow measurement. Also used to increase the friction losses in pipelines. In this case, it is known as a restriction orifice.

Overhead accumulator Horizontal holding tank containing overhead products which are produced in a fractioning column.

Overhead lines Vapor lines from distillation columns.

P

Parallel piping When several pieces of similar equipment have their inlet nozzles connected to one header and their outlet connected to another, the piping is in parallel.

Petrochemical Chemicals derived from petroleum.

Pipe Hollow cylinder used to carry fluids or gases.

Pipe alley Main bank of headers.

Pipe bend Directional change in pipeline obtained by bending the pipe. Normal bend radius is five pipe diameters.

Pipe bending Forming the pipe sections into predetermined radii by hot or cold bending procedures. Pipe sections are often bent to a five-diameter radius, although sharper radii of three pipe diameters are sometimes required. Bending to the radius of 6, 10, or 15 times the diameter of the pipe can be accomplished.

Pipe coupling Device for joining end of screwed pipe.

Pipe fabrication Production of sections, configurations, or assemblies of pipe and various pipe components such as valves and fittings. Fabrication includes a variety of processing: welding, forming, shaping, heating, cleaning, machining.

Pipe rack Structural steel framework used to carry pipe runs through a mill. The pipes are usually located above ground enough so that access is not obstructed under the pipe rack.

Pipe rack bent Structure consisting of horizontal connecting members and two vertical columns (stanchions). The horizontal member is referred to as a strut when it connects two bents.

Pipe scale Hard, flakelike material frequently found in new pipe caused by heating during fabrication.

Pipe strap Device used to hold lightweight pipe to wall or ceiling.

Pipe support Devices which support piping.

Pipe support system Complete arrangement of pipe supports (hangers, anchors, guides, snubbers) that (1) hold the weight of the system with a minimum safety factor of 3; (2) permit thermal and seismic movement; (3) dampen vibration caused by mechanical equipment; and (4) maintain a safe stress limit.

Pipe thimble Sleeve used where pipes pass through walls or floors or are embedded in concrete or masonry.

Pipe volume Total interior volume, expressed in cubic measurement, which a cylindrical object can hold. The volume can be found by multiplying pi times the inside radius squared by the length of pipe (area of cross section times length).

Piping, drains Small valved pipes to drain main piping system on shutdown.

Piping, vents Valved piping which is used for venting system on startup and shutdown.

Plug Screwed fitting used for shutting off a tapped opening.

Plumbing Piping system of small-diameter threaded pipe used for utility purposes.

Pocket Low point in a pipe where liquids can be trapped, or high point where vapors may become trapped.

GLOSSARY

Pontoon Floating roof used on storage tanks.

Port, inlet Opening connected to the upstream side of a fluid system.

Pressure, atmospheric Pressure exerted in every direction upon a body by the atmosphere; equivalent to 14.7 psi at sea level (1 pascal).

Pressure, back Pressure on the upstream side of valve seats, or pressure surge in the downstream piping system.

Pressure alarm Pressure switch attached by a coupling or flange to equipment or pipe; when pressure becomes excessive, the switch turns on an alarm.

Pressure controller Instrument that regulates pressure by means of a control valve connected to a measurement point on the pipe or vessel.

Pressure differential Difference in pressure between any two points of a piping system.

Pressure differential controller Instrument that controls pressure differential between two pipes or vessels.

Pressure differential indicator Dial which indicates pressure differential between two pipes or vessels.

Pressure gauge Pressure in excess of atmospheric pressure.

Pressure indicating controller Valve that has an indicator transmitter (or a remote mounted indicator) to control line or vessel pressure.

Pressure recorder Instrument that makes a permanent record of line or equipment pressure. The recorder can be local or board-mounted.

Pressure recording controller Instrument similar to a pressure recorder but has a control valve with a pneumatic signal hookup.

Pressure (differential) recording controller Instrument which controls pressure differential between two pipes or vessels by means of a valve.

Pressure switch Switch activated by pressure of the material.

Pressure vessel Any vessel (container) designed to withstand the internal pressure of gases or liquids.

Protection saddles Saddles used on insulated pipes that require free longitudinal movement.

Pump Piece of equipment, normally electrically powered, that draws a liquid from a source and propels it through pipe to a receiving point.

Pyrometer Device which measures high temperatures; usually associated with petrochemical processes.

Q

Quenching Interjecting a cooler material into the discharge line, thereby lowering its temperature.

Quick opening valve Valve which can be opened quickly, usually by a pull chain.

R

Rated capacity (working load limit) Maximum allowable working load established by the sling manufacturer.

Rebar Steel rod used for reinforcing concrete.

Reducer Coupling with a smaller opening at one end for reducing the size of the pipe opening (see Increaser).

Reducer Pipe fitting to reduce from one pipe size to a smaller one.

Reducing tee Pipe fitting in which the side outlet is smaller than other outlets.

Reducing valve Automatic valve to reduce pressure.

Refinery Petrochemical complex that utilizes crude oil in its original form as the primary process medium, converting it to a variety of products, such as gasoline, tar, propane, fuel, oil, asphalt, and gas.

Refrigeration Reduction of temperature by means of mechanically driven thermodynamic processes in fluids.

Regulator Automatic valve which controls pressure of flow rate in a pipeline.

Reinforcing pad Metal collar that is welded around a nozzle on a vessel in order to reinforce the opening area.

Relief valve Automatic valve to relieve excess pressure.

Restraint Device used to restrain pipe from lateral and axial movement; there are one-, two-, and three-way types.

Return bend U-shaped fitting used to reverse the direction of a pipe run.

Reynolds number Value of a mathematical equation used in the determination of the flow characteristics of liquids at various velocities.

Ring joint Type of flange joint for pipe.

Riser Vertical pipe from a header. Also the vertical distance between stair treads.

Rising stem valve Valve in which the stem rises when the valve is opened.

Rod hanger Type of hanger used extensively to carry vertical downward loads. It allows limited horizontal pipe movement in all directions.

Rollers Long pieces of round hardwood, or long pieces of pipe, used under heavy pieces of equipment to facilitate rolling along flat surface.

Rubber lined Vessel or pipe lined with rubber to prevent attack by process materials.

Run Main line of piping that has branch connections and side outlets.

Rupture disk Diaphragm placed between flanges on pipe. The diaphragm has a definite breaking pressure so that it will break before the safety limit of the piping or vessels has been reached.

S

Saddle U-shaped piece of metal which provides support or reinforcement to insulated pipelines. It can also be used to establish a sloped run. Horizontal vessels or other pieces of equipment are supported on pedestals by means of thick steel saddles welded to the vessel which allow for the anchoring of equipment of the concrete pedestal. Saddles may also be used with guides.

Saddle flange Flange, usually curved and riveted or welded, to fit a boiler, tank, or other vessel and receive a threaded pipe. Also called a tank flange or boiler flange.

Safety valve Rapid-opening valve designed to relieve buildups of excess gas or pressure within a vessel.

Sample connection Valved spigot in a pipe or vessel to obtain process samples.

Saran lined Plastic lining to protect metal furnaces from chemical attack.

Sargol Joint in which a lip is provided for welding to make the joint fluid-tight; mechanical strength is provided by bolted flanges.

Sarlun Improvement of the sargol joint.

Scale Mineral deposits on boiler and heat exchanger tubes.

Schedule Measure of the relation of the wall thickness of pipe to the inside diameter.

Schedule number Approximate value of the expression $1000 \times p/s$, where p is the service pressure and s is the allowable stress (in psi).

Screwed end Pipe or fitting that is joined by threaded connections.

Screwed fittings Pipe fitting which is attached to pipes by means of threads.

Screwed flange Flange that is attached to a pipe by a screwed connection.

Seal leg Liquid-filled U-shaped leg of pipe which is used to prevent the flow of gases past this point.

Seamless Piercing and rolling of a solid billet or cupping.

Seizing Wrapping at the end of a rope with wire or cord to keep the strands and lay of the rope in place. Seizing should be applied before cutting the rope.

Series connection or piping When a group of similar equipment such that the outlet nozzle of one vessel is connected to the inlet of the next, this is called series piping.

Service fitting Street tee or street ell having a male thread at one end.

Shearing stress One that resists a force tending to make one layer of a body slide across another layer.

Shell Outer wall of a vessel, whether horizontal or vertical.

Shoe Metal piece attached (usually welded) to the underside of a pipe and resting on supporting steel. Used to reduce wear from sliding of lines subject to movement; also permits insulation to be applied to pipe and to provide elevation from the support to allow for a light slope of the pipeline.

Shop fabrication Fabrication completed in a shop as opposed to field construction. Shop fabrication has many advantages over field construction, offering a wider variety of production methods in a controlled environment.

Short radius Elbow pipe fitting whose radius equals one diameter of the pipe.

Short spring Heavier spring than the medium type, giving 50 percent of the deflection a medium spring would provide under a given load.

Side loading Load applied at an angle to the vertical plane of the boom.

Sight glass Glass plate suitably inserted in a vessel or pipe for internal observation.

Signal Hydraulic, pneumatic, or electric message that operates instruments and equipment.

Skelp Plate prepared by forming and bending and welding into a pipe.

Skirt Portion of a vertical vessel which extends from the vessel to the base and foundation.

Slip-on flange Flange which is slipped over the end of the pipe and welded in place.

Slope Angle from the horizontal at which a pipe is placed for drainage reasons.

Slurry Flow medium which has solid material suspended in a liquid.

Snubber (shock suppressor, pipe arrestor) Device, mechanical or hydraulic, which absorbs shock forces which could damage the pipe system. Snubbers do not resist slow thermal movement.

Socket fitting Fitting used to join pipe in which the pipe is inserted into the fitting. A fillet weld is then made around the edge of the fitting and the outside wall of the pipe.

Socket weld Type of weld performed when a piece of pipe is fitted into the socket of a fitting.

Socket-welded fitting or valve Socket-end fitting or valve for low pressures and small diameters.

Socket-welding fittings Pipe fittings in which the pipe is inserted in a socket.

Soldering Method of joining metals using fusable alloys, usually tin and lead, having melting points under 700°F.

Source nipple Short length of heavy-walled pipe between high-pressure mains and the first valve of bypass, drain, or instrument connection.

Spectacle blind Double-flange unit that can be swiveled to either a blind position (closed) or an open position.

Spool Assemblage of pipe and fittings normally welded in the fabricator's shop.

Spring hanger Support which allows variations in pipe position due to changes in temperature; often used for vertical lines.

Spring support (1) Variable load: spring hanger which responds only to the dead weight of a piping system and contents. This device allows both vertical and horizontal movement. Excellent for critical applications that need flexible support, this type of support has three variations based on travel distance. (2) Constant load: system of springs and lever arms used for high-temperature pipe installations where thermal movement may cause stress transfer of the system's load. The constant-load spring hanger balances the load and tension regardless of pipe movement (within certain established limits).

Spur gear hoist Most common type of chain hoist; utilizes spur gears to gain the mechanical advantage for ease of lifting.

Square knot Method of joining together two ropes of equal size. Sometimes called the reef knot, sailor's knot, or flat knot.

Standard Designation of cast iron flanges, fittings, and valves suitable for a maximum working steam pressure of 125 psig.

Standard weight Schedule of wrought iron pipe weights in common use.

Steam Vapor phase of water.

Steam, saturated Steam in contact with the water from which it was generated; may be either wet or dry.

Steam, superheated Steam heated above the temperature of the boiling point corresponding to the pressure.

Steam, traced Steam tubing that is wrapped around a pipe to keep line contents at a desired temperature.

Steam jacket Outer shell around a pipe or vessel for steam heating purposes.

Steam trap Device which permits the passage of condensate but not steam.

Street elbow Threaded elbow pipe fitting with male thread on one end and female on the other.

Steel, fireproofed Coating of special concrete placed on steel structures to prevent the steel from buckling from heat during a fire.

Steel strap Simple support used for stationary and noninsulated pipelines in heating and common housing plumbing.

Stem Part of valve trim which moves the disk on and off the valve seat.

Straight cross Cross pipe fitting in which all of the outlets are of the same diameter.

Straight tee Tee pipe fitting in which all of the outlets are of the same diameter.

Strain Change of shape or size of a body produced by the action of a stress.

Strength, minimum breaking Minimum load at which the sling will break when loaded to destruction in direct tension.

Strength, nominal breaking Load at which the sling could be expected to break when loaded to destruction in direct tension.

Stress Intensity of the internal, distributed forces which resist a change in the form of a body. When external forces act on a body they are resisted by reactions within the body which are termed stresses. The load on an object, usually pipe or structural members, measured in pressure units (psi).

Stub end Ends welded on a pipe or into a vessel in order to make a Van Stone joint.

Stub-in Connection or branch created when a small-diameter pipe is welded directly into a large one.

Subheader Branch line from a large header or pipe.

Sump Reservoir for the collection of spills and wastes.

Superstructure Rotating upper frame structure of the machine and the operating machinery mounted thereon.

Surge Violent movement of fluid in a piping system caused by quick opening of valves or collection of condensate in pockets in a steam line; usually associated with steam or water hammer. Surges can cause carryover of the line material to vapor lines (puking).

Swagging Process of reducing the ends of a pipe or tube section with rotating dies which are pressed intermittently or rotated against the pipe or tube end.

Swedged nipple Pipe similar to a reducer in which the ends may have male threads if desired.

Sweepolet Fitting that combines a stub-in and a saddle and offers a greater amount of reinforcing.

Swing Rotation of the superstructure for movement of loads in a horizontal direction about the axis of rotation.

Swing angle Rotation angle of the hanger rod caused by movement of pipe from cold to hot position.

Swing strut Pipe support component that can carry load in its axial direction while not restricting the pipe in any direction perpendicular to its axis within certain limits. A swing strut resists both tension and compression.

Symmetrical Similarity of arrangement on either side of a dividing line, axis, or plane; equal measure from a central point.

T

Tackle Assembly of ropes and sheaves arranged for hoisting and pulling.

Tank Basic process vessel used for the storage of raw materials, intermediates, and finished products. When modified, tanks can be used as dissolvers, precipitators, reactors, fermenters, and stills. Tanks are constructed in many sizes and shapes; the most common forms are vertical and horizontal right cylinders with end closures. Pressure tanks are fabricated as spheres or portions of spheres.

Tank, holding Any tank or vessel in which material is stored before it goes through processing.

Tank, horizontal Vessel, often cylindrical in shape, with closed ends. Head shapes are similar to those on vertical vessels or tanks. Usually supported below by saddles fastened to a foundation or floor or supported by slings hung from an overhead structure. Length of the horizontal tank should not be more than five times the diameter.

Tank, multicylinder Vessel with cylindrical shell segments and internal diaphragms with heads which may be partial multispheres.

Tank, multisphere Vessel used for storage of gas at pressures that would require prohibitive thickness in a single sphere.

Tank, pressure Vessel used for storage of volatile materials. Can be constructed to withstand the maximum pressure developed in a process. Many shapes other than the upright cylinder are possible. A full spherical shape is sometimes used which is capable of withstanding higher pressures; for gasoline, the Horton spheroid is used. Ellipsoidal pressure tanks are available in large capacities. Cylindrical tanks with outward dished heads can serve for storage at medium and low pressures.

Tank, settling Any vessel or tank which allows solids and liquids in the process material to separate gravitationally.

Tank, vertical (vertical vessel) Vertical right cylinder, the most common tank. Can be built to virtually any size and capacity required. Tanks used for storage in the open are covered. The roof may be self-supporting, but it is often supported by a central column and radial rafter. Tanks built at grade are of the flat-bottom type; outdoor elevated tanks may be flat-bottomed, supported by a suitable structure, or suspended from the sides. Suspended bottoms may be hemispherical, hemiellipsoidal, conspherical, or conical. Outdoor flat-bottom tanks at grade may rest on a foundation of sand or both sand and a curbing of concrete slightly larger than the tank diameter. Indoor storage tanks are usually flat-bottomed. Pressurized tanks may have dished heads.

Tank farm Any area of a refinery or plant that is primarily used for the storage of liquid or gaseous products; usually a group of vessels or holding tanks.

Tap Tool used for forming female (internal) threads.

Tee Three-way fitting shaped like the letter T.

Tee Fitting that provides a branch of line size or smaller in the run pipe.

Tee joint Joint between two members located at right angles to each other forming a T.

Temperature, absolute Temperature measured from absolute zero.

Temperature alarm Temperature-sensitive device with an alarm which warns against rise or drop in the line or vessel temperature.

Temperature control valve Control valve that is regulated by temperature fluctuations.

Temperature controller Instrument which regulates pipeline or vessel temperature by a control valve which is actuated by pneumatic signals from a transmitter.

Temperature differential controller Temperature control valve combined with a temperature-recording instrument for the controlled line or vessel which together control temperature differences between two vessels or pipelines.

Temperature element Instrument that is thermo-coupled without connections for measuring a line medium's temperature.

Temperature indicator Device used for measuring temperature: (1) a locally mounted dial; (2) remote mounted dial capillary tubes; or (3) electric thermocouple.

Temperature recorder Instrument which records a permanent continuous history of pipe or equipment temperature.

Temperature-recording controller Device which records and regulates the temperature of a vessel or line by a pneumatic signal to control valve and recorder.

Tensile strength Maximum tensile stress which a material will develop. The tensile strength is usually considered to be the load in pounds per square inch at which a test specimen ruptures.

Tensile stress One that resists a force tending to pull a body apart.

Thermal movement Any expansion or contraction of the piping system due to a change in temperature associated with the line material or external conditions.

Thermal shock Shock caused by sudden rise or drop in temperature of the fluid in contact with a valve.

Thermal stress Forces built up as anchored piping undergoes changes in temperature during operation. A rise in temperature causes expansion; as the pipe cools during down time or between operations, the pipeline contracts.

Thermocouple Temperature-sensing device consisting of two dissimilar metals joined together.

Threader Tool used for cutting male threads on the end of a pipe.

Threadolet A fitting that is welded into the run pipe to provide a threaded branch connection.

Tower Column or vertical vessel which increases the degree of separation that can be obtained during the fractionation and distillation of oil in a still.

Tower, forced-draft Cooling tower which forces air through the water being cooled by utilizing a horizontal-shaft fan on the side of the tower.

Tracing Using electrical or steam tubing to keep a pipeline or equipment at a constant predetermined temperature. Tracing methods usually require insulation of both tracer and pipeline.

Transmitter Device which transmits instrument readings or pneumatic signals for recording and control.

Trap Device that removes condensate, air, and gases from a steam line without releasing steam.

Trapeze Pipe hanger arrangement formed by two suspended hanger rods and a crossbar which supports the pipe.

Trim Certain internal valve parts such as seat rings, disks, stems, and repacking seat bushings.

Tube bundle Tubular part of a tubular heat exchanger.

Tubing System of small-diameter, lightweight (usually copper, brass, or plastic) pipes where OD always equals tubing size.

Turbine, single-flow Turbine which utilizes steam in a single path over the turbine blades.

Turbulence Nonparallel flow in a pipe; usually caused by obstructions, changes of direction, or other internal deviations from a straight run of pipe.

Turnbuckle Simple position-adjustment device for use on pipe supports.

U

Union Three-piece screwed fitting used to join lengths of pipe to permit easy opening of a line.

Unit fabrication Pipe components formed, welded, and produced by a variety of methods to create assemblies that can be shipped and installed in a complete section (unit) in the field.

Utility air Air line used in cleanup operations, to drive motors, and other service functions.

Utility piping Small-diameter threaded pipe used to convey water, instrument air, and sewerage.

Utility station Strategically located stations supplied with one or all of the following: drinking water, wash water, air, emergency showers, and eye wash fountain.

Utility steam Steam that is generated and used for service in a plant.

V

Valve Mechanical device used to interrupt or regulate flow in a piping system.

Valve, control Automatic valve.

Valve, flow-dividing Valve which divides the flow from a single source into two or more branches.

Valve body Main part of a valve into which the stem and other parts are installed.

Van stone Type of flanged pipe joint. Also known as a lapp joint.

Vapor lock Blocking of liquid flow by air or vapor trapped in a pipe.

Variable-spring hanger support Spring-support device (short, medium, or double spring). The load it exerts on the pipe depends on the amount of compression (loaded strength) in the spring (see Cold load, Variability).

Variability Change in the load of a spring hanger during operation of the piping system.

Velocity Time rate of motion in a given direction and sense, usually expressed in feet per second.

Vent Pipe to remove air or vapor from a pipe or vessel.

Venturi Nozzle design to increase velocities which causes a pressure drop. The suction due to the pressure drop is used to remove air and other materials.

Vessel Any container used in conjunction with a piping system to hold, transform, or store the medium.

Viscosity Internal friction in a fluid; its "thickness" or resistance to flow.

Viscosity index Measure of the viscosity-temperature characteristics of a fluid compared to other fluids.

Viscous material Material having a high resistance to flow.

Volume of a pipe Measurement of the space within the walls of the pipe. To find the volume of a pipe, multiply the length (or height) of the pipe by the product of the inside radius times the inside radius by 3.142.

W

Water hammer Shock or stress caused by liquid pressure waves that move through piping and come in contact with a change-in-direction fitting or closed piece of equipment. Can be caused by rapidly closing valves or (in the case of stem lines) condensate trapped in various portions of a valve body or in a line which is not sloped properly.

Wedge Disk shape for certain types of gate valves.

Weir port Particular shape of a valve body interior.

Welded fitting Forged or wrought steel elbow, tee, or similar piece for connection by welding to each other or to a pipe.

Welded joint Union of two or more members produced by the application of a welding process.

Welding Process of joining metals by heating until they are fused together, or by heating and applying pressure until there is a plastic joining action. Filler metal may or may not be used.

Welding end End of a fitting, pipe, or valve that is joined to other piping elements by welding.

Weld-neck flange Flange with integral extended neck for welding to pipe.

Weldolet Small fitting welded directly to a pipe, creating a branch or connection for welded pipe smaller than the run.

Winterizing Insulation, jacketing, or heat tracing of piping systems, including equipment and components.

Wrought pipe Pipe worked as in the process of forming furnace-welded pipe from skelp; distinguished from cast pipe.

X

X-ray test Test performed by x-ray to determine the soundness of welds.

Y

Y-strainer Y-shaped strainer for insertion into pipelines.

Y valve Y-shaped valve.

Yield strength Stress at which a material exhibits a specified limiting permanent set.

22

STANDARDS

Fig. 22-1

STANDARDS

American Gas Association AGA, 1515 Wilson Boulevard, Arlington, VA

American Iron and Steel Institute AISI, 350 5th Avenue, New York, NY

American National Standards Institute ANSI (formerly American Standards Association/ASA/), 1430 Broadway, New York, NY

American Society of Mechanical Engineers ASME, 29 West 39 Street, New York, NY

American Society for Testing Materials ASTM, 1916 Race Street, Philadelphia, PA

American Petroleum Institute, Dallas, TX API, Division of Production

American Water Works Association AWWA, 666 West Quincy Avenue, Denver, CO

American Welding Society AWS

Commodity of Standards Division, CS, U.S. Department of Commerce, Washington, DC

Department of the Navy, U.S. Government,* Department of the Navy (docks or ships)

Federal Specifications FED, U.S. Government,* Department of Commerce

Instrument Society of America ISA

Joint Army-Navy Specifications JAN, U.S. Government,* Department of Defense

Manufacturers Standardization Society of the Valve and Fitting Industry MSS, 1815 North Fort Myer Drive, Arlington, VA

Military Specifications MIL, U.S. Government,* Department of Defense

NATIONAL BUREAU OF STANDARDS: Superintendent of Documents, U.S. Government Printing Office, Washington, DC

Pipe Fabrication Institute PFI

STEEL AND IRON WROUGHT PIPE

Standard Number   Title ASA B36.1-1956   Welded and Seamless Steel Pipe (ASTM a-53-57t)

ASA B36.2-1956   Welded Wrought Iron Pipe (ASTM a72)

ASA B36.3-1956   Seamless Carbon-Steel Pipe for High Temperature Service (ASTM a106)

ASA B36.4-1956   Electric Fusion (arc) Welded Size 16 in. and Over (ASTM a134)

ASA B36.5-1956   Electric-Resistance-Welded Steel Pipe (ASTM A135-57)

ASA B36.9-1956   Electric-Fusion-Welded Steel (sizes 4 in. and over) (ASTM A139-55)

ASA B36.10-1950   Wrought Steel and Wrought-Iron Pipe (ASME)

ASA B36.11-1956   Electric-Fusion-Welded Steel Pipe for High-Temperature Service (ASTM A155-55)

ASA B36.16-1956   Spiral-Welded Steel or Iron Pipe (ASTM A211-54)

ASA B36.19-1957   Stainless Steel Pipe

ASA B36.20-1951   Black and Hot-dipped Zinc-Coated (Galvanized) Welded and Seamless Steel for Ordinary Uses (ASTM A120-57)

ASA B36.23-1956   Welded and Seamless Open-Hearth Iron (ASTM A253-55t)

ASA B36.26-1956   Seamless and Welded Austenitic Stainless Steel (ASTM A312-57t)

ASA B36.36-1956   Seamless and Welded Ferritic Stainless Steel Tubing for General Service (ASTM A268-55)

---

*Index of Government Specifications information available at U.S. Government Printing Office, Washington, DC.

ASA B36.1956   Austenitic Stainless Steel Tubing Seamless and Welded, for General Service (a269-57 ASTM)

ASA B36.38-1956   Austenitic Stainless Steel Sanitary Tubing Seamless and Welded (a270-57 ASTM)

ASA B36.40-1956   Low-Temperature Service, Welded and Seamless Pipe (ASTM A333)

ASA B36.41-1956   Low-Temperature Service, Welded and Seamless Tubes (ASTM A334)

ASA B36.42-1956   Seamless Ferritic Alloy Steel Pipe, for High-Temperature Service, (ASTM A335)

ASTM A376-55t   Austenitic, Seamless, for High-Temperature Central-Station Service

ASTM A139-55   Electric-Fusion (arc) Welded, Sizes 4 in. and Over

ASTM A358-55t   Electric-Fusion-Welded Austenitic Chromium-Nickel Alloy for High-Temperature Service, Specs

ASTM A369-55t   Alloy Steel, Ferritic, for High-Temperature Service

ASTM A381-54t   Metal-Arc Welded for High-Pressure Transmission Service

ASTM A419-57t   Wrought Iron Plate pipe, Fusion Arc Welded

AWA C201-50   Electric-Fusion Welded Steel Water Pipe, Sizes 30 in. and Over

AWWA C202-49   Steel Water Pipe, sizes up to but not including 30 in.

API STD. 5A Mar. 54   Casing, Tubing, and Drill Pipe and Supplement 1, April 1955

API BUL 5A July 1955   Dimensional Data for Sharp-Thread Casing and Tubing

API STD. 5L March 1958   Line Pipe

API STD. 5LX   High Test Line Pipe

Fed WW-P-40A   Pipe, Steel, Seamless and Welded, Black and Zinc-Coated

Fed WW-P-406A   Pipe, Steel, Seamless and Welded (for ordinary use)

Fed WW-P-441b(1)   Pipe; Wrought Iron (Welded, Black or Zinc-Coated)

MIL-P-3107   Steel Drive Pipe with Couplings

MIL-P-1144a (ships)   Pipe, Stainless Steel, Seamless or Welded, Corrosion Resisting

MIL-T-7880(1)   Tubing, Flexible, Corrosion Resistant and Heat Resistant Steel

MIL-P-15692C   Pipe, Steel Seamless and Welded Black and Zinc Coated

MIL-P-18501   Pipe and Fittings, Steel, Lightweight, Spiral and Longitudinal Welded Seam

MIL-T-20155   Tubing, Steel, Alloy, Molybdenum, Seamless

MIL-T-20157   Tubing, Steel, for Oil, Steam, or Water

MIL-T-20160   Tubing, Steel, Seamless or Welded

MIL-T-20162   Tubing, Steel, Welded (including resistance welded) for Oil, Steam, or Water

MIL-T-20169   Tubing, Steel, Resistance-Welded (for Oil, Steam, or Water)

MIL-P-20220   Pipe, lead, lining for iron pipe and steel tubing.

CAST IRON PRESSURE PIPE

Standard Number   Title ASA A21.1-1957 Computation of Strength and Thickness of Cast-Iron Pipe (AWWA C101-57)

ASA A21.2-1953   Cast Iron Pit Cast Pipe for Water or Other Liquids (AWA C102)

ASA A21.3-1953   Cast Iron Pit Cast Pipe for Gas

ASA A21.4-1953   Cement Mortar Lining for Cast Iron Pipe and Fittings (AWAA C104)

ASA A21.6-1954   Cast Iron Pipe Centrifugally Cast in Metal Molds for Water or Other Liquids (AWA C106)

ASA A21.7-1953   Cast Iron Pipe Centrifugally Cast in Metal Molds for Gas

ASA A21.8-1953 Cast Iron Pipe Centrifugally Cast in Sand Lined Molds for Water or Other Liquids (AWWA C108)

ASA A21.9-1953 Cast Iron Pipe Centrifugally Cast in Sand Lined Molds for Gas

ASA A21.11-1953 Mechanical Joint for Cast Iron Pressure Pipe and Fittings, Specifications for (AWWA C111)

ASTM A377.57T Cast Iron Pressure Pipe

Fed WW-P-421a Pipe; Water, Cast-Iron (Bell and Spigot)

Fed WW-P-00421 Pipe; Water, Cast Iron (Bell and Spigot) (Navy-Docks)

IRON DRAIN PIPE

Standard Number Title ASA A40.1-1935 Cast Iron Soil Pipe and Fittings (ASME)

ASTM A74-42 Cast Iron Soil Pipe and Fittings

ASA G26.1-1942 Cast Iron Culvert Pipe (ASTM A142-38)

CS 188-53 Service-Weight Cast Iron Soil Pipe and Fittings

Fed. WW-P-356 Pipe, Cast Iron; Drainage, and Waste (Threaded)

Fed WW-P-401(3) Pipe and Pipe Fittings; Soil Cast-Iron; including Amendment 30

MIL-P-236b(2) Iron or Steel Culvert pipe, nestable

MIL-P-21624 (docks) Pipe, drainage, zinc bituminous coated corrugated, iron or steel

NONFERROUS METALLIC PIPE AND TUBE

Standard Number Title ASA H23.1 1956 Seamless Copper Water Tube (ASTM B88-55)

ASA H26.1-1956 Seamless Copper Pipe, Standard Sizes (ASTM B42-57)

ASA H27.1-1956 Seamless Red Brass Pipe, Standard Sizes (ASTM B43-57)

ASA H34.1-1955 Nickel Seamless Pipe and Tubing (ASTM B161)

ASA H34.3-1955 Nickel-Chromium-Iron Alloy Seamless Pipe and Tubing (ASTM B167)

ASA H34.2-1955 Nickel-Copper Alloy Seamless Pipe and Tubing (ASTM B165)

ASTM B68-55 Seamless Copper Tube, Bright Annealed

ASTM B75-55 Seamless Copper Pipe

ASTM B135-55 Brass Tube Seamless

ASTM B241-55t Aluminum-Alloy Pipe

ASTM B251-56T General Requirements for Wrought Seamless Copper and Copper-Alloy Pipe and Tube

ASTM B280-55T Copper Tube for Refrigeration Field Service, Seamless

ASTM B302-57 Threadless Copper Pipe

ASTM B306-58 Copper Drainage Tube (DWV) CS 95-41 Lead Pipe

Fed WW-P-325 Pipe, Bends and Traps; Lead (for) Plumbing and Water Distribution

Fed WW-P-351(1) Pipe; Brass, Seamless; Iron-Pipe, Size, Standard and Extra Strong

Fed-WW-P-377b Pipe Copper, Seamless, Standard

Fed WW-T-783b Tubing, Aluminum Alloy (25) Round, Square, Rectangular and Other Shapes, Seamless Drawn

Fed WW-T-785a Tubing, Aluminum Alloy 24S, Round, Square, Rectangular and Other Shapes, Seamless Drawn

Fed WW-T-787a Tubing, Aluminum Alloy (AL-52) (Aluminum-Magnesium-Chromium); Round, Seamless

Fed WW-T-788b(2) Tubing, Aluminum Alloy 3S, Round, Square, Rectangular and Other Shapes, Seamless, Drawn

Fed WW-T-789a(2) Tubing, Aluminum Alloy 61S, Round, Square, Rectangular and Other Shapes, Seamless, Drawn

Fed WW-T-791(1) Tubing, Brass, Seamless

Fed WW-T-797(1) Tubing; Copper, Seamless (for general use with IPS Flanged Fittings)

Fed WW-T-799a(1) Tubing, Copper, Seamless (for use with solder-joint or flared-tube fittings)

Fed WW-T-816 Tubing, Flexible, Aluminum Alloy (changed from R.R-T-791)

Fed WW-T-825 Tubing-Magnesium-Alloy, Round, Extruded

MIL-T-6945 Tubing, Brass, Seamless

MIL-T-873 (ships) Tubing, Copper, Seamless, for Pressures up to 4,000 lbs. per sq. inch

MIL-T-1368a Tubing, Nickel, Copper Alloy, Seamless and Welded

MIL-T-3595a Tubing, Phosphor Bronze, Round, Seamless

MIL-T-6013 Tubing, Copper-Silicon-Manganese, Round, Seamless

MIL-T-20168 Tubing, Brass, Seamless for Pressures from 451 to 4,000 pounds per square inch

NONMETALLIC PIPE
(Clay, Concrete, Fiber, Plastic) Standard Number Title

ASTM C4-55 Drain, Tile

ASTM C12-54 Clay Sewer Pipe, Rec. Practice For

ASTM C13-57 Sewer Pipe Standard Strength, Clay

ASTM C14-57 Concrete Sewer Pipe

ASTM C75-56 Concrete Sewer Pipe Reinforced

ASTM C76-57 Concrete Culvert Pipe

ASTM C118-56 Concrete Irrigation Pipe

ASTM C200-57 Extra Strength Clay Pipe

ASTM C211-57T Standard Strength Clay Pipe, Perforated

ASTM C261-57T Ceramic Glazed Sewer, Standard Strength Pipe

ASTM C278-57T Ceramic Glazed Pipe, Extra Strength

ASTM C296-55T Asbestos-Cement Pressure Pipe

ASTM C301-54 Testing of Clay Pipe

ASTM C361-57T Low Head Pressure Reinforced Concrete Pipe

ASTM C362-55T Low Head Internal Pressure Reinforced Concrete Sewer Pipe

ASTM C700-78A Vitrified Clay Pipe, Extra Strength, Standard Strength & Perforate

AWWA C300-52 Reinforced Concrete Water Pipe Steel Cylinder Type, Not prestressed

AWWA C301-55T Reinforced Concrete Water Pipe-Steel Cylinder Type, Prestressed

AWWA C302-53 Reinforced Concrete Water Pipe-Non-cylinder Type, Not Prestressed

AWWA C400-53T Tentative Standard Asbestos-Cement Water Pipe

AWWA C900-75 Polyvinyl Chloride (PVC) Pressure Pipe 4 in. through 12 in. for Water

CS 143-47 Standard Strength and Extra Strength Perforated Clay Pipe

CS 116-54 Bituminized-Fibre Drain and Sewer Pipe

CS 143-47 Standard Strength and Extra Strength Perforated Clay Pipe

CS 197-54 Dimensions & Tolerances for Standard Wall Polyethylene Pipe

Fed SS-P-331a Pipe; Asbestos-Cement, Sewer, Non-Pressure

Fed SS-P-351a Pipe; Asbestos-Cement

Fed SS-P-356 Pipe, Bituminous Fibre, Sewer and Fittings for same

Fed SS-P-361b(1) Pipe; Clay, Sewer

Fed SS-P-371a Pipe; Concrete, Sewer, Non-reinforced

Fed SS-P-375 Pipe; Concrete, Sewer, Reinforced

Fed SS-P-381 (GSA-FSS) Pipe; Pressure, Reinforced Concrete, Pretensioned Reinforcement (Steel Cylinder Type)

Fed WW-C-633 Couplings, Hose; Pneumatic, Universal Type

ASTM D702-56 Cast Methacrylate tubing

ASTM D701-49 Cellulose Nitrate (Pyroxylin) Tubing

ASTM D922-54T Nonrigid Polyvinyl Tubing

ASTM D1503-57T Cellulose Acetate Butyrate SW

Fed HH-T-791a Tubing Flexible, Non-metallic

MIL-T-16407a Tubing, Rubber, Hard, Rigid

MIL-P-19119A Pipe, Plastic, Rigid Unplasticized High Impact Polyvinyl Chloride

MIL-T-20537 Tubing, Rubber, Synthetic

CS 197-57 Flexible Polyethylene Plastic Pipe

CS 206-57 Dimensions and Tolerances for Solvent Welded (SWP size) Cellulose-Acetate Butyrate Pipe

CS 207-57 Dimensions and Tolerances for Rigid Polyvinyl Chloride Pipe

STEEL AND IRON FITTINGS

Standard Number Title ASA B16.b-1944 R1953 Cast-Iron Pipe Flanges and Flanged Fittings, Class 250 (ASME)

ASA B16.b1-1931 R1952 Cast-Iron Pipe Flanges and Flanged Fittings (for 800 pound Hydraulic Pressure) (ASME)

ASA B16.b2-1931 R1952 Cast-Iron Pipe Flanges and Flanged Fittings (for Maximum WSP of 25 pounds) (ASME)

ASA B16.1-1948 R1953 Cast-Iron Pipe Flanges and Flanged Fittings, Class 125 (ASME)

ASA B16.20-1956 Ring Joint Gaskets and Grooves for Steel Pipe Flanges (ASME)

ASA B16.3-1951 Malleable-Iron Screwed Fitting, 150 pounds (ASME)

ASA B16.4-1949 R1953 Cast-Iron Screwed Fittings, 125 and 250 pounds (ASME)

ASA B16.5-1957 Steel Pipe Flanges and Flanged Fittings (ASME)

ASA B16.9-1951 Steel Butt-Welding Fittings (ASME)

ASA B16.10-1957 Face-to-Face and End-to-End Dimensions of Ferrous Valves (ASME)

ASA B16.11-1946 R1952 Steel Socket-Welding Fittings (ASME)

ASA B16.12-1953 Cast-Iron Screwed Drainage Fittings (ASME)

ASA B16.14-1949 R1953 Ferrous Plugs, Bushings, and Locknuts with Pipe Threads (ASME)

ASA B16.16-1948 R1952 Cast-Iron Flanges and Flanged Fittings for Refrigerant Piping, Class 300 (ASME)

ASA B16.19-1951 Malleable-Iron Screwed Fittings, 300 pounds (ASME)

ASA B16.20-1956 Ring-Joint Gaskets and Grooves for Steel Pipe Flanges (ASME)

ASA B16.25-1955 Butt-Welding Ends

ASA A21.10-1952 Short Body, Cast-Iron Fittings, 3 inches to 12 inches, for 250 psi, Water Pressure plus Water Hammer (AWWA C110)

ASA G38.1-1957 Carbon and Alloy Steel Nuts for Bolts for High-Temperature Service (ASTM A194-53)

ASA G46.1-1956 Forged or Rolled Steel Pipe Flanges, Forged Fittings, and Valves and Parts for General Service (ASTM A181-55T)

ASTM A182-53T Alloy, Pipe Flanges, Forged Fittings, and Valves and Parts for High-Temperature Service

ASTM A350-55 Rolled Carbon and Alloy Pipe Flanges, Forged Fittings, Valves and Parts for Low-Temperature Service

ASTM A181-57T Pipe Flanges, Forged Fittings, and Valves and Parts for General Service

ASA G17.3-1947 Forged or Rolled Steel Pipe Flanges, Forged Fittings, and Valves and Parts for High-Temperature Service (ASTM A105)

ASTM A234-57T Welded Fittings, Carbon Steel and Ferritic Alloy Steel, Factory-Made

ASTM A403-56T Welded Fittings, Wrought Austenitic Steel, Factory Made

AWWA C100-55 Cast-Iron Pressure Fittings

AWWA C207-55 Steel Pipe Flanges

AWWA C208-55T Dimensions of Steel Water Pipe Fittings

AWWA C502-54 Specifications for Fire Hydrants or Ordinary Water Works Service

API STD 6B Sept. 1953 Specifications for Ring-Joint Flanges for Drilling and Production Service

API STD 6BX Ring-Joint Flanges for Drilling and Production Services

CS 5-46 Pipe Nipples; Brass, Copper, Steel, and Wrought Iron

CS 7-29 Standard Weight Malleable Iron or Steel Screwed Unions

MSS SP-43-1957 Stainless Steel Butt-Welding Fittings

MSS SP-44-1955 MSS Steel Pipe Line Flanges

MSS SP-46-1955 Assembly of Steel Raised Face Flanges to Cast Iron, Brass, Bronze or Stainless Steel Flanges

MSS SP-48-56 Steel Butt Welding Fittings (26 in. and larger)

MSS SP-49-56 Forged Steel Screwed Fittings, 2000, 3000, and 6000 Pound

MSS SP-50-56 Forged Steel Plugs and Bushings

MSS SP-51-57 150 lb. Corrosion Resistant Cast Flanges and Flanged Fittings

Fed WW-N-351a Nipples, Pipe, Threaded

Fed WW-P-471(1) Pipe Fittings (Bushings, Plugs, and Locknuts); Bronze and Ferrous (Screwed)

Fed WW-P-491a(1) Pipe Fittings; Cast Iron, Drainage

Fed WW-P-501c(1) Pipe Fittings; Cast Iron (Screwed) 125–250 pounds

Fed WW-P-521c Pipe Fittings; Malleable Iron, Wrought Iron and Steel (Screwed) 150 pounds

Fed WW-U-531a Unions; Malleable Iron or Steel, 250 pounds

Fed WW-P-541b(1) Plumbing Fixtures (for) Land Use

Fed WW-U-536(1) Unions; Malleable Iron or Steel, 300 pounds

Navy 45U6b Dec., 1945 Unions, Iron (Malleable) or Steel, 300 pounds, including Amendment 1

MIL-P-10388B(3) Pipe Fittings, Malleable Iron, Steel, or Aluminum Alloy, Grooved Type

MIL-U-16365 (ships) August 1951 Unions, Iron (Malleable) or Steel 250 psi Saturated Steam 500 psi Cold Water, including Amendment 1

MIL-P-18174 Pipe Fittings, Screwed, Steel

MIL-P-18178 Pipe Bends, Fabricated, for Shore Use

MIL-F-18180 Flanges and Flanged Fittings, Pipe, Steel (150, 300, 400, 600, 900, 1500 and 2500 pounds)

MIL-U-18181 Docks Union, Pipe, Flanged, Bronze and Iron (125, 150, 250, and 300 pounds WSP)

MIL-U-18182 Union, Pipe, Forged Steel, Ground Joint, Water, Oil, Gas (2000, 3000 and 6000 psi)

MIL-U-18183 Union, Pipe, Flanged, Steel

MIL-U-18249 Unions Pipe Bronze or Naval Brass, threaded pipe connection 250 lbs. WSP

MIL-U-18250 Unions, Pipe, Steel or Malleable Iron, threaded pipe connection, 300 lbs. WSP

NONFERROUS METALLIC FITTINGS

Standard Number Title ASA B16.15-1947 R1952 Brass or Bronze Screwed Fittings, 125 pounds (ASME)

ASA B16.17-1949 R1953 Brass or Bronze Screwed Fittings, 250 pounds (ASME)

ASA B16.18-1950 Cast-Brass Solder-Joint Fittings (ASME)

ASA B16.22-1951 Wrought Copper and Bronze Solder-Joint Fittings (ASME)

ASA B16.23-1955 Cast-Brass Solder-Joint Drainage Fittings (ASME)

ASA B16.24-1953 Brass or Bronze Flanges and Flanged Fittings (ASME)

ASA B16.26-1958 Brass Fittings for Flared Copper Tubes

CS 5-46 Pipe Nipples; Brass, Copper, Steel, and Wrought Iron

CS 96-41 Lead Traps and Bends

Fed WW 325 Lead Bends & Traps for Plumbing and Water Distribution

Fed WW-P-460 Pipe Fittings; Bronze (Screwed) 125 and 250 pounds

Fed WW-T-696a Traps; Steam, Thermostatic (for land use)

Fed WW-U-516 Unions; Brass or Bronze, 250 pounds

Navy 45u5a Unions, Brass or Bronze, 250 pounds

MIL-P-10388B Pipe Fittings, Malleable Iron, Steel or Aluminum Alloy, Grooved Type, including Amendment 2

VALVES

Standard Number Title API 6C Specifications for flanged steel gate and plug valves for drilling and production service

API 6D Specifications for Iron and Steel gate, plug and check valves for pipeline service

MSS SP-3-44 Roughing-in Dimensions for Low Pressure Radiator Valves, Union Elbows and Return-Line Vacuum Valves

MSS SP-6-55 Finishes for Contact Faces of Connecting End Flanges of Ferrous Valves and Fittings

MSS SP-9-55 MSS Spot-Facing Standard

MSS SP-20-46 MSS Specification for Steam-Bronze Castings for Valves, Flanges and Pipe Fittings

MSS SP-37-49 MSS 125 lb. Bronze Gate Reaffirmed 53 Valves

MSS SP-38-55 MSS 100 lb. Bronze Gate Valves

MSS SP-40-46 MSS Specifications for Leaded Red Brass, and Leaded Semi-Red Brass Castings for Valves and Pipe Fittings

MSS SP-41-53 MSS Specification for Stainless Steel Castings for Valves, Flanges and Pipe Fittings

MSS SP-42-57 MSS 150 lb. Corrosion Resistant Cast Flanged Valves

MSS SP-45-57 MSS Bypass and Drain Connection Standard (formerly SP-5 and SP-28)

MSS SP-52-57 Cast Iron Pipe Line Valves

Fed WW-V-51a(2) Valves, Bronze; Angle, Check and Globe, 125 and 150 pounds, Screwed and Flanged (for land use)

Fed WW-V-54 (2) Valves, Bronze, Gate; 125 and 150 pounds, Screwed and Flanged (for land use)

Fed WW-V-151 (1) Radiator, Air, Thermostatic (Gravity, Steam Heating Systems)

Fed WW-V-58(1) Valves; Cast Iron, Gate; 125 and 250 pounds, Screwed and Flanged (for land use)

MIL-V-1202A Valve, angle: valve, cross: valve, globe: bronze 100 WSP

MIL-V-1187B Valves, angle: valves, globe-bronze threaded 300 lbs. WSP

MIL-V-20063A (ships) Valve, plug, lubricant sealed.

MIL-V-18826 Valves, globe and angle: cast iron.

MIL-V-18827 Valves gate bronze 300 psi

MIL-V-18436 Valves, check

MIL-V-17547N Valves, check bronze

MIL-V-12003A Valves, plug lubricated and lift plug, nonlubricated, C.I. and steel

PIPE AND FITTING THREADS

Standard Number Title ASA B2.1-1945 Pipe Threads (ASME)

ASA B26-1935 R1953 Fire Hose Couplings Screw Thread (ASME)

ASA B33.1-1935 R1947 Hose Coupling Screw Threads (ASME)

API STD 5B Sept., 1954 Specifications for Inspection of External and Internal Pipe Threads

API STD 6A May 1957 Specifications for Threads in Valves, Fittings and Flanges Navy 44T2C Threads, Standard, for Pipe and Pipe Fittings

AWWA C800-55 AWWA Standard for Threads for Underground Service Line Fittings

MIL-P-7105 Aeronautical National Form Taper Pipe Threads

CODES AND INSTALLATION STANDARDS

Standard Number Title ASA A40.4-1942 Air Gaps in Plumbing Systems (ASME)

ASA A40.6-1943 Backflow Preventers in Plumbing Systems (ASME)

ASA A40.8-1955 National Plumbing Code (ASME)

ASA B31.1-1955 Code for Pressure Piping (Sections 1-7) (ASME)

ASA B31.1.8-1955 Gas Transmission and Section 8 of 1951 Distribution Piping Code Systems (ASME)

ASA Z21.30-1954 Installation of Gas Piping and Gas Appliances in Buildings (Not Applicable to Undiluted Liquefied Petroleum Gas)

AWWA C206-50 Standard Specifications for Field Welding of Steel Water Pipe

AWWA C600-54T Tentative Standard Specifications for Installation of C.I. Water Mains

API STD 1104 1958 Standard for Field Welding of Pipe Lines

ASTM C12-54 Installing Clay Sewer Pipe, Rec. Practice

SYMBOLS AND MARKINGS

Standard Number Title ASA A13.1-1956 Scheme for Identification of Piping Systems (ASME)

ASA Z32.2.1-1949 Welding, Graphical Symbols for (ASME)

ASA Y32.4-1955 Plumbing, Graphical Symbols for (ASME)

ASA Z32.2.3-1949 Pipe Fittings, Valves and Piping, Graphical Symbols for (ASME)

ASA Z32.2.4-1949 Heating, Ventilating, and Air Conditioning, Graphical Symbols for (ASME)

MSS 25-1958 Standard Marking System for Valves, Fittings, Flanges and Unions

ES STANDARDS AVAILABLE FROM PIPE FABRICATION INSTITUTE

ES1 End Preparation and Machined Booking Rings for Butt Welds (1974)

ES2 Method of Dimensioning Piping Assemblies (1974)

ES3 Fabricating Tolerances (1974)

ES4 Hydrostatic Testing of Fabricated Piping (1975)

ES5 Cleaning of Fabricated Piping (1975)

ES6 Recommended Practice for Heat Treatment of Pipe Bends and Formed Components of Carbon and Ferritic Alloy Steels (1977)

ES7 Minimum Length and Spacing for Welded Nozzles (1962) (Reaffirmed 1975)

ES11 Permanent Marking on Piping Materials (1975)

ES16 Access Holes and Plugs for Radiographic Inspection of Pipe Welds (1974)

ES18 Ultrasonic Examination of Tubular Products (1975)

ES19 Preheat and Postheat Treatment of Welds (1972) (reaffirmed 1975)

ES20 Wall Thickness Measurement of Pipe Bends by Ultrasonic Examination (1975)

ES21 Manual Gas Tungsten-Arc Root Pass Welding End Preparation and Dimensional Joint and Fit Up Tolerances (1974)

ES22 Recommended Practice for Color Coding of Piping Materials (1969) (Reaffirmed 1975)

ES24 Pipe Bending Tolerances—Minimum Bending Radii—Minimum Tangents (1975)

ES25 Random Radiography of Pressure Retaining Girth Butt Welds (1973)

ES26 Welded Load Bearing Attachments to Pressure Retaining Piping Materials (1976)

ES27 Visual Examination—The Purpose, Meaning and Limitation of the Term (1974)

ES28 Recommended Practice for Welding of Transition Joints Between Dissimilar Steel Combinations (1976)

I

Insulation, 137-43
   block, 143
   85% magnesia, 143
   properties, 142
   thickness, 142

L

Leaky joints, 52

M

Miters (*see* Pipe miters), 111-15

N

Non-right triangles, 172-73

O

Offsets (*see* Pipe offsets), 88-96

P

Parabola, 162
Patterns (*see* Developments and patterns; Templates), 115-22
Pipe, 2-22
   expansion, 22
   heat loss, 20, 22
   schedule number, 5-6, 10-15
   sizes, 5-15
   sizing, 8
   specifications, 8, 9
   steam, 8
   weight, 22
Pipe alignment, 83
Pipe jig, 83
Pipe joints, 52-57
Pipe materials, 2-5
   aluminum, 4
   brass, 3, 17
   cast iron, 20
   clay, 5
   copper, 3, 17
   glass, 4
   iron, 2, 6
   lead, 4, 17
   plastic, 4, 19
   titanium, 4
   wood, 5
Pipe miters, 111-15
   four miter bend, 115
   90 degree miter, 112
   one miter bend, 113
   three miter bend, 114
   two miter bend, 113
   two piece, 112
Pipe offsets, 88-96
   45 degree offsets, 89-90, 94-96
   45 degree rolling offsets, 90-92
   simple offsets, 91
   30 degree offsets, 92
Pipe setup, 82-87
   cutback, 85
   elbows, 84
   elbows (45 degree)-to-pipe, 83
   elbows (90 degree)-to-pipe, 83
   flange-to-pipe, 83
   header and branch layout, 83
   jig, 83
   pipe-to-pipe, 83
   reinforcing pads, 87
   tee-to-pipe, 83
Pipe supports, 144-49
   spacing, 145-47, 149
   stresses, 149
   thermal expansion, 146
Polygons, 160-61
Prisms, 162
Pyramids, 163
Pythagorean theorem, 171

R

Reinforcing pads, 87
Relief valves, 72
Resistance welding, 124
Rigging, 150-56
Right triangles, 171-73
Rod swing angle, 147
Rope, 153-54
   knots, 153-54
   safe working load, 155

S

Safety valves, 71
Solids, 162
Specific gravity, 135
Spheres, 164-65
Spherical, 165
Standard organizations:
   ANSI, 2
   ASA, 2
   ASTME, 2
Steam traps, 71-72
Structural bracing, 148

Symbols, 74-81
   fittings, 76-78
   flow diagram, 79-81
   instrumentation, 80-81
   pipeline, 74
   special, 75
   utilities, 74

T

Templates, 116-22
   blunt and development, 122
   branch and header, 119
   lateral development, 120-21
   multipiece, 117
   90 degree branch, 177
   90 degree multipiece turn, 119
   90 degree three-piece turn, 118
   90 degree two-piece turn, 118
   orange peel, 122
   Thermal expansion, 148
   wraparound, 177
Thermal expansion, 146
Thermal insulation, 138-40
Threads, 52-57
   American National Standard Taper, 53
   American Standard Threads, 57
   British Standard Taper Threads, 58
   engagement, 52, 54
   making up joints, 55
   tap drill sizes, 58
   tools, 56
Torus, 165
Triangles, 159, 170-77
   formula, 171
   non-right triangles, 172-73
   right triangles, 171-72
Trigonometry, 171-76
   functions, 171, 173-76

V

Valves, 59-72
   abbreviations, 62
   bonnets, 63-64
   end connections, 66
   flanged joints, 67
   flow, 67
   functions, 60-61
   materials, 64-65, 68
   placement, 63
   resistance, 61
   service conditions, 62
   stems, 63-64
   types, 69-70
Volume, 166-69

W

Water pressure, 135
Welding, 123-31
   arc welding, 124-26
   electrodes, 127-29
   gas welding, 124
   joint types, 129
   materials, 126
   resistance welding, 124
   symbols, 126-27
   temperature, 127, 130
   trouble shooting, 130
Welding symbols, 126-27
Wire rope, 151-53
   clamp fasteners, 153
   handling, 151-52
   lubrication, 151
   safe working load, 152
   sheaves, 155